FROM THE

TWENTY-THIRD ANNUAL REPORT

ON THE

NEW YORK STATE CABINET OF NATURAL HISTORY,

For the Year 1869.

APPENDIX D

ENTOMOLOGICAL CONTRIBUTIONS.

By J. A. LINTNER.

APPENDIX E.

On Cucullia intermedia nov. spec., etc.: By A. Speyer, M. D.

PRINTED IN ADVANCE OF THE REPORT.

ALBANY:
WEED, PARSONS AND COMPANY, PRINTERS.
1872.

CONTENTS.

(D.)

ENTOMOLOGICAL CONTRIBUTIONS.

BY J. A. LINTNER.

I. BIOGRAPHY OF HEMILEUCA MAIA (DRURY).

On the 11th of May (1869) there was found at Center, Albany county, N. Y., a belt of eggs encircling a small twig of *Quercus ilicifolia*, which was evidently the deposit of some species of Heterocera, but which, at the time, could only be referred problematically to the Bombycidæ.

The Egg. — Its form is obovate, laterally and apically compressed; its transverse diameter is .075 of an inch, its greater conjugate .06 of an inch, and its lesser conjugate .045 of an inch. Its color is reddish-brown on so much of its apical portion as is visible when observed in the belt, and yellowish-white on the remaining part. When examined with a high magnifying power, minute blackish dots are seen sparsely sprinkled over it, in each of which is a white central point, apparently papilliform.

During the ensuing two months, five additional clusters of egg-shells (the larvæ having emerged) were found at the same locality, from which the following description is drawn:

Egg-belt (Plate 8, fig. 3). — The belts vary in length from .25 to .60 of an inch, and surround twigs of *Quercus ilicifolia* and *Q. prinoides* which range in diameter from .08 to .16 of an inch. The smallest number of eggs in a belt is seventy — the greatest number about one hundred and seventy. While resembling, in their arrangement, those of the Lackey Moth of Europe (*Clisiocampa neustria* Linn.), unlike these, they are not deposited in a continuous spiral, but in somewhat irregular rings. Those of the first ring (the first in order of deposit) which is invariably placed the lowest on the twig, are attached to the bark by their sides; in the next ring they are somewhat inclined; those of the succeeding rings, including the last, are placed on end, perpendicularly, with their transverse diameter at right angles with the twig. A quincuncial order of arrangement is readily traceable, *i. e.*, any one of the interior eggs is central to four others, of which two are in the ring above and two below; to these it is cemented at points midway to the apices of the lesser and greater conjugates. When the regularity of deposit has been interrupted, the quincuncial order is not maintained: but, through the crowding together of the members of a ring, the assumed central egg may be united to one or two of the lateral ones, thus giving it five or six points of union with those surrounding it. From the above-described arrangement, it follows that an extension over all the rings, of the lines joining the eggs of one ring to those of another, will give us a succession of spiral bands crossing the stem at an angle approaching 45°, and making about one circuit around it. The belt is entirely free from any covering of down or other protecting material.

Egg Cement. — The eggs are firmly united to one another and to the twig by a glutinous matter,* in which they are enveloped as they pass from the oviduct. When, by the drying of the twig and its consequent shrinkage, the belt is loosened and capable of being withdrawn, this matter is seen as rings of a black substance surrounding the points of attachment of the egg to the twig; but around the points of union of the eggs to one another it assumes a reddish color.

For this difference of color, no ready explanation presents itself. Were its black hue owing to a thickening of the glutinous matter by its gravitative flow down the egg while in a semi-fluid state, then upon the lower side of a horizontal or inclined twig, it would be found either surrounding the points of union of the egg to the adjoining ones, or collected at the apex. If it were a special secretion, deposited upon the base of the egg on its expulsion from the oviduct, different from that enveloping it, then it should not occur on the sides of the horizontally-placed eggs of the first ring, as required by their exceptional position. That it is not a chemical change resulting from contact with the bark, appears from the fact that it assumes the same character in the oviposition of several of the Bombycidæ occurring in confinement within paper, wooden and metallic boxes.

---

* In many of the Lepidoptera this matter imparts to the eggs their color. Among a large brood of *Telea Polyphemus* (Linn.) reared by me, one individual, after having been pinned and poisoned as were all the others, was found to have strewn her eggs loosely about her; they were without the slightest degree of adhesiveness, and all were of an abnormal uniform white color.

Experiments instituted upon the solubility of this cement, show it o be unaffected by immersion for twenty-four hours in cold water, lcohol, ether or chloroform. Subjected to boiling water in a state f rapid ebullition for the space of one hour, it became softened suf ciently to allow the eggs which were still cohering to be separated by he point of a knife. Upon the subsequent drying of the belt, it was ound to be as firmly bound together as at first.

Oviposition.—The eggs are deposited in the fall, probably in the nonth of October, soon after the emergence from the pupa, as in the eighboring sub-family of Attacinæ, in which the female comes from the ocoon with her abdomen distended with fully matured eggs, finds her nate the same evening if there be one in the vicinity, and deposits er eggs very soon thereafter,* or in some instances a portion even efore mating.† While engaged in ovipositing, the head of the moth s directed toward the tip of the twig, and she probably performs uccessive circuits about the stem, corresponding in number to that f the egg-rings.

Hybernation.—How it is possible for eggs, wholly unprotected as re these, to endure the rigors of winter, is still a mystery. Their ilm-like shell, no thicker than a sheet of delicate note paper, seems vholly insufficient to preserve unharmed the principle of life which it s destined to protect against a temperature of zero, and even several legrees lower. It might justly be regarded as a wise provision in nature that so very few of the Lepidoptera are subject to the severe ordeal of hybernating in the egg; and in the few instances where this occurs, the coverings which we find thrown over the eggs by the nstinct and ofttimes cunning skill of the parent moth, appear so admirably adapted to afford them the required protection, that we night be excusable if, from partial observation, we educed a law admiting of no exceptions. Thus, among the European moths the following instances occur: The Gipsy Moth (*Ocneria dispar* Linn.) has the extremity of the body of the female thickly clothed with downy hairs, which she employs for bedding, for singly enveloping, and for exteriorly coating, in beautiful regularity, her deposit of eggs intended to survive the winter. The *Cnethocampa processionea* (Linn.) deposits her eggs n July on the trunks of trees, first coating the bark with a gummy natter extruded from her abdomen, which she covers with hairs plucked from her terminal tuft, and upon these places, in regular order, her eggs, completing the operation by spreading over them additional hairs of a color so resembling that of the bark as to serve the additional pur-

* A *Telea Polyphemus* which emerged with me on the 18th June, coupled during the night, oviposited the following day, and died on the 22d June.

† Trouvelot, in the *American Naturalist*, vol. 1, page 36.

pose of eluding observation. *Ocneria salacis* (Linn.) employs for a protection a thick covering of a wool-like substance. *Dicranura verbasci* (Godt.) and *D. furcula* (Linn.) which place their eggs singly or in two's and three's, also cover them with hairs from their body. *Lozotænia rosana* (Stephens) whose eggs occur in oval clusters on trees or neighboring objects, guards them with a yellowish cement.

Of our moths, we have, unfortunately, but a very limited knowledge of their early history. Of two species, their hybernating clusters are familiar to us, viz., those of *Orgyia leucostigma* Harris, a moth which has abounded in Albany during the last and several preceding years to such an extent as to prove a serious nuisance, even defoliating some of our shade trees; and those of *Clisiocampa decipiens* Walker, the imago of the common apple-tree caterpillar, allied to the European *C. neustria*. Both of these seem admirably protected from cold and other exposure; the former — an oblong mass — by a bedding upon the cocoon from which the female emerged, and a thick covering of a tough, white, frothy secretion, as may be seen in numerous specimens on the trunks of the trees along our streets; the latter by a thick, compact, gummy substance entirely coating the belt and binding it to the twig of wild cherry or apple. We may venture to ascribe the same method of oviposition and protection to *Clisiocampa sylvatica* Harris, the eggs of which have never been observed by me, nor am I able to find any record of their observation.*

The eggs of the Catocalas also hybernate, as observed in Europe by

* Since the above was written the egg-belt of this moth has been described and figured in the *American Entomologist*, vol. ii, p. 261. Mr. Riley states that "with each egg is secreted a brown varnish which firmly fastens it to the twig and to its neighbor, and which, upon becoming dry, forms a carinated net-work of brown over the pale egg-shell." In this thin covering of varnish over the eggs of this species, we may note a very near approach to the naked eggs of *C. neustria*, to which species our American moth is so closely allied in appearance, that were it not for marked larval difference, we might believe the two identical.

Mr. Riley has also communicated to me the interesting fact, that the egg-belts of *C. decipiens* occurring in Missouri, are covered with a coating of gum much thinner than are those which occur in New York, specimens of which were shown him. In this State, the eggs are entirely hidden beneath their smooth, thick coating, while in those of Missouri they are distinctly visible. A figure of one of these belts, from Missouri, is given in the *American Entomologist*, vol. i, p. 208, where the moth is referred to under its familiar synonymical name of *C. Americana* Harris. The eggs are represented as so entirely destitute of any coating, that it may be questioned whether they were not figured from the shells after the larvæ had emerged and consumed most of the gummy matter, as they are accustomed to do, before commencing to feed upon the leaves.

This variation in the covering of these egg-belts presents the interesting subject of inquiry, whether it may not be the result of climatic causes, and an adaptation to the degree of protection required. If it be so, we shall find the covering diminishing in thickness as we follow the moth southward in its range from Maine to Georgia

Dr. Speyer, and in this country by Mr. Riley. They have been found by Mr. Riley, in clusters, beneath the bark of the trees upon which the larvæ feed, appressed closely together and partly overlapping, with no protection beyond that afforded by the shelter of the bark.

But that these coverings are not indispensable to a safe hybernation is shown by the entirely naked and exposed egg-belts of *H. Maia*, and of the Lackey Moth (*Clisiocampa neustria*) of Europe. The eggs of the former moth, from which emerged the larvæ which form the subject of this paper, endured a winter in which the thermometer in their immediate vicinity fell to 8° below zero of Fahrenheit.*

Some experiments which have been made to ascertain the effects of cold upon the eggs of insects have disclosed the fact that they possess a remarkable power of retaining their vitality under very low temperatures. The eggs of a Vanessa and of several other Lepidoptera, were exposed by Spallanzani for five hours to a freezing mixture indicating —22°, without the least injury resulting to them, for they all subsequently disclosed their larvæ.

It is probable that the eggs are in reality not frozen, but that their contents continue fluid under the greatest cold to which they may be subjected in their natural exposure. Some eggs of *O. leucostigma* before referred to, which I had divested of their covering and left exposed during a portion of the winter, were examined by me under a temperature of 18° below zero, and were found to be in their natural fluid condition.

For this extraordinary capability of resisting cold, no satisfactory reason has been assigned. Among other conjectures, it has been suggested that a spirituous or an oleaginous element might possibly enter into the composition of the eggs.

Hatching. — The larvæ emerged on the 27th of May, sixteen days after they had been brought within doors, and deposited in a glass-covered box. Their escape from the shell was, without a single exception, from the apex, through a regular elliptical opening eaten by the larvæ, corresponding in outline to a depression previously existing. A very few of the eggs did not develop.†

Egg-shell. — The shells are translucent, of a brownish-yellow color superiorly, and whitish on their basal half. No portion of the shells, beyond that above mentioned, was eaten by the larvæ.

---

* The eggs of our November Moth (*Oporabia dilutata* Albin) — pronounced by Fitch identical with the European species — are doubtless deposited in the month of November, nakedly, upon the twigs or buds, as are those of its European allies.

† In the belts subsequently collected, there were found quite a number of undeveloped ova. In one belt, but one of the eggs composing the first ring had produced its inclosed larvæ.

Young Larvæ.—When first escaped from the shell, they are of a pale reddish-brown color, which gradually changes in the course of a few hours to fuscous. Their length is .13 of an inch. Under a high magnifying power, the larva appears as follows: The head is shining black and bears a few whitish hairs of the length of about one-half its diameter; its dimensions taken from the case after its molt are, length .032 of an inch, breadth .03 of an inch. The body is glossy black with minute granulations dorsally and laterally, reddish-brown and smooth ventrally. Its armature consists of eight rows of spines on the anterior segments (varying from five to eight rows on the other segments as stated hereafter) which are glossy black, subcylindrical, muricate, in length nearly equaling the diameter of the larva, giving out at the apex four diverging spinules and a curved ciliated bristle of nearly double the length of the spine. The spines of the three anterior segments in the six superior rows are forked at about their apical fourth, with each fork bearing its spinules and bristle as above. The anal plates of the larva have on them a few black hairs.

There being no food at hand, the larvæ collected in a body in an angle of the box in which they were confined. The following day the colony was several times observed in motion, probably in quest of food, in regular procession of two or three abreast. On the 30th, after their refusal of a half dozen species of leaves, oak-leaves were given them, and upon these they immediately commenced to feed vigorously. Beginning at the margin of a leaf, they ate inwardly, consuming in their progress even the larger veinlets, but not the midrib except at its tip. On the 2d of June, they had collected in three clusters—two on separate leaves and the third on a twig—and had assumed their position for molting on a slight web spun beneath them, in which the hooklets of their anal legs were firmly implanted.

First molt.—On the morning of the 4th of June, they were found to have undergone this change during the night. Their cast skins (exuviæ) were adhering to the surface to which the larvæ had been attached, and had not been fed upon by them; the detached head-cases were strewn over the bottom of the box. They now measure three-tenths of an inch in length. The head is shining black, with a few black hairs. The body is blackish, and when highly magnified, granulated. The spines are glossy black, with three or more branches, in the truncated tips of which are inserted a bristle of about one-third the length of the spine. The spines of the two anterior segments are longer than the others; on segments three to seven inclusive in the two superior rows, the main spine (which is trifid on segment three, bifid on segment four and simple on segments five to seven) is acutely terminated, while its principal branches have the termination and armature of those elsewhere.

At this age, the larva was observed to carry a thread with it in all its movements over the leaf.

On the morning of the 9th, the larvæ had taken position for another molting, and were of the length of .45 of an inch.

Second molting.—Two and a quarter days from the time above noted, or at noon of the 11th, the first of the colony molted for the second time, and the entire number before the close of the day.* A half hour after the change, the head of the larva was of a rufous color. The body, fuscous with obscure red stigmatal spots, and sprinkled with rufous granulations, of which there is a larger one laterally on each segment below the subdorsal row of spines. Spines, rufous, with fulvous branches; those of the two superior rows tipped with a bristle, often black and numerously branched; the lateral spines less branching, and tipped with longer fulvous bristles. The legs and prolegs, rufous; the former with the tarsus black. Three hours after the molt the larvæ had changed to a uniform reddish-brown color.

During this stage of development, the larvæ disclosed characters in accordance with descriptions and representations of *H. Maia*, which enabled me to refer them to that species.

Third molting.—June 20th.† Length of larva, one inch. Color, fuscous. Spines of the two superior rows on segments 3 – 10 and the mesial one on segment 11, red with their terminal third black, simple, fasciculate, thickly radiating from a black, slightly elevated tubercle; the other spines are glossy black, sparsely branched, the branches cylindrical, whitish near their tips, and having inserted in them a delicate, acute black bristle. Stigmata linear, tawny colored.

Sting of larva.—Some experiments made with the larva subsequent to the molting above recorded, in examination of its stinging powers, which were first noticed at this stage of its growth, indicate its possession of this means of defense in a degree considerably exceeding that of the closely allied species, *Hyperchiria Io* (Fabr.)

A larva was dropped three or four times from a height of about ten inches upon the back of the first joint of the thumb. The sensation did not differ materially in kind from the sting of the nettle, but was more acute. In a few minutes the surface became reddened, and in a short time numerous slightly elevated whitish blotches made their appearance, accompanied with a burning and itching. The following

---

* Of their exuviæ, a small number had the head case attached; of those which were separated, nearly half had the collar, with its spines, united to the head-case.

† About one-half of the cast skins were separated from their head-cases, and nearly all of the cases had attached to them the collar, bearing its four spines.

day the thumb could not be bent without experiencing a sensible degree of pain, which was materially increased by an attempt to bring the joints to a right angle. This stiffness of the joint continued for four days. When the blotches subsided, small purplish spots of coagulated blood appeared in their place, which by degrees became more circumscribed, until after the lapse of a few days, when they presented an appearance similar to that of grains of gunpowder burned beneath the skin. These gradually disappeared; those nearer the surface by a scaling of the skin above them; those deeper, removed by the slower process of absorption, were visible at least two weeks.

When the larva was permitted to fall upon the thicker skin of the palm of the hand, a slight stinging sensation was experienced, and minute purple dots were developed, continuing a shorter time than the above.

The sting is doubtless the result, not of broken tips of the spines remaining in the flesh — for none such could be observed by careful scrutiny with a lens — but of a poison secreted by the larva, and probably injected through a minute aperture in the tip of the spine. Whether its excretion is voluntary or involuntary was not determined, it not having occurred to institute the simple experiment by which that point could readily have been ascertained. A slight motion of the larva, apparently a contractile one, was frequently observed to accompany the sting; but this may have been either defensive, or simply the consequence of alarm at being rudely touched.

Some tips of the spines clipped off and placed between slides under a high magnifying power, showed, under varying pressure, a motion of a fluid within them; but no apical opening could be discovered for its escape.

The ability to inflict a sting does not belong to all the spines of the larva, but only to those of the two subdorsal rows on segments three to ten, and the dorsal spine on segment eleven. These differ from those elsewhere on the body in their fascicular arrangement, their shorter length, the regular taper of the branches, and their tawny color, as appears in detail in the description given of the mature larva. With this interesting structural peculiarity in mind, the larva may be handled with impunity, as was repeatedly done with the fifty or more individuals composing the colony from which these notes were drawn, in the frequent transfers which they required as they approached maturity, to fresh food and cleansed quarters. With proper care, the thumb and fingers could safely be passed along their sides and beneath them, slowly raising them from the leaf or stem to which they were attached; but if attempted too hastily, the larva throws itself in a circle, projects its defensive armor, and inflicts a sting which effectually releases it from the grasp.

Fourth molting — June 28 – 29. Length, one inch and one-tenth; ameter, two-tenths of an inch; including bristles, four-tenths of an ch. Upon first emerging from the old skin, the head and all the ines are pale red. After a few hours, the head becomes reddish-own, and the spines of their normal color; the body is black with merous small, whitish, oval papillæ, from each of which a short black ir proceeds. The stigmata are acutely ellipsoidal and white.

The head-cases of the larvæ with the spined collar adhering by one its corners, with few exceptions remained attached to the cast skin, this and in the following molting.

On transferring the larvæ to fresh leaves on the afternoon of the 9th of ly, five were found in position for molting, attached by their prolegs the twigs, with their heads downward. On the morning of the th, nearly all had taken their position. Length, 2.5 inches; diameter, 3 of an inch.

Fifth molting — July 11 – 12. Length, one inch and one-half; ameter, one-fourth of an inch. The superior spines on segment 2, ve short, fascicular, tawny, black-tipped spinules, like those of the two rsal rows, encircling their base; on segments 3 – 11, the lateral row spines (second from above) have similar fascicular spinules bor-ring the upper half of their base, as have also the two ante-superior ines of segment 12 at their anterior basal half. This interesting partation of the characteristic features of the seventeen dorsal fascicles each one of the surrounding spines, to a greater or less extent, encircl-g some and bisecting others, is a feature developed at this molting.

At no stage of its growth, is the exuvia eaten by the larva, as is done *Hyperchiria Io* at its first and second moltings, but probably not those later.

On the 17th of July several of the larvæ descended from the branches which they had been feeding, and gave indications of having attained aturity.

Mature larva. — Length from one and three-fourths to two and one-lf inches; diameter, three-tenths, or, inclusive of lateral spines, six-nths of an inch. The head is round, depressed at the clypeus, with few short, brown hairs. The collar is glossy and is bordered in front ith four of the eight spines, directed somewhat anteriorly, with which e first segment is armed. The body tapers moderately at the tremities, is fuscous, sometimes black, sprinkled with numerous hitish ovoid spots bearing each a short fuscous hair, which are con-uent along the stigmata. The two superior rows of spines on seg-ents three to ten inclusive are fascicular, spreading over the anterior alf of the segment, thirty to forty in each fascicle, cylindrical, tawny-lored, tapering near their apex to an acute black tip; the spines of the lateral row, the substigmatal and the ventral (the latter interrupted on the proleg-bearing and penultimate segments) are twice the length of the superior spines, of a glossy black color, with a tapering trunk, which gives off laterally and apically about twelve cylindrical branches of nearly equal length with the trunk; of these branches the lateral ones are white, translucent and mucronate, having the terminal spinule, black, slender, acute and of about one-half the length of the branch from which it proceeds. The number of spines borne by the several segments is from five to nine, as appears in the following formula: $\frac{1,2,3,4,5,}{8}$ $\frac{6,7,8,9,}{6}$ $\frac{10,}{8}$ $\frac{11,}{5}$ $\frac{12.}{7+2.}$ The stigmata are of the color of the abdominal spots, and acutely ellipsoidal in outline. The caudal plates and shield are deep red, with pitted surfaces and short hairs. The legs are glossy black, with black hairs. The prolegs are red, of the shade of the head, with black hairs exteriorly, granulated interiorly, a smooth glossy spot externally, and with fuscous terminal hooklets. Beneath, a median line of round red spots, of which there is one on the middle of each segment from five to eleven.

Food-plants.— The larva probably feeds on most, if not all, of our oaks. My colony, during its progress to maturity, partook of five species and was readily changed from one to another. Mr. Walsh states* that "the eggs of the moth are deposited, out west, on the scrub willow and different species of oak;" from which it may be inferred that the former (unknown to us under its local name) is the plant on which it more frequently occurs in that region. It has also been reported to Mr. Walsh as occurring on the wild cherry and on black walnut.

Parasites.— Although so enveloped in spines as scarcely to leave a space sufficiently exposed for other than a random thrust of an ovipositor, our larva does not enjoy entire immunity from parasitic attack. Of a colony of about thirty individuals found after their second molt, eight of the number proved to have been ichneumonized, and during the months of June and July, two species of parasites were obtained from them. Several days after the larvæ were collected, two of them, which had meanwhile increased very little in size, and had rested frequently from feeding, were found apparently affixed to a stem by their anterior and posterior legs, with the central portion of the body raised up and enfolding in its curve, a parasitic cocoon lying between it and the stem and closely clasped on each side by the prolegs. The cocoons disclosed their imagines on the 26th of June and 3d of July. They were submitted to Mr. E. T. Cresson for determination and were found to be the *Limneria fugitiva* (Say), which Mr. C. states "seems also to

---

* *American Entomologist,* 1868, vol. i, page 186.

be parasitic on *Dryocampa stigma*."* An examination of the remains of the larva showed that only the skin was left; the parasite through its consumption of the entire interior portions had attained such a size that in the contraction of its pupal change, it had broken through the larval skin ventrally, disclosing one-half of the cocoon, while the other half occupied all of the skin except the extremities. The cocoon is regularly oval in form, and measures .35 of an inch in length.

Of the second parasite twenty-five individuals were obtained from six larvæ between the 5th and 12th of July, the number from each larva varying from one to ten. Their small, white, elongate-oval cocoons were spun upon the outside of the larva, and after a few days of pupation (the period was not noted) gave out through their apical lid the imago, which, according to Mr. Cresson, is an undetermined species of Microgaster.

The editors of the *American Entomologist* have had sent to them a Maia larva with its body covered with the egg-like cocoons of some Hymenopterous Ichneumon, the species of which had not been ascertained by them, but which was probably the above Microgaster.

Pupation.—After about a day passed by the larvæ in traveling around the box in which they were inclosed, among and beneath the leaves with which the ground was strewn, they selected their positions for pupation. The larger number prepared for themselves a simple cell, by scooping out the ground from beneath a leaf, to the depth of the diameter of their body; a few buried themselves just beneath the surface, barely covered by a mantle of grains of earth, loosely spun together.

On the 22d of July occurred the first transformation to the pupa, and the last about August 1st. Several of the larvæ died without entering their pupal state, which may have been the result of transferring them too soon to the pupa-box. No change of color was perceived, indicating their having arrived at maturity;† but they were presumed to have attained that stage when they descended from the leaves, and seemed disinclined to remain upon them when replaced, or to partake of food. About forty of the colony passed safely through their moltings, and assumed their pupal form.

Pupa (Plate 8, fig. 2).—Color, black; cephalic and thoracic region uniformly rounded, with the exception of a slight sutural elevation between the first and second segments; antennæ-cases and leg-cases

* Some parasites, bred by Mr. Bassett of Waterbury, Ct., from larvæ which were identified by Mr. Riley as those of *Dryocampa senatoria*, were, upon examination, found to be identical with these. In the larval remains from which they had been procured, a similar cocoon had been formed, and occupying the same position in relation to the larva as those above described.

†Among the Sphingidæ, the readiness for pupation is unerringly indicated by a material change in color to duller and diffused shades.

showing no relief; their surface, as also that of the wing-cases, rough or crape-like; the abdominal portion, conical in the male, and slightly ovoid in the female. The margins of the three sutures pertaining to segments 4 – 7 have a plaited appearance; the next three sutures are regularly striated, as if milled on their anterior margin, as appears more distinctly when the pupa-case has been extended by the escape of the imago; the anterior margin of the eleventh segment is prominently plaited. The terminal spine is triangular, flattened, and ends in a number of short, brown, bristly fibres. Length of the male pupa, .80 to .95 of an inch; of the female, from .95 to 1.10 of an inch; diameter of male, .31 to .38 of an inch; of female, .38 to .42 of an inch.

The pupa may be readily separated from that of *Hyperchiria Io* (Fabr.) by the absence of the short ferruginous hairs which characterize the latter.

Imago.—The moth has been described and figured by various authors from the time of Fabricius to Harris, and its description at the present would, therefore, be quite unnecessary. In the last edition of Harris' *Insects Injurious to Vegetation*, an excellent figure is given of it. Its translucent crape-like wings, the white mesial band traversing the black wings, the conspicuous red anal tuft of the male—combine to render it one of the most beautiful of our moths, and a valued addition to entomological collections.

The first imago from the above pupæ emerged on September 18th, a second on the 24th, a third on the 29th, a fourth on the 30th, and the last for the season on October 4th. Of these six individuals, only one was a female.

Metamorphoses.— For convenience of reference, the periods required for the several changes embraced in the transformation of *H. Maia* is herewith tabulated:

From hatching to first molt	8	days.
" first molt to second molt	7	"
" second molt to third molt	9	"
" third molt to fourth molt	8	"
" fourth molt to fifth molt	13	"
" fifth molt to maturity	6	"
" maturity to pupa	5	"
" pupa to imago	58	"
Duration of larval state	56	"
Duration of pupa state	58	"
From the egg to the imago	114	"

Discrepancies.—The published accounts of this moth are so uncertain and conflicting as to show a very imperfect knowledge of its larval state, and of the method and duration of its pupation. The figures given by Abbot of the larvæ are quite inaccurate; and he also ttes that they enter the ground for their transformation. Harris ys that the moth has been reported to him as occurring in July and e beginning of August. The description of the larva given in orris' Synopsis is also inapplicable to it. In a notice of a commu-cation read by Mr. Wood before the Entomological Society of iladelphia,* it is asserted that the larvæ "went into the ground the st of August, 1859, and became perfect insects on October 10th, 60." We believe this statement to be an error, which may have curred either in inserting in the report the supposed years not men-oned in the communication, or the transformation of the August 859) larvæ into imagines in October of the same year may have caped the observation of Mr. Wood "near the sea-shore;" and he ould, therefore, naturally refer the moths observed by him in Octo-r, 1860, to the larvæ of the preceding year. It is very improbable at a transformation, requiring less than two months of a cool, within-oor temperature during the latter part of summer, would, under tural conditions, be extended over an entire summer and prolonged fourteen months. (See note appended.)

Mr. Walsh writes of it:† "The larvæ are at first entirely black. hen full grown they have a yellow band, variegated with short black ies on each side of the body; the head and collar are chestnut-brown. uring the month of August they descend into the ground, where ey change to chestnut-brown chrysalids." Our larvæ were without e band and short black lines, and the pupæ were chestnut-brown ly in the brief interval between the casting of the larva-skin and eir assumption of their normal color.

Rarity.— The moth is quite rare in the State of New York. It d never been taken by me during fifteen years of collecting; and have heard of its capture but once in this State, an individual hav-g been caught a few years since in the vicinity of Albany. The imber of clusters of its eggs found at Center, without search having en made for them, would indicate a greater frequency of its occur-nce.

Its rarity may find an explanation in the social habit of the larvæ iring the first half of their existence. Unfortunately possessing a lor in marked contrast with the leaves on which they feed, even a litary individual would be but illy fitted to escape the searching eye bird or parasite that preys upon it; but assembled in a compact ass, and feeding without the slightest attempt at concealment, it is mply impossible for it to elude detection. Its formidable array of ines undoubtedly induces many of the insectiverous birds to pass it by

* *Proc. Ent. Soc. Phil.*, 1867, vol. 1, p. 46.

† *American Entomologist*, 1868, vol. 1, p. 186.

unmolested. A more courageous bird, venturing an experimental taste, may find in the stinging bristles, as it passes down its throat, no inducement to repeat the experiment. But as our cross-bills (*Curvirostra Americana* and *C. leucoptera*) are furnished with a mandibular structure peculiarly adapted to opening the cones of hemlock and pines and extracting their seeds, so there are probably birds specially fitted by formation of beak or method of feeding to find in our repellant *Maia* a harmless and attractive morsel. To such an one, the discovery of a colony of the larvæ would be equivalent to the destruction of each individual member. From a belt of eggs now in my possession, a brood of larvæ, numbering over one hundred, had emerged at Center, and were feeding socially on a small bush of *Q. prinoides*. Desirous of instituting a comparison between their development under the conditions there existing, and my colony being reared in confinement, I observed them on several occasions until after their second molting. Returning after an interval of three days, not an individual remained on the bush, nor was I able by a rigid search to discover a single one on any of the several oaks surrounding it. They had not scattered, as they probably do when further advanced; but the entire colony had without doubt been destroyed.*

Habits of the Imago.— The moths reared by me manifested a great degree of restlessness upon their emergence from pupa, and an apparent disinclination to accept the provision made for the suspended position assumed by them during the expansion of their wings, viz., a thin, coarse-threaded muslin covering of the pupa-box, which had been found well adapted to the wants of large numbers of newly-emerged Lepidoptera. A small branch of oak placed in the box

* From an observation subsequently made, we have reason to believe that very efficient destroyers of these colonies of larvæ are to be found among the "bugs" of the order of Hemiptera, especially in the family of Pentatomidæ. In the early part of June, a small number of these larvæ were discovered on their usual food-plant, and near them was the egg-belt, of about the ordinary size, from which they had emerged. Thinking that the colony might, from some cause, have separated, the bush and the adjoining ones were examined in search of the remainder, without finding any trace of them. Returning to the larvæ to secure them for rearing, the explanation of their reduced number was disclosed in the discovery of an Arma in proximity to them, with one of them impaled upon his beak (rostrum of Fabricius). Finding the locality a favorable feeding ground, he had no doubt selected it for his abode, taking one from the company as often as his appetite demanded, until their original number of one hundred and ten, as indicated by the egg-shells of the belt, had been reduced to twenty-two (they were at this time between their second and third moltings). In a few days the last one of the brood would doubtless have been appropriated by the intrusive guest. To the kindness of Mr. Uhler, I am indebted for its determination as the *Arma modesta* of Dallas — congeneric with a valued ally, as shown by Riley, in our contests with *Doryphora* 10-*lineata* Say, for the preservation of our potato-vines, viz., *A spinosa* Dallas, and also nearly allied to another friend, *Podisus placidus* Uhler, which preys upon the current-worm.

roved no more acceptable; nor were they satisfied with any of the everal objects or surfaces offered them. Their singular activity at his stage suggested the suspicion that the expenditure of a liberal mount of physical energy must necessarily precede their perfect evelopment.

The same disposition was again manifested very soon after the full xpansion of their wings. A large proportion of the Bombycidæ, and ll of the subfamily of Ceratocampinæ with which *Maia* is grouped, o far as we are acquainted with their habits, are characterized by a rolonged state of inactivity, or lethargic condition, following their naginal development, from which they are not readily aroused in the rocess of pinning, or even by two or three unsuccessful efforts proprly to insert the pin. But so marked is the contrast presented by *Maia* that, unless its apparition be carefully watched, and the individual inned within an hour after attaining its full development, a perfect pecimen can rarely be obtained. It resumes its movements over the ides of the breeding cage, and the continual fluttering of its wings, lthough comparatively moderate, suffices very soon to divest it of nany of its slightly attached wing-scales; the attempt to pin it while n this excited condition is rarely accomplished without seriously marring its beauty.

The flight of the moth has been observed by Mr. William Calverley, f Utica, N. Y. He informs me that numbers were seen by him from he 18th to the 25th of October, in oak openings, at Kankakee, Indiana, where they were commonly known as the "deer moth." Their flight was short, and usually terminated by their suddenly dropping in the rass, where they could very seldom be found, although the spot of heir fall was carefully noted, and immediate search instituted. Several were taken by him, by striking them down during flight with his hat.

Geographical range.—Our knowledge upon this point is not very ull. It is known to extend from Maine through each of the sea-board States to Georgia. In its western distribution, we have it reported rom Indiana, Illinois,* Missouri † and Iowa.‡

Synonymy.—In the following table the principal authors only, who ave described or written of this species, are cited, omitting several ninor references occurring in our scientific journals and other publications:

Phalæna (Bombyx) Maia Drury. Illus. Nat. Hist., 1773, II, p. 42, pl. 24, f. 3, ♂.
Bombyx Proserpina Fabr. Syst. Ent., 1775, p. 561, n. 17.
Phalæna Maja Cramer. Pap. Exot., 1776, II, pl. 98, p. 3, f. A, ♂.
Bombyx Proserpina Gmel. Linn. Syst. Nat., 1788–93, n. 2407.
Bombyx Proserpina Fabr. Spec. Ins., 1781, II, p. 173, n. 31.

* The *American Entomologist*, vol. i, p. 186. † Ib. p. 247. ‡ Ib. p. 252.

Bombyx Proserpina Fabr. Mant. Ins., 1787, II, p. 110, n. 35.
Bombyx Proserpina Fabr. Ent. Syst. em., 1792, III, p. 419, n. 40.
Bombyx Proserpina Pal. Bauv. Ins. Af. Amer. Lep., 1786–97, pl. 24, figs. 2, 3.
Bombyx Proserpina Olivier: in Enc. Method. Hist. Nat. Ins., 1789, V, p. 37, n. 48.
Phalæna Proserpina Smith. Sm.-Abb. Lep. Ins. Geor., 1797, II, p. 99, pl. 50, ♂, ♀.
Saturnia Maia Hübner. Verz. Schm., 1816, p. 157.
Saturnia Proserpina F. Harr. Cat. An. and Pl. Mass., 1835, p. 72.
Saturnia Maia Westw.-Drur. Illus. Exot. Ent., 1837, II, p. 45, pl. 24, f. 3, ♂.
Saturnia Maia Drur. Harr. Rep. Ins. Mass., 1841, p. 285.
Saturnia Maia Duncan. Nat. Lib., 1845, XXXII, p. 154, pl. 16, f. 1.
Saturnia Maia Drur. Harr. Treat. Ins. New Eng., 1852, p. 305.
Saturnia Maia Emm. Agric. N. Y., 1854, V, p. 231, pl. 39, figs. 2, 3, ♂, ♀.
Hemileuca Maia Walker. Cat. Lep. Br. Mus., 1855, v. VI.
Saturnia Maia Drur. Morr. Synop. Lep. N. A., 1862, p. 221.
Saturnia Maia Drur. Harr. Ins. Inj. Veg., 1862, p. 396, f. 193, ♂.
Euchronia Maia Packard: in Proc. Ent. Soc. Ph., 1864, III, p. 383.
Hemileuca Maia Gr. and Rob: in Ann. Lyc. Nat. Hist. N. Y., VIII, p. 376.

Note.—Since the above was written, I observe that Mr. P. S. Sprague reports * that from a brood of *H. Maia* which he reared from the larvæ, a portion emerged in October, and one deformed specimen in the following May, and that according to Miss C. Guild, "of the same brood of larvæ all going into the chrysalis at the same time, part came out in October and others not until the following October, some lying in chrysalis one year longer than others." This statement of Miss Guild conforms to that of Mr. Wood, previously cited, but it seems so remarkable that a moth should have three distinct periods of emergence, viz., September-October, May-June and October of the following year, that we are not prepared to receive it as an accepted portion of the history of our insect, without additional confirmatory evidence. It is not very rare among the Sphingidæ for a pupa to pass over one spring and emerge the year thereafter, but in these instances, there is not, to our knowledge, a regular previous late-summer or early-fall brood, as with *H. Maia.*

From my observations it appears that a limited number of the moths emerge after a pupation of about two months. Much the larger portion of the pupæ of my colony survived the winter. On June 4th, one produced the moth; thence to July 4th, five additional ones emerged. An examination of the pupæ a month or two thereafter, showed the remaining ones to be dead. While the number of moths produced as above in the summer was the same as the number in the fall (six in each instance), yet the fact of most of the pupæ continuing alive during the winter, would indicate the summer as the regular period of appearance. This, however, seems to be at variance with Mr. Sprague's observations.

* *Canadian Entomologist*, 1859, vol. i, p. 41.

I embrace the present opportunity to record the additional fact, that, at Center on Sept. 19, 1870, I saw four specimens of *H. Maia* in flight at midday; and, at a distance not permitting of positive identification, what I believed to be three others. Its occurrence was in a portion of the Center locality which I visited on this occasion only, and am therefore without knowledge of its appearance at this place at other seasons or of its abundance. The flight was rapid, in a direct line, and disclosed the same phenomenon of sudden disappearance as noticed by Mr. Calverly, so that I was unable to effect a capture. Its midday flight is undoubtedly voluntary, for in most of the above instances the moth was first observed so remote from me as to exclude the idea of its being disturbed at my approach.

Of the larvæ referred to in the foot note of page 150, a small number were successfully carried through their several transformations to their perfect state. On the 30th of September, a moth presenting a remarkable and beautiful variety was disclosed, which will be found faithfully represented on Plate 8, fig. 1.

The anterior wings above are wholly black, with the exception of some orange scales on the discal cross-vein and submedian nervules. Beneath is a faint whitish band, one-half line in diameter, consisting of rather long and delicate hairs implanted among the black scales; if the surface of the wing be brought in plane with the eye and observed from its outer margin, the white hairs then appear as a distinct white band. In normal specimens of the species, this band on the lower surface is usually about double the width of that of the upper.

On the posterior wings, the band is quite narrow and only extends to the subcostal nervure, instead of reaching the costal margin as ordinarily. The black of the basal region unites with the discal spot, and the white band is not elbowed behind the cell. Beneath, the band corresponds in outline with that of the upper surface, and at a point on the costa, which it would reach if extended, are a few white hairs. Expanse of wings, 2.6 inches.

II. OBSERVATIONS ON MELITÆA PHAETON (FABR).

Two males and one female of the above species were taken at Center, July 6th, which, from their worn appearance, had evidently been abroad for several days. Owing to frequent rains and unusually cold weather, the locality had not been visited during the preceding four weeks, except on the 15th and 22d of June — the latter a very unfavorable day for the flight of diurnals; the first apparition of the species, therefore, for the present year (1869) could not be noticed. In a record kept by Mr. Otto Meske, of Albany — an enthusiastic student of entomology, who for several preceding years had diligently collected the Lepidoptera of this vicinity — the first capture of this butterfly is noted on June 30, 1868, and June 19, 1867.

Observing the abdomen of the above female to be much distended, apparently with eggs, she was pinned (otherwise uninjured) in a box. Upon opening the box on the 9th of July, a cluster of eggs was found deposited therein, numbering about one hundred.

The eggs were of a pale orange-color, smooth, moderately pyriform, with a slight apical concavity. On the 12th they were observed to have changed to a brownish color. On the 13th, they were of a reddish-brown, and had developed some coarse ribs. By the 17th, they had passed into a purple shade, and were flattened apically. On the 28th, they had assumed a grey shade, and were marked with a black spot at the apex, indicating the position of the head of the inclosed larva.

The larva emerged from the eggs July 29th. Their length was six hundredths of an inch. The head was round and of a glossy black; the body of a dull, pale green, bearing some short, whitish hairs. Their motions were very sprightly. Showing a disposition to leave the stem of *Chelone glabra* on which they had been placed and commenced to feed, they were inclosed in a small tin box, with a few of the tender, terminal leaves.

On the 6th of August occurred their first molting. The larvæ now measured one-tenth of an inch in length. The head was shining black, and bilobed, and the collar was also black; the body of a pale brown shade, with rows of short black spines, and with scattered whitish hairs. Of the large number of eggs deposited, but twenty larvæ remained at this date, and the dead bodies of several others were lying in the box.

The larvæ fed on the lower portion of the leaf (leaving the upper cuticle), within a web which they had spun beneath it and extending thence a short distance to the side of the box, within which they could be seen at rest or diligently employed in adding to the web.

The second molting commenced on the 12th of August, and was completed on the 14th. The length of the larvæ was now one-fourth of an inch. Head black, subcordate, with black hairs; the body yellow-brown, darker at the extremities, with seven rows of black spines having conical trunks and numerous fine branches.

When about in readiness for their third molting, they had attained a length of four-tenths of an inch, and a diameter of seven-hundredths of an inch centrally, with attenuated extremities. The anal and the two anterior segments are now black, and the third partially so; the rest of the body is of a clear brown, with the incisures and two narrow bands on each segment, black. The spines of the dorsal row are small; in the next row (subdorsal) they are large and placed on an extended black base; in the next (lateral) they are slightly larger than the dorsal, and like them, are located on the anterior of the segment; those of the next (substigmatal) row are of the size of the subdorsal and correspondingly located on the posterior of the segment. Some of the spines of the posterior extremity of the body are branchless.

August 20th, the larvæ were transferred to stems of *Chelone* within a glass jar, when they at once commenced spinning a web between some leaves, but not drawing them together. On the 22d, some of their exuviæ were observed in the web, indicating a molting (the third) of some of their number. Although fresh leaves were several times given them, they partook of no food after their removal to the jar, nor, judging from the condition of the leaves, had they eaten for a few days prior to their removal — perhaps not after they had attained the period just preceding their third molt. Toward the last of the month, some of the larvæ died, and by the middle of September none remained alive.

From the above observations, and from similar ones on *M. Nycteis*, we may conclude that these two Melitæas, at least, hybernate without attaining any considerable size, and probably after their third molting. It would seem, that, in preparing for their hybernation, they collect within a slight web spun between a couple of leaves, in which shelter they remain, and fall with the leaves to the ground. An additional covering of leaves above them would secure the amount of moisture necessary for their preservation, and serve to shield them from the winds and frosts of autumn, until the snows of winter throw over them a protecting mantle. About the 20th of May (inferring from larval observations on *M. Nycteis*) the larva awakens from its protracted repose of from eight to nine months duration, and seeks its food. As its proper food-plant can hardly have made its appearance thus early in the season, some other plant probably serves its purpose.*

Very many of the larvæ inevitably perish during the long period of their lethargic condition and consequent abstinence from food, in which interval they are necessarily subjected to trying vicissitudes of temperature, to various hygrometrical conditions, and to destruction from many foes. When to these we add the increased danger to which they are exposed during the earlier stages of their existence, it will not be a matter of surprise, that, of the eggs deposited, so small a proportion — estimated at from one to four per cent, as the seasons may prove more or less favorable — attain the perfect state.

In localities where the butterfly occurs in any considerable numbers, the young larvæ could doubtless be easily found upon its food-plant. On some stems of *Chelone glabra* gathered on the 28th of July, at Center, from the ditch beside the New York Central Railroad, and placed beneath an inverted jar to retain them in condition for food, eight days thereafter six of the larvæ, in readiness for their first molting, were discovered feeding on the lower surface of a leaf where it rested against the glass.

This species presents a notable illustration of the localization of certain insects, several other examples of which, in connection with the Center locality, although less marked, will be referred to in future notes. While this prolific collecting field, as limited by almost fruitless explorations of adjacent territory, embraces a tract of about three-fourths of a mile square, the Phaeton habitat proper has a radius of but one-eighth of a mile, with an occasional elliptical extension to one-fourth of a mile in diameter. Its central point is the extension of a swamp over a seldom traveled road, where a few inches of water is found throughout the summer. Upon the wet sands adjacent, the butterfly can usually be met with during its season, often in little companies, sipping the moisture from the ground, after the habit of *C. Philodice*, and, when alarmed, flying for a short distance and then alighting in the surrounding bushes. Beyond these limits the species

---

* It has been reared by Mr. Scudder on the black currant, and has been taken by him on the barberry (*Berberis Canadensis*). He also records it as having been seen by Dr. A. S. Packard, Jr., feeding on *Aster, Viburnum dentalium* [dentatum?] and *Corylus Americana.*

It is also reported in the *Canadian Entomologist*, vol. ii, p. 36, as having been found in West Virginia by Mr. J. L. Mead, within close webs, attached to Iron weed (*Veronica*) [Vernonia?] and a species of *Solidago* — in one instance a web being attached to the two plants.

The English Melitæas feed on various species of germander (*Teucrium*), speedwell (*Veronica*), cow-wheat (*Melampyrum*), plaintain (*Plantago*), and other low-growing plants; most of them are confined to one food-plant, but others are equally common on three or four different species. — Newman's *British Butterflies.*

has not been observed, during five years of frequent visits to this locality by Mr. Peck and Mr. Meske. One specimen has been taken by the former in Sandlake, N. Y. It has also been captured near Utica, N. Y. I have not met with it at Schoharie, N. Y., nor in the neighborhood of Albany except at Center, although its favorite food-plant is of common occurrence in this vicinity.

While the genus Melitæa, as recently restricted by Edwards,* is only represented east of the Mississippi river by the single species *Phaeton* (our smaller Melitæas being placed by him in the genus Phyciodes of Hübner), it is interesting to note, as illustrative of faunal distribution, its full representation in our western States and territories. Of the seventeen other species recorded as congeneric, one is from Texas, one is credited to Colorado only, one to Oregon only, one to Alaska, and thirteen occur in California (of which number three are also found in Colorado, in Nevada and in Oregon).

---

* *Synopsis of North American Butterflies*, pp. 15, 16.

III. NOTES ON MELITÆA NYCTEIS (DOUBL).

On June 15th two larvæ of this species were collected in different localities at Center feeding on *Helianthus divaricatus* L. They were of the length respectively of .75 and .80 of an inch. The head was cordate, broadest at the summit, of a glossy black shade, with numerous blackish hairs of unequal length. The body was reddish-brown, with a few gray dots; laterally with a broad testaceous stigmatal stripe, embracing the stigmata and the substigmatal row of spines. The spines were rufous, black-tipped, thickly verticillated with black bristles; on the first segment are three dorsal granulations, a substigmatal spine, and some clusters of short hairs projecting over the head; on segments 4 to 11, a dorsal row of spines, the latter segment bearing two dorsal spines; on segments 2 to 11, a subdorsal, a lateral and a substigmatal row of similar spines; on segment 12, two subdorsal spines and an anal, ovate, blackish tubercle; above the legs and prolegs, a row of short spines, with sparse tawny hairs. Stigmata oval, blackish. Legs black; prolegs rufous.

On the morning of the 18th one of the larvæ was found resting on the upper surface of a leaf in position for molting. The following morning (19th) it was observed to have molted during the night, and had resumed its feeding. Instead of the broad stigmatal stripe, there was now a narrow substigmatal one, embracing the substigmatal row of spines; above, in range with the stigmata, some scattered rufous dots. Body superiorly fuscous; spines reddish-brown. Prolegs rufescent; legs black.

The mature larva measures .95 of an inch. Its color is fuscous, with white dots on the annulets. The bases of the spines are glossy black. The substigmatal stripe is rufous; and the stigmatal rufous spots, with the adjacent papillæ, form an interrupted stripe. The stigmata are black and broadly oval. On the 21st of June, the more advanced of the two larvæ (the other was sent to a correspondent that a drawing might be made of it) was found changed to a chrysalis, attached by its anal spine to the margin of the under side of a leaf.

The chrysalis (Plate 8, fig. 14) was .44 of an inch in length. Color fuscous, with white markings on the wing-cases, especially at their bases, and at the bases of the spines posteriorly, covering most of the 4th and 5th segments; on the thoracic projection which is rounded and moderately elevated, are five subtriangular white spots (of which two mark protuberances), radiating from its center; stigmatal region rufescent. The head-case is slightly excavated in front. The spines are short, not acute, three each on the 6th, 9th, 10th and 11th segments, and five on the 7th and 8th; terminal segment recurved, so that the body of the chrysalis forms a right angle with the short spine by which it is suspended.

The imago from the above emerged July 1st, after a pupal period of ten days.

On the 20th of August some young larvæ (a colony) were found at Center clustered on a leaf of *H. divaricatus*, of which they had eaten the upper portion. From their appearance, they doubtless pertained to Melitæa, and may have been of the above species. When alarmed by a sudden motion of the plant, they immediately loosened their hold to the leaf and suffered themselves to fall to the ground, with their bodies bent in a circle. After lying motionless for a few minutes, if not again disturbed, they would arouse themselves and travel rapidly away to some place of concealment.

When about to transfer them to fresh leaves on the 23d, they were found to have molted, and to have left their exuviæ with attached head-cases, within a slight web which they had spun on a leaf. From a comparison of the size of the head-cases with those of *M. Phaeton* which were being reared at the same time, this was their second molting. They subsequently fed moderately on the surface of the leaves, which becoming dried were removed and pinned to a fresh stem of the *H. annuus*. Without leaving their position, and after several days' cessation from feeding, they underwent their third molting on the 29th.

Although fresh leaves of *H. divaricatus* were provided for them, they could not be induced to resume feeding, but one after another died and fell from the plant. It is probable that at this stage of their growth the larvæ habitually cease from feeding, and assume the lethargic condition in which they pass the winter — which some of this colony would doubtless have done, had they been favored with the conditions to which they are ordinarily subjected.

The butterfly has occurred abundantly at Center during the two years of my collecting there, appearing from the middle of June until about the 20th of July, its period of greatest abundance being about the 1st of July. At Bethlehem (an excellent collecting locality three miles south of Albany) it has not been seen. It is quite rare at Schoharie, where only two individuals have been taken by me.

IV. NOTES ON PIERIS OLERACEA (HARRIS).

Fifty eggs of this butterfly were collected at Schoharie, N. Y., on the 22d August from a small patch of turnips of perhaps twenty square feet. A small bed of cabbage plants adjoining, yielded none of the eggs, indicating a marked preference by the insect for the former plant.

With few exceptions, the eggs were placed on the under side of a leaf, so near the edge as to render it probable that the butterfly in ovipositing alights on the margin of the upper surface, and bends her body over its edge to place her egg on the less exposed under surface. Usually but one occurs on a leaf, but occasionally two or three are found so near together, as to indicate their having been deposited at the same time.

The eggs are ovoid in form, corresponding with the familiar representation of those of *Pieris brassica* of Europe; they are of a yellow-green color, and measure .047 of an inch in length (average of three), with a diameter of about one-third their length. They are fluted longitudinally, presenting fourteen ribs in two specimens examined and sixteen in a third, which unite in about half the number near the apex. The ribs are sharp-edged, while the intervening flutings show about forty transverse lines.

Ten eggs disclosed their larvæ during the night of August 24 – 25. The larvæ were pale green, cylindrical, with some short, whitish hairs, and measured .075 of an inch long.

The emergence of a larva from the shell was observed, and for a half hour it was seen to be vigorously plying its black-tipped mandibles on the interior surface near the apex, before it effected an opening. The opening made with so much labor was rapidly enlarged by the larva eating a sufficient portion of the shell to permit its egress. Immediately upon having wholly withdrawn itself, it resumed its feeding upon the shell, nearly all of which it consumed. It increased rapidly in size; on the morning of the 26th (twenty-four hours after its disclosure) it measured .13 of an inch in length.

On the 27th P. M. four larvæ had molted for the first time, and four more on the 28th A. M. (two were missing.)

On the morning of the 30th the larvæ were found to have molted for the second time; probably some of the number underwent their change the preceding afternoon when they were not observed. The length of the smallest was .23 of an inch. They now showed numerous dark colored dots, which were more contiguous on the dorsal region, where they define a vascular line of pale green.

August 31st, A. M., the first of the six larvæ now remaining, molted for the third time; its length, after its change was .31 of an inch. A second one molted in the afternoon of the same day, and the four others, by noon of the day following, September 1st.

On September 3d, A. M., the first of the larvæ molted for the fourth and last time, when it measured .45 of an inch; three were in position for molting, which occurred on the following day.

On the 7th two of the larvæ suspended themselves for their pupal change, and the other three (one having died) on the morning of the 8th. Later in the day they had all changed to pupæ. On the 15th the first imago emerged, a second on the 18th, and of the others no note was made.

The transformations of the butterflies of this brood are completed in remarkably short periods of time. The intervals between the hatching of the egg and the first molting, and that between the two following moltings, are each but two days, and a period of three days carries it to its last molt. Thence to its pupal change requires but five days, and the brief space of seven days suffices for converting the pupa into the perfect insect. The entire conversion of the egg into the imago is effected in the space of three weeks. I have elsewhere shown (*Proc. Ent. Soc. Ph.*, vol. iii, p. 52), that there are at least three annual broods of *P. oleracea*, viz., the last of April, the early part of July, and last of August. In favorable seasons, a fourth apparition may be added to the above, in the latter part of September. From the observations recorded below, it is possible that this last appearance is but a portion of the spring brood brought forth by an unusually warm autumn.

On the 19th of September I obtained, from the same small bed of turnips from which the former collection of eggs was made, fifty-three *P. oleracea* larvæ, most of which were nearly of full size. By the 25th, all had transformed to pupæ, and on the 27th one made its appearance as an imago. During the following two weeks several others emerged. Some weeks having passed without further developments, the box containing the pupæ was removed to a cold room for the winter. In early March it was returned to a warm apartment, and on the 6th of April an imago was disclosed, and others continued to appear during the remainder of the month. It was observed that all those given out during the first week were of the male sex.

V. DESCRIPTION OF NEW SPECIES OF NISONIADES.

Nisoniades Icelus nov. sp.* Plate 7, figs. 5, 6, ♂.

Head and palpi dark brown, the latter lighter beneath, and inter spersed with gray or gray-tipped hairs. Antennæ brown, annulated with white obscurely above, with the club orange-tipped. Thorax dark brown, with scattered scales of lighter brown. Abdomen dark brown, with some gray scales, especially at the posterior margin of the segments.

Anterior wings above dark brown, basally mottled with umber, and sprinkled with yellow-brown and bluish-gray scales. A continuous dark brown discal band (interrupted or much constricted below the cell in *N. Martialis*, Plate 7, fig. 7) crosses the cell from the end of the costal fold in the ♂ to the submedian nervure, with fuscous borders usually obscure, and having on its superior half some bluish hairs; in *N. Brizo* (Plate 7, fig. 9) the borders are well-defined, black, and the bluish hairs are continued over the entire length of the band. The submarginal band, consisting of bluish hairs, is regularly curved, parallel to the hind margin, or sometimes, as in the figure, slightly receding from it as it approaches the internal margin; its borders are well-defined in fuscous, the anterior one but moderately sinuate on its superior half, the posterior one with six sagittate spots superiorly (the second and third apical ones more elongate than in *N. Brizo*), thence reaching the submedian nervure in three curves similar to the corresponding ones of the anterior border; upon the band, between the subcostal nervules, an indistinct elongated whitish spot (not in *N. Brizo*). Intermediate to the two bands, resting on the costa and extending to the second median nervule, a patch of bluish scales, interspersed with umber-colored ones; thence to the inner margin, the space is umber brown, similar to the shade of the posterior wings. Along the hinder margin is a series of umber spots, usually crescentic in the females, surrounded by bluish scales; behind these, a narrow dark brown marginal line. Fringe, of the color of the preceding spots, with short basal bluish hairs.

Posterior wings above, umber-brown, with two marginal rows of brownish-yellow spots, usually eight in each, and two contiguous smaller discal ones (not existing in *N. Brizo*), separated by the cellular fold; the first costal spot of each row is nearly as distinct as the others (in *N. Brizo*, obsolete).

---

* A description of the male genital armature of this and the following species, has been published by Messrs. Scudder and Burgess in their paper on "Asymmetry in the Appendages of Hexapod Insects" (*Proc. Bost. Soc. N. H.*, 1870, vol. xiii, pp. 287, 288).

Beneath (Plate 7, fig. 6), on the superiors, touching the discal cross-vein and separated by the disco-central nervule, are two elongate, sometimes indistinct, yellow-brown spots. Between the subcostal nervules, resting on elongate dark-brown spots, of which the second and third are usually forked posteriorly, are three grayish quadrangular spots; behind the cell are two smaller ones (sometimes obsolete) surrounded by brown; between the median nervules are two quadrangular gray spots of larger size than the preceding, and shaded behind with brown: the above seven spots which form a less regular curve than in *N. Brizo* ♀ beneath (Plate 7, fig. 10), with the spots less conspicuous and not so uniform in size, correspond in position to the submarginal band of the upper surface. Posterior to these is a row of grayish spots running from an apical patch of pale blue scales; there is also a row of elongate whitish spots resting anteriorly on a black terminal line.

On the inferiors are two discal, yellow-brown, rounded spots, and the two rows of similar spots of the hind margin, the anterior one of which is preceded by a row of elongate dark-brown spots. Cilia of the color of the ground of the wings, with a few of the shorter scales grayish. Described from 11 ♂, 6 ♀.

N. Icelus is readily distinguished from *N. Brizo*, to which it is closely related, by its uniformly smaller size, its expanse varying from 1.20 to 1.40 of an inch, while the smallest *Brizo* in my collection measures 1.50 of an inch. A marked characteristic feature is the costal patch of bluish scales between the bands.

The egg is of a pale green color. In shape it is a semi-ellipsoid; its base is flat and its apex depressed between the tips of the ribs which terminate exterior to the depression. It is distinctly fluted even to the naked eye, and with an one inch lens, the ribs may be seen of the number usually of eleven, but not uniformly, for of nine specimens examined, one was observed with ten ribs and one with twelve. Connecting the ribs are from thirty to thirty-five transverse striæ. The diameter of the egg is .031 of an inch, and its height .028 of an inch. The larva has not been observed by me.

The imago was captured for the first time the present year (1869), on the 25th of May. The second week of June — from the 9th to the 15th — it was found abundantly at Center, resting with outspread wings on damp sand in the road. A female Nisoniades, taken on the 7th of July, differing in the much greater width of the submarginal band at the inner margin, I have referred to the same species. I have taken it at Schoharie, N. Y., on the 14th of June.

Icelus (Ic′-e-lus), in mythology, was a son of Somnus, the god of sleep, associated with his brothers, Morpheus and Phantasos, in the government of the palace of sleep, their principal duty being to inspire dreams in mortals. To Morpheus was committed dreams relating to men; to Phantasos those concerning inanimate objects; while Icelus was charged with such as relate to animate objects, through a personation of bird, insect, or other form. Brizo was a divinity of the island of Delos, where she was worshipped as the goddess of sleep.

The above described butterfly having heretofore been confounded with *N. Brizo*, the mythological name selected for it is deemed so appropriate as to warrant a departure from the established custom of naming the species of this genus after celebrated Roman poets.

Nisoniades Lucilius nov. sp. Plate 7, fig. 1, ♀; 2, ♂.

Thorax, abdomen and palpi dark brown with a red lustrous reflection, the latter tipped beneath with gray. Antennæ reddish brown, with white annulations which are obsolete superiorly.

Anterior wings of a lighter shade of brown than the body, and giving the same reflection. Discal band interrupted, fuscous, obscurely defined except at its hind margin where it crosses the cell; its course, as in the other species of the genus; in the ♀, and occasionally in the ♂, resting upon the outer one of the two cellular teeth formed by this band, is a white hyaline spot, sometimes obsolete. The submarginal band consists of interspaceal sagittate fuscous spots, which are somewhat squarely truncated anteriorly, and have umber-colored scales centrally; its course is direct from the submedian nervure to the subcostal nervules, whence it is broadly reflected anteriorly to the costal margin, embracing in this portion four interspaceal minute white hyaline spots, of which the first, third and fourth are nearly in line, the second and largest lying behind (in one specimen but three spots are seen); between the median nervules there are two hyaline spots, of which the inner one is sometimes obsolete in the ♂, or wholly absent. Between the bands the ground is umber-brown, with a few bluish-gray scales toward the submarginal band, and a larger number between the subcostal nervules. The sagittate spots of the submarginal are bordered behind with gray (not in *N. Persius*), followed by a series of rounded umber spots, having a few gray scales resting on obscure yellowish spots (these spots not in *N. Persius*) between them and the brown marginal line. The cilia are umber-colored with a very few basal gray scales.

Posterior wings, of a more uniform brown than the anterior, and more shaded with red. The two marginal rows of spots are usually obscure, and of a yellow-brown; the discal spots, which in *Persius* are ordinarily visible as a transverse line across the extremity of the cell, are barely seen. The cilia are grayish-brown.

Beneath, reddish-brown; the anterior wings conspicuously so at the apex; the posteriors are darker and lustrous; the terminal margins are but a shade lighter than the rest of the wings. On the anteriors, in some specimens, is a small white cellular spot. The four subapical hyaline spots are constant. The median spots are larger than the subapical, and are subquadrangular in form; rarely, the inner spot is obsolete. The secondaries are without discal spots. Of the two rows of the hind margin, the outer one is the most distinct; in some of the darker-colored males these spots are scarcely visible. The cilia are of the color of the wings, with their base of a paler brown.

The coloring and markings of the sexes are very nearly alike. Expanse of wings from 1.10 to 1.25 of an inch. Length of body of ♂, .44; of ♀, .50 of an inch.

The female of this species has so strong a resemblance to *N. Persius* female, of Scudder, that the two are not readily separated by those not familiar with them. It is of a smaller size (the figure is from an individual of extreme size), and in addition to the differences above referred to, the following may be noted:

It is without the bluish-gray hairs which sprinkle the upper surface of the primaries of *N. Persius* (and the bands of *Brizo* and *Icelus*), and is also destitute of the bluish-gray scales of the hind margin above, and apex beneath, of that species. The submarginal band, in its course toward the inner margin, recedes less from the hind margin than in *Persius;* hence it follows, that, while the fourth apical hyaline spot and the first median one are equidistant from the hind margin, or the latter is the nearer in this species, in *Persius* the latter is always relatively the furthest removed. The sagittate spots of the submarginal band in this are shorter and less acute.

The males of the two species are not liable to be confounded. In *Persius* the anterior wings are of almost an uniform fuliginous hue, and consequently much less conspicuously marked than those of *Lucilius.* The hyaline spots are smaller and less constant. Very rarely are there two of these spots present between the median nervules; often the apical ones only appear, and occasionally these are obsolete. The male of *Persius* is figured for comparison in fig. 3 of Plate 7, and in fig. 4 the upper and lower wing surfaces of the ♀ are given. Fig. 3 is from a very distinctly marked individual, but is represented with too light a shade upon the terminal half of the anterior wings.

N. Lucilius would appear to be the common Nisoniades at Schoharie, as among my collections made several years since at that place, a number of specimens of the species occur, with two or three each of *Juvenalis* and *Icelus*, but none of *Persius.* At Center, where the other species of the genus are abundant, it has not been seen. A single individual of it was taken by me last season, in Bethlehem, on the 30th of July. The Schoharie specimens, as they appear in my collection, differ from this, in being less shaded with red; but it is probable that their colors have become somewhat impaired, through a partial exposure to the light during my earlier collections. The colors, as above given, are from the perfectly fresh Bethlehem specimen.*

Nisoniades Ausonius nov. sp. Plate 7, figs. 11, 12, ♂.

Head, palpi, thorax and abdomen reddish-brown; the latter with a few grayish scales at the margins of the segments, and with yellow-brown hairs bordering the genital organs, less conspicuously so than in *N. Martialis;* antennæ red at tip, annulated with a clearer white than in the other species, having the joints beneath almost entirely white.

Anterior wings above, pale umber-brown with grayish scales sprinkled over most of their surface (more diffused than in the other species) except on the fuscous bands, showing especially behind the submarginal band. There are two brown basilar spots resting on the subcostal and median nervures, not so dark as those of the disc. The discal band usually continuous in this genus, here consists of three elongate fuscous dashes (appearing to the unaided eye as a single spot) resting on the subcostal near the discal cross-vein, extending nearly half-way to the median, the intervening space having merely an indication of the spot which appears distinctly in most of the species as the inner cellular tooth of the discal band; following this is an obscure fuscous spot at the fork of the first and second median nervules, and beyond, the usual hour-glass shaped spot extending from the second median nervule to the submedian with its constriction on the interspaceal fold. The discal cross-vein is quite curved and is conspicuously marked in brown. The submarginal band of fuscous spots is doubly curved, being convex toward the hind margin, from the costa to the third median nervule, thence concave to its termination at the submedian. It consists of four acutely ellipsoidal fuscous spots between the subcostal nervules, which are wholly destitute of the usual hyaline spots, followed by three others of similar form but of greater breadth, the next subacute posteriorly, and the last, similar in outline to the corresponding one of the discal band. There is a marginal row of interspaceal brown spots, the first four of which are surrounded with gray scales and lie near the margin, and the remaining four more remote from it than in *N. Martialis;* also, an obscure row of brown spots resting on the tips

* The delay in the publication of this report has permitted a revision of the description of this species, from the inspection of thirty specimens subsequently collected, and of a few reared from the larvæ. Notes upon the earlier stages of the insect (egg, larva and chrysalis) have been made, and will be given in a future paper.

of the nervules and extending on the cilia. The cilia are of the color of the ground of the wings, with a few of the basilar scales gray.

Posterior wings above, of a darker ground than the anterior, sprinkled with blackish scales, darker basally, and with pale yellow-brown spots, of which the discal spot (conspicuous in *N. Persius* ♀), is obsolete; the spots of the submarginal row are crescentic in form; those of the marginal row are obsolete; between these two series, and nearly inclosed by them, is a range of oval fuscous spots, and anterior to the submarginal row is a similar range of sub-connected spots. Cilia light brown, with dark brown basilar scales.

Beneath (Plate 7, fig. 12), reddish-brown with the terminal margin gray. The anterior wings have the fuscous spots of the submarginal band and marginal row as on the upper surface; of the discal band, the cellular spots are alone obscurely visible; the basal ones are lost in the color of the ground; the marginal interspaceal brown spots below the subcostal nervules rest centrally on elliptical gray patches, while those of the posterior wings approach a semi-oval form, and are preceded by conspicuous gray crescents which nearly inclose them by uniting with some marginal gray scales; at the tips of these crescents, a submarginal row of fuscous spots is obscurely seen; the discal spots, so distinct on the secondaries of *N. Martialis*, are here obsolete. Cilia of the wings, reddish brown; those of the anteriors are somewhat encroached upon by the gray of the margin.

Expanse of wings 1.06 of an inch. Length of body, .45 of an inch.

This interesting species was taken at Center on the 12th of May, 1871.* It may be known by its small size, being the smallest yet discovered of the genus; by the entire absence of the usual white apical spots pertaining to all the other known species except *N. Brizo;* by the quite curved submarginal band of elongate black dashes; by the peculiar cellular spot and the brown scales covering the discal cross-vein. In its markings it approaches *N. Martialis* more nearly than any other of our species.

Only a single individual was obtained. The time of its appearance another season will be awaited with no little anxiety, in the hope that it will prove another instance of a solitary capture being the precursor of many others the ensuing year. Thus, it had excited much surprise

---

* If an apology is due for embodying in a "report for 1869" a few observations made during the two following years, it may be found in the temptation to embrace the earliest favorable opportunity for publication, in consideration of the unavoidable delays which sometimes occur in the issue of the State Cabinet reports, as in the case of the present one, which, when nearly all in type and within, perhaps, two weeks of its completion, was destroyed by fire, in the burning of Weed, Parsons & Co.'s printing house in April of the present year (1871).

that *N. Brizo* during several years of thorough collecting had not occurred at Center where *N. Icelus* was found so abundantly, the two being associated in about equal numbers in New England, and elsewhere. In 1870, an individual, believed to be *Brizo*, was inclosed in the net, but escaped therefrom before it could be positively identified. The following spring the species was not at all rare, and several pairs were taken *in coitu*. In 1869, a single ♂ *Hesperia Logan* was secured at Center, followed the ensuing season by several of each sex. Previous to the present year (1871), Messrs. Tepper and Graef of Brooklyn had obtained but one specimen of the rare *H. Massasoit*, while this season they report it as not uncommon on Long Island.

VI. DESCRIPTION OF A NEW SPHINX.

Ellema pineum nov. sp. Plate 8, figs. 12, ♂, 13, ♀.

Male. — Head and collar, umber; palpi brown; thorax umber at the sides, and brownish-cinereous on the middle. Abdomen immaculate, brownish-cinereous. Legs brown, with white scales on the femora and at the joints. Anterior wings as long as the body, umber colored, dusted with grayish at the base, along the terminal margin and on the principal nervures and their branches; within the cell is a subquadrangular blackish-brown spot; an umber-brown shade is placed over the base of the nervules, filling the lower half of the post-apical interspace * half way to the hinder margin, entirely filling the disco-central interspace within one-third of the margin, the middle portion of the medio-superior, the base of the central and posterior interspaces; the outer margin of this shade is doubly curved, convex toward the hinder margin, becoming concave from the medio-superior nervule; the inner margin of the wing beneath the submedian nervure, is brownish from the base to its middle; the tips of the nervules are touched with umber-brown. Cilia umber-brown, spotted with white on the interspaces. Posterior wings above and beneath, ochreous-gray, lighter at the base. Expanse of wings 1.75 of an inch; length of body .80 of an inch.

Female. — Head and thorax umber-brown, the latter grayish at the sides and in the middle, with a short white line on the upper edge of the wing-covers. Anterior wings broader than in the male, and longer than the body; color umber-brown, with a darker brown costo-basal spot, another on the internal margin near the base, which is continued in a dark shade along the internal margin; a similar colored spot occupies most of the apical interspace, and there are two within the post-apical; within the cell, a subquadrangular blackish-brown spot; of the umber-brown shade which in the male rests on the base of the nervules, scarcely more than its hinder margin is visible, and that indistinctly; middle of the wing at the base dusted with grayish scales, and the nervules are also more or less dusted with grayish, especially the branches of the subcostal vein. Posterior wings above,

---

* For the sake of better comparison, the names by which the veins and interspaces are designated in this description are those used by Dr. Clemens in his *Synopsis of North American Sphingidæ.*

darker upon the apex and upon the hind margin than in the male; and beneath, without the obscure band which crosses the middle of the nervules in *E. Harrisii.* Cilia white, spotted with dark umber on the ends of the nervules. Expanse of wings, 2.10 inches; length of body .90 of an inch.

The species is readily distinguished from *E. Harrisii* (Plate 8, figs. 10, ♂, 11, ♀), by the darker ground of its wings, the absence of the gray shades, and its much less distinct markings.

Larva. — Length two inches. Color, grass-green. Head subtriangular, green, bordered with bright yellow, within which at the apex is a ∧ of black. Body subcylindrical, tapering at the extremities, and without a caudal horn. Dorsally, a reddish-brown line interrupted on the hinder portion of each segment by a square of green traversed by diagonal lines; a subdorsal yellow line borders the above; lateral stripe yellow; substigmatal stripe white, interrupted at the sutures by light green; ventral stripe and prolegs, rose-red. Feeds on the white pine, and matures about the middle of September, when it enters the ground and forms a cell for pupation.

Several of the larvæ were taken by me at Schoharie, N. Y., in the years 1858 and 1859, but, unfortunately, I succeeded in rearing but a single individual of each sex. The larva, with the exception of its characteristic feature — the dorsal row of squares — resembles so closely that of *E. Harrisii* (Plate 8, fig. 8), that, meeting with the latter for the first time in 1860, I believed the two identical, and accordingly appended to my notes of the former, "In those taken this year the dorsal squares are not visible." Since then I have not met with it, although the same locality has continued to give me *Harrisii*, and in another favorable locality recently found at Bath, Rensselaer county, N. Y., I was able to procure of the latter species between the 7th and 24th of September of last year (1869), from the trunks and branches of *Pinus strobus*, twenty larvæ, while as many more were taken at the same time by Mr. Meske.

E. Harrisii is, by Mr. Grote (*Trans. Amer. Ent. Soc.* vol. II, p. 115), referred to the genus Sphinx, and subgenus Hyloicus of Hübner, of which *S.* (*Hyloicus*) *pinastri* of Europe is cited as typical. In the absence of reasons advanced for such a reference, it is not easy to surmise why it has been made. The style of ornamentation in *E. Harrisii* and *E. pineum* differs very materially from that of *S. pinastri*, especially in their immaculate abdomens. In the earlier stages of the insects the differences are still more marked, and would seem effectually to remove them from a generic relationship with the species to which Mr. Grote would ally them. The pupæ in Ellema

(Plate 8, fig. 9, pupa of *E. Harrisii*) are without an exserted tongue case, for the short tongue, not exceeding the length of the palpi, does not require such a provision; in *pinastri*, the tongue-case is long and elevated in its middle above the thorax. * The larval states are very dissimilar, the former being without a caudal horn, and the latter provided with a conspicuous one. The larvæ also differ in several particulars in their habits.

Mr. Grote (loc. cit.) expresses his opinion that the *Lapara bombycoides* of Walker will prove to be identical with *E. Harrisii*. It has also been suggested to me that *E. pineum* might be equivalent to Walker's species. Accepting his diagnosis as, at least, approximately correct, I cannot believe that the species described in this paper has been anticipated, or that *L. bombycoides* will be found to be a synonym of *E. Harrisii*. Walker states that his species "has much outward resemblance to the Bombycidæ," but neither of our Ellemas would even remotely suggest such a comparison.

The statement of Mr. Grote, that Walker's description was drawn "from a Canadian specimen in Mr. Saunders' collection," led me to communicate with Mr. Wm. Saunders of London, Ontario, with a view of ascertaining what the specimen really was. He informed me that it had never been in his possession, but that he had ascertained, after considerable effort, that the insect in question had been received by Mr. Saunders of London, England, from a correspondent in Canada, from whom it was procured by Walker for description.

From Mr. Wm. Saunders I have received specimens of the larva and imago of *E. Harrisii*, taken in his vicinity — the larva within five days of its pupation at the date of its transmission, September 6th.

It is probable that *E. pineum* occurs also in Canada, for having (in 1864) transmitted my specimens to Mr. W. H. Edwards for his inspection, he informed me that he had just received, in a box from Mr. Saunders, an individual seemingly identical with those sent by me.

Harris, in his *Insects injurious to Vegetation*, p. 328, speaks of "the curiously checkered caterpillar of *Sphinx coniferarum* on pines." It is possible that this may have been the larva of *E. pineum* which may have fallen under his observation, associated in the Eastern States with *E. Harrisii* (his *S. coniferarum*), yet it is more probable that it is simply a reference to the representation as given by Abbot of the *coniferarum* larva of the Southern States.

---

* Each of the species, of which the pupation is known to me, which Mr. Grote refers to the genus Sphinx proper, viz.: *chersis, drupiferarum, kalmiæ, Gordius, eremitus* (*luscitiosa* and *lugens* unknown), has an exserted tongue-case for the tongue, which latter is nearly or quite as long as the body.

VII. LIST OF SPHINGIDÆ OCCURRING IN THE STATE OF NEW YORK.

MACROGLOSSINÆ.

Sesia Fabricius.

1. **diffinis** Harris. Id. Walk., Clem. *Macroglossa diffinis* Boisd. *Sphinx fuciformis* Smith.
2. **gracilis** Gr. & Rob.; in Proceed. Ent. Soc. Ph., 1865, V, p. 174, pl. 3, figs. 1, 2, ♂ (*Hæmorrhagia gracilis*).
3. **Buffaloensis** Gr. & Rob.; in Ann. Lyc. Nat. Hist. N. Y., 1867, VIII, p. 437, pl. 16, fig. 18 ♂, 19 ♀.
4. **Thysbe** Fabr. Id. Clem., Gr. & Rob. (in List). *Sphinx pelasgus* Cram. *Sesia pelasgus* Harr. *Cephonodes pelas.* Hübn. *Sesia cimbiciformis* Steph. *Hæmorrhagia thysbe* Gr. & Rob.
5. **uniformis** Gr. & Rob. Trans. Am. Ent. Soc., 1868, II, p. 181. *Sesia ruficaudis* Walk.

Thyreus Swainson.

6. **Abbotii** Swainson. Id. Harris, Walker, Clemens.

Amphion Hübner.

7. **Nessus** (Cram.) Hübn. Id. Gr. & Rob. (in Cat.) *Sphinx Nessus* Cram., Fabr. *Thyreus Nessus* Harr., Walk., Clem. *Macroglossa Nessus* Harr. (in Cat.)

Deidamia Clemens.

8. **inscripta** (Harr.) Clem. *Pterogon ? inscriptum* Harr. *Thyreus ? inscriptus* Walk.

CHŒROCAMPINÆ.

Darapsa Walker.

9. **Chœrilus** (Cram.) Walk. Id. Clemens. *Sphinx Chœrilus* Cram. *Chœrocampa Chœrilus* Harr. *Sphinx Azaleæ* Smith. *Otus Chœrilus* Hübn., Grote.
10. **Myron** (Cram.) Walk. Id. Clem., Gr. & Rob. (in List). *Sphinx myron* Cram. *Sp. Pampinatrix* Smith. *Chœrocampa Pamp.* Harr. *Otus myron* Hübn., Grote. *Otus Cnotus* Hübn.
11. **versicolor** (Harr.) Clem. Id. Gr. & Rob. *Chœrocampa versicolor* Harr., Walk. *Otus versicolor* Grote.

Chœrocampa Duponchel.

12. **tersa** (Linn.) Harr. Id. Walk., Clem., Her.-Sch., Grote. *Sphinx t.* Linn., Drury, Fabr., Cram., Smith. *Deilephila t.* Westw.-Drur. *Metopsilis t.* Duncan. *Philampelus t.* Burmeister.

Deilephila Ochsenheimer.

13. **chamænerii** Harris. Id. Grote. *Sphinx epilobii* Harr. (in Cat.) *D. Galii* Clem. ? *D. intermedia* Kirby.

14. **lineata** (Fabr.) Harris. Id. Clem., Grote. *Sphinx lineata* Fabr. *Sphinx daucus* Cram. *D. daucus* Steph., Wood, Walk., Her.-Sch.

Philampelus Harris.

15. **Pandorus** (Hübn.) Walk. Id. Gr. & Rob. *Sphinx Satellitia* ? Fabr., ? Drury. *Philampelus Satellitia* Harr., Clem., Fitch, Grote. *Daphne Pandorus* Hübn.

16. **achemon** (Drury) Harris. Id. Clem., Fitch. *Sphinx achemon* Drury. *Sp. Crantor* Cram., Smith. *Pholus Crantor* Hübn.

17. **vitis** (Linn.) Harris. *Sphinx vitis* Drury, Fabr., Cram., Smith. *Dupo jussieuæ* Hübn. *Philampelus fasciatus* Her.-Sch.

SMERINTHINÆ.

Smerinthus Latreille.

18. **geminatus** Say. Id. Walk., Clem., Gr. & Rob. *Smerinthus geminata* Harris. *Sphinx ocellatus Jamaicensis* Drury, Gr. & Rob. *Smerinthus Cerisyi* Kirby.

19. **excæcatus** (Smith) Walk. Id. Clem., Fitch. *Sphinx excæcata* Smith. *Smerinthus excæcata* Harr. *Paonias excæcatus* Hübn.

20. **myops** (Smith) Harr. Id. Walk., Clem., Fitch. *Sphinx myops* Smith. *Paonias myops* Hübn. *Sm. rosacearum* Boisd.

21. **Astylus** (Drury) Westw. Id. Harr., Walk., Clem. *Sphinx Astylus* Drury. *Sphinx Io* Boisd. *Smerinthus Io* Wilson. *Sm. integerrima* Harr. (in Cat.)

22. **modestus** Harr. Id. Walk., Clem. *Smerinthus princeps* Walk.

Cressonia Gr. & Rob.

23. **juglandis** (Smith) Gr. & Rob. *Sphinx Juglandis* Smith. *Amorpha dentata Juglandis* Hübn. *Smerinthus Juglandis* Harr., Walk., Clem., Fitch.

SPHINGINÆ.

Ceratomia Harris.

24. **Amyntor** (Hübn.) Gr. & Rob. *Agrius Amyntor* Hübn. *Ceratomia quadricornis* Harr., Walk., Clem.

Daremma Walker.

25. **undulosa** Walk. Id. Gr. & Rob. (in List). *Sphinx Brontes* Boisd. *Macrosila Brontes* Walk. *Ceratomia repentinus* Clem. *Daremma repentinus* Gr. & Rob.

Diludia Gr. & Rob.

26. **jasminearum** (Boisd.) Gr. & Rob. *Sphinx jasminearum* Boisd., Wilson, Clem.

Macrosila Walker.

27. **Carolina** (Linn.) Clem. Id. Gr. & Rob. *Sphinx C.* Linn., Drur., Fabr., Smith, Steph., Harr., Her.-Sch., Fitch. *Manduca obscura C.* Hübn. *Phlegethontius C.* Hübn.

28. **quinquemaculata** (Steph.) Clem. *Sphinx Carolina* Donovan, Harr. (in Sill. Jour.) *Phlegethontius Celeus* Hübn. *Sphinx quinquemaculata* Steph., Wood, Walk., Harr., Fitch.

29. **cingulata** (Fabr.) Clem. Id. Gr. & Rob. (in List). *Sphinx cingulata* Fabr., Drur., Harr., Walk., Burm., Grote. *Sphinx convolvuli* Drur., Cram., Smith. *Sphinx Druræi* Donov., Steph., Wood. *Agrius cingulata* Drur.

Sphinx Linnæus.

30. **chersis** (Hübn.) Gr. & Rob. *Lethia chersis* Hübn. *Sphinx cinerea* Harr., Walk., Clem.

31. **drupiferarum** Smith. Id. Harr., Walk., Clem., Fitch. *Lethia drupiferarum* Hübn.

32. **kalmiæ** Smith. Id. Harr., Walk., Clem., Fitch. *Lethia kalmiæ* Hübn.

33. **Gordius** Cram. Id. Harr., Walk., Clem., Fitch. *Lethia Gordius* Hübn. *Sphinx poecila* Steph., Wood.

34. **luscitiosa** Clem. Id. Gr. & Rob.

35. **plebeia** Fabr. Id. Steph., Harr., Clem. *Anceryx plebeia* Walk. *Sphinx* (*Hyloicus*) *plebeia* Gr. & Rob.

Agrius Hübner.

36. **eremitus** Hübn. *Sphinx eremitus* Gr. & Rob. *Sphinx sordida* Harr., Walk., Clem. *Sphinx* —— *?* (larva) Lint., Proc. Ent. Soc. Ph. III, p. 652.

Ellema Clemens.

37. **Harrisii** Clem. *Sphinx coniferarum* Harr., Fitch. *Anceryx coniferarum* Walk. *Sphinx* (*Hyloicus*) *Harrisii* Grote.

38. **pineum** Lintner.

Dolba Walker.

39. Hylæus (Drury) Walk. Id. Clem., Gr. & Rob. *Sphinx Hylæus* Drur., Cram., Fabr., Harr. *Sphinx Prini* Smith. *Hyloicus Hylæus* Hübn.

Dilophonota Burmeister.

40. ello (Linn.) Burm. Id. Gr. & Rob. (in List). *Sphinx Ello* Linn., Drur., Fabr., Cram., Harr. *Erynnis Ello* Hübn., Grote. *Anceryx Ello* Walk., Clem.

VIII. LIST OF BUTTERFLIES OCCURRING IN THE STATE OF NEW YORK.

PAPILIONIDÆ.

Papilio Linn.

Turnus Linn.
Glaucus Linn.
Ajax Linn.[1]
Marcellus Cramer.
Asterias Fabr.
Calverleyi Grote.[2]
Troilus Linn.
Philenor Linn.

PIERIDÆ.

Pieris Schrank.

oleracea (Harris).
rapæ (Linn.).
protodice Boisd.-Lec.

Anthocaris Boisd.

Genutia (Fabr.).[3]

Callidryas Boisd.

Marcellina (Cramer).
Eubule (Linn.).[4]

Colias Fabr.

Philodice Godart.
Keewaydin Edw.

Terias Swainson.

lisa (Boisd.-Lec.).[4]
Nicippe (Cramer).[4]

NYMPHALIDÆ.

Danais Latreille.

Plexippus (Linn.).
misippus (Fabr.).

Euptoieta Doubl.

Claudia (Cramer).
Argynnis columbina Fabr.

Argynnis Fabr.

Aphrodite Fabr.
Cybele Fabr.
Atlantis Edw.
Idalia (Drury).
Bellona Fabr.
Myrina (Cramer).

Melitæa Fabr.

tharos (Drury).
Selenis Kirby.
Marcia Edw.[5]
Batesii (Reakirt).
Harrisii Scudd.
Nycteis Doubl.
Phaeton (Drury).

Grapta Kirby.

Progne (Cramer).
C-argenteum Kirby.
comma (Harris).
Dryas Edw.[6]
Faunus Edw.

---

[1] Mr. Graef of Brooklyn, reports this species as having been taken on Long Island.

[2] This is by many supposed to be merely a suffused variety of *Asterias*. The capture of another individual in Florida — a female — (Mr. Grote's type was a male) reported by Mr. T. L. Mead in the *American Naturalist*, vol. iii, p. 332, is favorable to its specific distinction.

[3] Taken by Mr. Edwards in Newburgh.

[4] Taken by Mr. Tepper on Long Island, who reports *T. lisa* as occasionally occurring there in abundance.

[5] The validity of this species is not fully established. Its author now deems it possible that it may be but a variety of *tharos*. It has not, however, been found to occur in New England associated with *tharos*.

[6] It is thought that this will prove to be a dimorphic form of *comma*.

interrogationis (Fabr.?, Godt.).[1]
J-album (Boisd.-Lec.) Lintn.

Vanessa Fabr.

Milbertii Godart.
Antiopa (Linn.).

Pyrameis Hübn.

huntera (Fabr.).
cardui (Linn.).
Atalanta (Linn.).

Junonia Hübn.

Lavinia (Cramer).[2]
Orythia (Smith).
cœnia (Boisd.-Lec.).

Limenitis Fabr.

misippus (Fabr.).
Disippe (Godart).
Arthemis (Drury).
Proserpina Edw.
Astyanax (Fabr., 1775).
Ursula (Fabr., 1793).

Apatura Fabr.

clyton Boisd.-Lec.[2]
Proserpina Scudd.

SATYRIDÆ.

Neonympha Hübn.

Canthus (Linn.).
Hip. Boisduvalii Harris.
Eurytus (Fabr.).

Satyrus Latr.

Portlandia (Fabr.).
Oreas mar. Andromacha Hübn.
Alope (Fabr.).
Nephele (Kirby).

LIBYTHEIDÆ.

Libythea Fabr.

Bachmannii Kirtland.[3]

ERYCINIDÆ.

Charis Hübn.

borealis Gr.-Rob.

LYCÆNIDÆ.

Thecla Fabr.

niphon (Hübn.).
Irus (Godart).
Arsace Boisd.-Lec.
Henricii Gr.-Rob.
Augustus Kirby.
Calanus (Hübn.) Westw.
Falacer Godt., B.-L., Gr.-Rob.
inorata Gr.-Rob.
Edwardsii Saunders.
Falacer Harris.
Calanus Gr.-Rob.
Acadica Edw.
Læta Edw.
Clothilde Edw. ♀
liparops Boisd.-Lec.
strigosa Harris.
Auburniana Harris.
smilacis Boisd.-Lec.[4]
Mopsus (Hübn.).
Melinus (Hübn.).
Favonius Boisd.-Lec.
Hyperici Boisd.-Lec.
Humuli Harris.

Lycæna Fabr.

Scudderii Edw.
Pembina Edw.[5]
violacea Edw.
neglecta Edw.
pseudargiolus Harris.
pseudargiolus Bois.-Lec., Edw.
argiolus (Smith).
Lucia (Kirby).
comyntas (Godart).

Chrysophanus Hübn.

Hyllus (Cramer).
Thoe (Boisd.-Lec.).
Americana (Harris).
phlæas (Boisd.-Lec.).
epixanthe (Boisd.-Lec.).

---

[1] The two forms of this species, viz., that designated as *Fabricii* by Mr. Edwards (Trans. Am. Ent. Soc., vol. iii, p. 5), and that described by me as *umbrosa* (Trans. Am. Ent. Soc., vol. ii, p. 313), have recently been ascertained by Mr. Edwards to be dimorphic forms of equal value of the same species. They are faithfully figured in "The Butterflies of North America" as *G. interrogationis* var. *Fabricii*, and *G. interrogationis* var. *umbrosa*.

[2] Taken by Mr. Edwards in Newburgh.

[3] Mr. Graef reports this species as having been once captured on Long Island by Mr. Grote.

[4] Edwards, in his Synopsis of N. A. Butterflies, p. 30, cites this as a distinct species.

[5] This species may prove to be identical with *Lygdamus* Doubleday.

Feniseca Grote.

Tarquinius (Fabr.).[1]
Pol. cratægi Boisd.-Lec.
Pol. Porsenna Scudd.

HESPERIDÆ.

Eudamus Swains.

Tityrus (Fabr.).
Lycidas (Smith).[2]
Bathyllus (Smith).
Pylades Scudd.[3]

Nisoniades Hübn.

Persius Scudd.
Lucilius Lintner.
Brizo (Boisd.-Lec., Harr.).
Icelus Lintner.
Martialis Scudd.
Ausonius Lintner.
Juvenalis (Fabr.).[4]
Catullus (Fabr.).

Thymelicus Hübn.

Numitor (Fabr.).
puer Hübn.
Het. marginatus Harris.

Hesperia Fabr.

Centaureæ Ramb.
Wyandot Edw.
Mandan Edw.[5]
? Mesapano Scudd.
vialis Edw.
Metea Scudd.
Samoset Scudd.[5]
alternata Gr.-Rob.
nemoris Edw.
Massasoit Scudd.
Logan Edw.
Delaware Edw.
conspicua Edw.
Zabulon Boisd.-Lec.[6]
Hobomok Harris.
Pocahontas Scudd.
Quadaquina Scudd.
Phylæus Drury.
Sassacus Harris.
Huron Edw.[7]
Leonardus Harris.
Peckius Kirby.
Wamsutta Harris.
Olynthus (Boisd.-Lec.).[7]
maculata Edw.[8]
Hianna Scudd.
Metacomet Harris.
rurea Edw.
verna (Edw.).
Ætna Boisd.
Egeremet Scudd.
Mystic Edw.
bimacula Gr.-Rob.
Acanootus Scudd.
Manataaqua Scudd.
Taumas (Fabr.).
Ahaton Harris.
cernes Boisd.-Lec.

The above list, embracing one hundred and thirteen species, can only be regarded as a preliminary one, although surpassing in number the list of New England butterflies, published by Mr. Scudder, in 1868,

---

[1] Of this rare species, two individuals were taken a few years since by Mr. C. H. Peck, at Bath, opposite to Albany. It has also been observed by him at Sandlake, Rensselaer county, and at Elizabethtown, Essex county; in each instance flying about bushes of Alder (*Alnus serrulata* Ait) indicating it as the food-plant of the larva.

[2] Taken by Mr. Edwards at Newburgh, and by Mr. Tepper on Long Island.

[3] Mr. Scudder has found this to be a distinct species from *Bathyllus* of the Southern States (figured and described by Abbot and Smith), with which it has been hitherto confounded.

[4] *Ennius*, of Scudder and Burgess, regarded by them as the northern representative of *Juvenalis*, has not occurred among my collections.

[5] Included in this list upon the authority of Mr. Edwards.

[6] *Pocahontas* and *Quadaquina* are dimorphic forms of this species.

[7] Included in this list upon the authority of Mr. Scudder.

[8] One specimen of this southern species (Louisiana to Florida) was taken by Mr. Meske at Center, Albany county, in 1866.

in which ninety-four species are catalogued, inclusive of two (perhaps three) which are cited above as synonyms, and two southern forms (*H. Oneko* and *H. Panoquin*) erroneously reported from New England. The Scudder list contains several species which have not yet been detected within the limits of the State of New York, most of which, together with others occurring in neighboring States and in Canada, will undoubtedly be found among us, when more faithful and general explorations shall have been made. *Grapta gracilis* Gr.-Rob.,[1] and *Argynnis Montinus* Scudd., will probably be taken in the Adirondack mountains. We may also expect to have the following species included in our future lists:

Pieris vernalis Edw. New Jersey and Pennsylvania.
Pieris Virginiensis Edw. West Virginia and Ontario.
Colias Eurytheme Boisd. Vermont and Connecticut.
Terias Delia Boisd. Southern New England (Scudder).
Thecla Ontario Edw. Port Stanley, Ontario.
Eudamus Proteus (Linn.). Eastern N. Amer. to Conn. (Scudder).
Nisoniades Horatius Scudd.-Burg. New England.
Nisoniades Virgilius Scudd.-Burg. New England.
Hesperia viator Edw. Illinois and Massachusetts.
Hesperia Wingina Scudd. Southern New England to Florida.
Hesperia Monaco Scudd. Connecticut and Massachusetts.
Hesperia punctella Gr.-Rob. Connecticut to Louisiana (Scudder.)
Hesperia Uncas Edw. Pennsylvania and Ohio.

Information of the capture of any of the above, within the limits of New York, or of any other diurnals not included in the list, would be gratefully received.

[1] Said by Mr. Scudder to be a dimorphic form of *Faunus*, thus giving two forms (a dark-winged form having been found by him in *Progne*) to each of our Graptæ.

IX. CALENDAR OF BUTTERFLIES FOR THE YEAR 1869.

Thecla Irus (*Godart*).—April 27th, one ♂ taken at Center, N. Y., — a remarkable locality for this species, usually so rare elsewhere. On the 11th of May, at the next visit made, both sexes were found abundantly, most of them somewhat worn. Sixty individuals were taken in about three hours' collecting. Previous to 11 o'clock, much the larger proportion of captures consisted of females; subsequent to that hour, the males were the more numerous. On the 25th, they were still abundant. June 7th and 9th, a few much worn were seen; on the 15th it was observed for the last time for the season, it being single-brooded. Its flight is short and rapid, frequently alighting on the hot sands in the roads. The ♂ was often taken while resting on bushes by the roadside. This species has not been observed at Bethlehem, nor at Schoharie.

Pieris oleracea (*Harris*).—May 9th, a couple were observed by me at Schoharie, N. Y., where they were reported as having occurred about two weeks previously. August 30th, they were abundant at the same place, and their eggs were found on turnips. September 13th they were still abundant.

Thecla Melinus (*Hübn.*).—May 11th, two males taken; on the 21st six of ♂ and ♀. On the 15th of July, two of a second brood were obtained, and on the 23d six of ♂ and ♀ in good condition; on the 28th two were taken. August 6th, it occurred for the last time. At Center only.

Thecla niphon (*Hübn.*).—May 11th, four were taken, associated with *Irus;* on the 14th seven were taken at Bethlehem in a grove of varied timber.

Thecla Augustus *Kirby*.—May 11th, two specimens somewhat worn, associated with *Irus*, at Center; none others were found.

Nisoniades Juvenalis (*Fabr.*).—May 11th, first observed; on the 15th, several males collected. June 9th, two males, and on the 11th the first ♀; on the 15th, worn specimens only were abroad. Taken resting on wet sand in the roads.

Nisoniades Persius *Scudder*.—May 11th, the ♂ occurred, and on the 21st the ♀; was taken for the last time on the 27th. Abundant at Center, alighting on the wet sands.

Colias Philodice *Godart.* — Observed for the first May 11th, and thence continuously to October 1st. The white variety of the ♀ was taken August 24th.

Chrysophanus Americana (*Harris*). — May 13th, ♂, and on the 27th ♀. Observed frequently up to September 30th. On the 8th of September fresh specimens were obtained, and on the 14th it was seen in greater abundance than at any time during the year.

Vanessa Antiopa (*Linn.*).— May 13th, several of this butterfly were seen, which, from their worn appearance, had evidently hybernated. June 8th, one emerged from its chrysalis. On the 15th, a colony of the larvæ, after the second molting, showing the dorsal row of red spots, was observed on willow. A number of them were removed and placed on elm, upon which they readily fed. July 2d, two imagines emerged from chrysalis, the larvæ of which had suspended for their pupal change on the 20th ult., and transformed on the 21st. July 20th, the butterfly was abundant. September 8th, it was observed, and for the last time on the 30th.

Grapta comma (*Harris*). — A worn specimen (hybernated) taken at Bethlehem, May 13th. It occurred with me but once subsequently during the year, on August 23d.

Lycæna neglecta *Edwards.* — May 21st, the ♂ abundant at Center, collecting in companies on damp places, and on excremental matter in the roads; three of the ♀ were taken at this date. June 1st, the ♀ was still infrequent; on the 7th of June, the butterfly occurred in flocks, and several of the ♀ were obtained; on the 9th, it was very abundant, and four pairs were captured *in coitu;* on the 22d, the ♀ was of frequent occurrence. By the 7th of July, only two or three battered specimens were seen. On the 30th of July, one ♀ was taken in Bethlehem, where, on the 17th of June, a very few of the ♂ had been obtained. This species appears to have but one brood.

Lycæna comyntas (*Godart*). — May 21st, a few ♂, and one ♀. June 9th, none were found. July 7th to 23d, it again appeared, and again from August 20th to September 8th, unless different species are included under this appellation, which, from the variations in the specimens secured, seems quite probable.

Papilio Turnus *Linn.* — May 21st, reported by Mr. Peck; on the 25th, it occurred on lilac blossoms. June 1st, five males were taken at one cast of the net, on a damp patch of earth by the road side; on the 7th of June several were observed, also on the 15th and 17th. July

12th, two were seen; August 20th, the larva was found on the wild cherry (*Prunus Pennsylvanica.*)

Hesperia Metea *Scudd.*—May 21st, two males, the first Hesperian observed this season; on the 25th two females were obtained, and on June 3d, additional ones of the same sex. Not rare; occurring usually among shrubbery.

Papilio Troilus *Linn.*—May 25th, on lilac blossoms, ♂; June 1st, the ♀ appeared; on the 15th and 17th, several were seen. July 7th, fresh specimens were taken; 15th and 20th, several; on the 30th, only worn specimens were abroad.

Nisoniades Icelus *Lintn.*—May 25th, 1 ♂. June 1st, several of ♂ and 1 ♀; from the 9th to the 15th, it was abundant at Center. On the 7th of July, it was observed for the last time. *N. Brizo*, with which this species has hitherto been confounded, and which occurs in New England associated with *N. Icelus*, has not been taken here.

Hesperia vialis *Edw.*—May 25th, 2 of ♂ and 1 ♀; additional males on the 25th and 27th. On June 9th, some females were collected. July 7th, one torn ♀ was taken.

Lycæna Scudderii *Edw.*—May 27th, 1 ♂; June 1st, males abundant, and three females; on the 6th, six females; on the 9th, still abundant; on the 15th, diminishing; on the 22d, several of each sex were seen. On the 15th of July, a fresh ♂ of the second brood was captured. On the 23d of July, males abounded in flocks, with a very few females among them. August 6th, both sexes were abundant; no other Lycæna abroad at this time. August 20th, but a few of each sex were observed, as also on the 27th. This butterfly sits at rest, with its wings partly open over the back. It was met with only at Center this season, and does not occur at Schoharie.

Lycæna violacea *Edw.*—May 27th, 1 ♀ at Center, and the only individual of the species taken.

Melitæa tharos (*Drury*).—June 1st, three males; 7th, seven males and one female; on the 15th, abundant. July 7th, none were seen at Center, but on the 15th both sexes occurred. On the 30th of July, at Bethlehem, eighteen males and two females, all apparently just from chrysalis, were taken on the damp stones of a little stream of water in a pasture. August 6th, 20th, 24th, 27th, September 1st and 8th, captures recorded.

Eudamus Tityrus (*Fabr.*).—June 1st, both sexes, about lilac blossoms; on the 7th, three were taken, flying over flowers in a wood.

July 12th, they were abundant; on the 20th several were seen, and on the 30th a few, they having continued without intermission for two months. On the 20th of August and 1st of September, the larvæ well advanced toward maturity, were found abundantly at Center on *Lespedeza capitata.* September 14th, twenty larvæ, nearly full grown, were taken in a few minutes' search concealed between leaves of locust (*Robinia pseudacacia*) at Bethlehem.

Eudamus Pylades *Scudd.*—June 1st, three taken of an unusually small size; on the 7th, a few seen, but too wild to capture; on the 9th, some were observed resting on excrement in the road; on the 25th they were quite abundant at Bethlehem. July 7th, diminishing in numbers; on the 12th, few were seen, and they were observed the latest on the 20th.

Pyrameis Atalanta (*Linn.*).—June 1st, both sexes on lilacs at Center; 3d and 9th at Bethlehem. August 23d, the larvæ were found abundantly on nettle (*Urtica gracilis*) at Schoharie, varying from their first molt to full size; thirty individuals were taken.* On the 30th of August others were collected at the same locality, from half-grown to mature size.

Nisoniades Martialis *Scudd.*—June 1st, ♂; on the 7th, seven of the ♂, but no ♀; on the 9th, the ♂ abundant and a few of the ♀; on the 15th, good specimens of each sex, but the ♀ rare. July 7th, a few old ones were seen, and, on the 15th, a worn ♀ was taken. On the 23d of July a fresh ♂ was captured, indicating a second brood. Another ♂, seemingly fresh from chrysalis, was taken August 20th. September 8th, a worn ♂ occurred.

Hesperia Zabulon *Boisd.-Lec.*—June 1st, five males were collected at Center, and several others were seen, which darted quickly from the damp earth, on which they were resting, into the neighboring bushes; on the 3d several males were taken; on the 7th observed, flitting about flowers in a wood with *E. Tityrus*; on the 15th a ♀ of the "*Pocahontas*" type was obtained; observed for the last time on the 17th. An abundant species at Center, Bethlehem and Schoharie.

---

* All the larvæ which had not attained their last molt were found concealed, singly, within a leaf spun together at its edges, of which the tip had, in most instances, been eaten away. The greater number of the nearly mature larvæ were hidden in a shelter made by spinning together several of the leaves at the tip of the plant, after the stalk had been partially eaten through at a suitable height so as to permit it to be readily bent downward among the leaves beneath, where a thicker shelter could be constructed. Notwithstanding these careful provisions for concealment, each one of the larvæ collected at this time proved to have been ichneumonized.

Hesperia Hianna *Scudd.* — June 1st, ♂; on the 9th, both sexes. Occurred only at Center, and rarely.

Hesperia Taumas (*Fabr.*). — June 3d, both sexes taken, and four specimens on the 25th. Bethlehem.

Hesperia Mystic *Edw.* — June 3d, ♂; 17th, not rare at Bethlehem; rare at Center. July 7th, last capture recorded. Occurs also at Schoharie.

Hesperia Peckius *Kirby.* — June 3d; on the 17th both sexes abundant. July 15th and August 27th, captures recorded. September 14th, both sexes occurred; an abundant species.

Neonympha Eurytus (*Fabr.*). — June 3d, several specimens, and on the 17th, common. Center and Bethlehem; usually in shady woods.

Limenitis misippus (*Fabr.*). — June 7th, a pair taken by Mr. Meske *in coitu* at Center; observed feeding on excrementitious matter in the road on the 9th; 15th, two females taken. July 7th, August 20th and 27th, September 1st and 8th, observed.

Melitæa Batesii (*Reakirt*). — June 7th, fifteen males and one female collected; on the 15th the female was not rare; last on the 22d at Center; found at Schoharie on the 14th of June; none at Bethlehem.

Argynnis Myrina (*Cram.*). — June 9th, 1 ♂ at Center; 17th, both sexes at Bethlehem; 25th, several. Fresh individuals observed July 30th. September 8th, fresh specimens again appeared; on the 14th several were seen.

Hesperia Sassacus *Harr.* — June 14th, two males were captured on the Western mountain at Schoharie; not observed elsewhere.

Thymelicus Numitor (*Fabr.*). — June 17th, sixteen males and one female, all perfectly fresh specimens, collected from a swamp at Bethlehem; on the 25th several good females and worn males were taken. July 30th, males of a second brood appeared. August 20th, a third (?) brood occurred at Center; on the 24th, abundant at Castleton; on the 27th, took a ♀ at Center, the last recorded. This species is not rare at Utica; it has not occurred at Schoharie.

Chrysophanus Hyllus (*Cramer*). — June 17th, 1 ♀ at Bethlehem, in a swamp, and on the 25th, two others in the same locality. This species has never occurred among my Schoharie collections.

Grapta interrogationis var. **umbrosa** — June 25th, a fresh individual was seen at Bethlehem.

Argynnis Aphrodite *Fabr.*—June 25th, the ♂ occurred and until July 28th, being most abundant July 15th. The ♀ was observed July 23d and 30th. Center, Bethlehem and Schoharie.

Argynnis Cybele *Fabr.*—June 25th to August 6th, males; females July 20th. Less numerous than the preceding species; occurs at the same localities. *A. Atlantis* was not taken.

Limenitis Astyanax (*Fabr.*).—June 28th, from pupa, after ten days of pupation; July 5th, a second from pupa after the same period of pupation; July 7th, it was observed at Center; on the 12th, ♂ and ♀ were taken in Bethlehem, and, on the 20th, several were seen at the same place; observed also on the 30th of July. The larvæ feeds on apple, and constructs for itself a hybernaculum similar to *misippus.**

Melitæa Nycteis *Doubl.*—July 1st, obtained an imago from a larva which had transformed to a pupa June 21st. July 7th at Center, took twelve males but no female; on the 15th four females were taken and a few fresh males, but most of the latter were worn; on the 23d, ♂ worn, ♀ fresh. This species occurs rarely at Schoharie.

Melitæa Phaeton (*Drury*).—July 7th, two of ♂ and one ♀ considerably worn; on the 15th, two additional females. Less abundant than reported in former seasons.

Argynnis Bellona *Fabr.*—July 12th a few seen; August 24th, observed at Castleton; September 9th, the ♀ at Center.

Satyrus Nephele (*Kirby*).—July 12th and 20th in Bethlehem on the borders of woods.

Thecla Calanus (*Hübn.*).—July 7th, one ♀; on the 12th, eleven males and ten females were collected, and on the 15th, it was abundant at the same place; on the 20th, some worn specimens were found at Bethlehem; 23d, a few at Center; 30th, worn specimens at Bethlehem.

Thecla Edwardsii *Saund.*—July 12th, at Bethlehem, both sexes; 15th, at Center, sixteen males and one female; 20th, at Bethlehem, only worn specimens; on the 23d, at Center, eleven males and two females in good condition, and again on the 28th, when a few good males and several females were secured. The above species of Thecla

---

* On the 19th of June, a Limenitis larva, nearly mature, was found feeding on *Quercus ilicifolia.* It suspended itself for its pupal transformation, but died while in the act of withdrawing itself from its larval skin; the species, therefore, could not be ascertained.

were usually captured when resting on bushes after a short and rapid flight in the warm sunshine.

Thecla liparops *Boisd.-Lec.* — July 12th, the ♂ and ♀ taken at Bethlehem, and on the 23d, one ♀ at Center. Occurs rarely.

Satyrus Alope (*Fabr.*). — July 12th, ♂ at Bethlehem, and on the 20th abundant, when a pair was taken *in coitu.*

Hesperia bimacula *Gr.-Rob.* — July 12th, a single ♀ taken at Bethlehem; no other capture of the species during the season.

Hesperia Metacomet *Harr.* — July 12th to 20th; the ♂ abundant at last date at Bethlehem, by the roadside on flowers of peppermint (*Menthis piperita*); from the 15th to the 30th, the ♀ occurred. Taken also at Schoharie.

Hesperia Ætna *Boisd.* — July 15th to 28th, the ♂ occurred, and from the 20th to the 30th, the ♀ at Bethlehem. Equally abundant with the preceding species, and associated with it.

Hesperia Manataaqua *Scudd.* — July 15th, taken at Center, and, on the 23d, at Bethlehem. A rare species.

Thecla Acadica *Edw.* — July 20th, a ♂ at Bethlehem; on the 23d, at Center; on the 28th, three worn individuals, one a ♀. August 6th, a few worn ones were collected.

Thecla Mopsus (*Hübn.*). — July 23d, three males; on the 28th, four males and two females, taken in company with *Acadica*, *Edwardsii* and *Calanus* on blossoms of Jersey-tea.

Hesperia verna (*Edw.*). — July 25th, a ♀ at Bethlehem, the only capture made of this species.

Hesperia Logan *Edw.* — July 28th, a ♂ taken at Center on damp sand in the road.

Grapta J-album (*Boisd.-Lec.*). — July 28th, one worn individual taken at Center.

Hesperia Leonardus *Harr.* — August 20th, three males of this late Hesperian were captured, and, on the 27th, six males and two females. September 1st, two additional females were obtained; on the 8th, four worn specimens were collected from asters. Only at Center.

Danais Plexippus (*Linn.*).—August 24th, several were seen flying about the dock at Castleton; September 8th, observed.

Pyrameis huntera (*Fabr.*). — Only one capture of this species was made, on the 24th of August at Castleton.

Limenitis Arthemis (*Drury*).—This butterfly, which in some years has appeared abundantly in the vicinity of Albany, was not once observed the present season.

Papilio Asterias *Fabr.*—Of this species, usually so common in most localities, not an individual was seen by me in Albany or Schoharie counties.

X. DATES OF COLLECTION OF NEW YORK HETEROCERA.

I. COLLECTIONS DURING THE YEAR 1869.

SPHINGIDÆ.

Sesia diffinis *Harris*		May 25.
Sesia uniformis *Gr.-Rob.*		May 25.
Sesia gracilis *Gr.-Rob.*		May 25.
Sesia Thysbe *Fabr.*		June 1.
Sesia Buffaloensis *Gr.-Rob.*,	larva, mature, on Viburnum	Sept. 19.
Amphion Nessus (*Cramer*)		May 25.
Darapsa Myron (*Cramer*) [1]	larva, last molt, grape	Sept. 7.
Philampelus Pandorus (*Hübn.*)	" " "	Sept. 9.
Smerinthus excæcatus (*Smith*)[2]	" " wild cherry	Sept. 9.
Ceratomia Amyntor (*Hübner*)	" linden	Sept. 15.
Sphinx chersis (*Hübner*)	" " ash	Sept. 7.
Sphinx drupiferarum *Smith*	" " plum	Aug. 30.
Sphinx drupiferarum *Smith*		June 14.
Sphinx kalmiæ *Smith*,	larva, last molt, ash	Sept. 16.
Ellema Harrisii *Clemens*	" " pine	Sept. 7.

THYRIDÆ.

Thyris maculata *Harris*	June 17.

ZYGÆNIDÆ.

Alypia octomaculata (*Fabr.*)	June 13.
Eudryas grata (*Fabr.*), larva, last molt, grape	Sept. 7.
Acoloithus falsarius (*Clem.*)	July 12.
Acoloithus Americanus (*Boisd.*)	July 12.
Ctenucha virginica (*Charp.*)	June 22.
Ctenucha fulvicollis (*Hübn.*)	June 22.

[1] Of the eight Darapsa larvæ taken from one vine at Bath, from the differences presented, some were thought to be *Chœrilus*. Unfortunately the moths were not obtained from any of the number, for each one had been ichneumonized; only one attained the pupa state.

[2] Four larvæ occurred at this time on the wild cherry (*Prunus Pennsylvanica*), and on the 24th of September two others on a species of Cratægus; the latter bore the marks of parasites inclosed within their bodies.

Bombycidæ.

Hyphantria textor *Harr.*	June	1.
Euchætes egle (*Drury*)	June	25.
Euchætes collaris (*Fitch*)	May	25.
Orgyia leucostigma (*Smith*)[1]	July	23.
Lagoa crispata *Pack.*, larva, last molt, on oak, etc	Sept.	8.
Lagoa crispata *Pack.*	Nov.	4.
Euclea querceti (*Her.-Sch.*)	June	14.
Limacodes scapha *Harr.*	July	7.
Lithacodes fasciola (*Boisd.*)	June	17.
Ichthyura albosigma (*Fitch*)	June	17.
Ichthyura inclusa *Hübn*,[2] larva, third molt, on aspen	Sept.	7.
Apatelodes Angelica *Grote*, larva, last molt, on ash	Sept.	9.
Datana ministra (*Drury*), larva, last molt	Sept.	16.
Edema albifrons (*Smith*), larva, last molt, on oak	Sept.	1.
Cerura borealis (*Boisd.*), larva, last molt, on aspen	June	22.
Telea Polyphemus (*Linn.*), larva, last molt, on maple	Sept.	9.
Actias Luna (*Linn.*)	June	17.
Callosamia Promethea (*Drury*), larva, last molt, ash and lilac,	Sept.	24.
Platysamia Cecropia (*Linn.*), larva, last molt	Sept.	9.
Hemileuca Maia (*Drury*), larva, 2d-3d molt, on oak	June	15.
Hemileuca Maia (*Drury*)	Sept.	18.
Hyperchiria Io (*Fabr.*)[3]	Sept.	23.
Hyperchiria Io (*Fabr.*), larva, last molt	Sept.	16.
Eacles imperialis (*Drury*), larva, last molt, on pine	Sept.	7.
Anisota senatoria (*Smith*), larva, last molt, on oak	Aug.	20.

[1] The larvæ of this species were so abundant in Albany during the summer as to prove a serious annoyance, by nearly defoliating many of the shade trees. At the time when they were seeking suitable places for their transformation, a person could scarcely walk a block without treading several of them under foot.

[2] Other colonies of this larva were found on willows within nests composed of several leaves spun together. The last of February, within a box which had been standing in a moderately warm room, the moths commenced emerging from the slight cocoons which they had constructed between the leaves at the bottom of the breeding cage. The insect seems unusually hardy (unlike *L. crispata*), for the number of imagines disclosed proved nearly equal to that of the larvæ which had been secured. They continued to emerge during the month of March, and all, it was observed, escaped from the cocoons in the day time.

[3] A colony of about twenty of these larvæ, measuring four-tenths of an inch in length, was found July 15th on *Populus tremuloides*. They were reared to pupæ, from which three imagines were disclosed between the 17th and 23d of September, and others the following Spring.

Anisota senatoria (*Smith*)[1]........................... June 15.
Anisota stigma (*Smith*), larva, last molt, on oak........... Sept. 7.
Xyleutes robiniæ (*Peck*)[2]........................... June 17.

Noctuidæ.

Thyatyra cymataphorides *Guen*..... .... May 25.
Acronycta Americana (*Harris*), larva, on oak............ Sept. 1.
Aplecta latex *Guen*.................................. June 18.
Chamyris cerintha (*Treits.*)............................ July 12.
Acontia candefacta (*Hübn.*)............................. May 27.
Xylina Bethunei *Gr.-Rob*............................... Sept. 17.
Eriopus monetifera *Guen*............................... June 7.
Eriopus mollissima *Guen*.............................. June 29.
Abrostola urentis *Guen*............................... Aug. 24.
Plusia festucæ (*Linn.*)................................ Aug. 4.
Plusia balluca *Hübn*.................................. July 14.
Plusia ampla *Walk*................................... July 14.
Deva purpurigera *Walk*................................ July 3.
Gonoptera libatrix (*Linn.*)............................ Aug. 24.
Syneda limbolaris (*Hübn.*)............................ July 12.
Catocala piatrix *Grote*................................ Sept. 9.
Catocala cara *Guen*................................... Sept. 7.
Catocala concumbens *Doubl*............................ Sept. 9.
Catocala amatrix *Hübn*................................ Sept. 12.
Catocala parta *Guen*.................................. Sept. 24.
Ophiusa bistriaris *Hübn*............................... June 17.
Drasteria erechtea (*Cram.*)............................ May 15.
Euclidea cuspidea *Hübn*............................... June 17.
Poaphila quadrifilaris (*Hubn.*)......................... May 21.

---

[1] In the capture of a male of this species on the wing at Center, the same sweep of the net inclosed a second male united to a female in copulation, and a specimen of *Sesia diffinis*, which were doubtless resting unobserved on some leaves against which the net had accidentally brushed. Three of *S. diffinis* had a short time before been observed flying together in the bright sunshine, one of which had been taken. The larvæ of *A. senatoria* occur so abundantly at Center as wholly to defoliate numbers of the smaller oaks. On the 7th of July the female moths were seen to have commenced the deposition of their eggs on the under side of oak leaves in patches often nearly covering the entire surface. On the 11th of July some newly-hatched larvæ were observed.

[2] This rare moth was found resting on the dust in the middle of a road near Albany over which vehicles were frequently passing. It was not easily alarmed, but gave ample opportunity for the observation of its appearance, attitude, etc., after having been covered by the net. In the only other instance in which I have met with it, the moth (a female, the other a male) was brought to me at Schoharie, with its abdomen partially crushed from being stepped upon on a sidewalk, where it was found.

Phalænidæ.

Angerona crocataria (*Fabr.*)........................... June 3.
Endropia marginata *Minot*........................... June 3.
Amphidasys cognataria *Guen.*........................... April 8.
Cleora pulchraria *Minot*[1]........................... Sept. 7.
Tephrosia cribrataria *Guen*........................... April 17.
Aplodes mimosaria *Guen.*........................... June 9.
Acidalia enucleata *Guen.*........................... July 12.
Corycia vestaliata *Guen*........................... May 25.
Lozogramma defluaria *Walk.*........................... May 25.
Numeria obfirmaria (*Hübn.*)........................... May 21.
Fidonia Faxonii *Minot*........................... May 27.
Fidonia bicoloraria *Minot*........................... June 1.
Aspilates dissimilaria (*Hübn.*)........................... July 7.
Zerene catenaria (*Cram.*)[2]........................... Sept. 8.
Cidaria diversilineata (*Hübn.*)........................... July 12.

Deltoidæ.

Hypena elegantalis *Fitch*........................... June 9.
Rivula propinqualis *Guen*........................... June 17.

Pyralidæ.

Pyralis olinalis *Guen*........................... July 15.
Ennychia octomaculalis (*Linn.*)........................... June 15.
Desmia maculalis *Westw.*........................... June 26.

II. COLLECTIONS DURING YEARS PRECEDING 1869.

Sphingidæ.

Sesia diffinis *Harris*........................... May 12.
Sesia Thysbe *Fabr*........................... May 25.
Sesia gracilis *Gr.-Rob.*........................... June 1.
Thyreus Abbotii *Swains*........................... May 25.

---

[1] Occurs abundantly on the trunks of pines during most of the month of September at Bath; abundant also at Schoharie.

[2] Very abundant at Center at this date, where at least a dozen could be seen from one point, resting on the upper surface of leaves of shrubbery. The conspicuously marked larvæ had, during the summer, been very common on Vaccinia and other plants. September 1st, its peculiar cocoon of strong and very open meshes, showing plainly the pupa suspended within, was found spun between leaves of willow (*Salix humuli*). The imago was last seen September 30th. It has not been observed at Schoharie.

Amphion Nessus (*Cram.*) .... June 17.
Darapsa Chœrilus (*Cram.*).......................... June 28.
Darapsa Myron (*Cram.*) .... June 18, Aug. 23.
Deilephila lineata (*Fabr.*), of second brood............... Sept. 2.
Deilephila lineata, larva[1] .... Oct. 6.
Deilephila chamænerii *Harr*.......................... May 25.
Deilephila chamænerii, larva.......................... Aug. 8.
Philampelus Pandorus (*Hübn.*) .... June 27.
Philampelus achemon (*Drury*), larva........... Aug. 14, Sept. 4.
Smerinthus geminatus *Say*.......................... Aug. 16.
Smerinthus excæcatus (*Smith*)........................ June 29.
Smerinthus myops (*Smith*)........................... June 18.
Cressonia juglandis (*Smith*) *Gr.-Rob*.................. July 3.
Ceratomia Amyntor (*Hübn.*)........... June 5, July 4, Aug. 17.
Ceratomia Amyntor, larva .... Aug. 24, Sept. 17.
Daremma undulosa *Walk*.......................... July 6.
Macrosila quinquemaculata (*Haw.*), of a second brood...... Sept. 2.
Sphinx chersis (*Hübn.*), larva, on lilac.................. Aug. 21.
Sphinx drupiferarum *Smith*, larva on plum.............. Aug. 26.
Sphinx kalmiæ *Smith* .... July 6, July 28.
Sphinx Gordius *Cram*.......................... June 27.
Agrius eremitus *Hübn*.......................... July 2.
Agrius eremitus, larva, on spearmint............ Aug. 5, Sept. 22.
Ellema Harrisii *Clem*.......................... June 22.
Ellema pineum *Lintn.*, larva, on pine.................. Sept. 14.

ÆGERIDÆ.

Trochilium marginitum *Harris*[2]........................ Aug. 21.
Trochilium tibiale *Harris*.......................... Aug. 16.
Ægeria caudata *Harris*.......................... Aug. 8.
Ægeria tipuliformis (*Linn.*) .... June 23.

THYRIDÆ.

Thyris maculata *Harris*.......................... Aug. 1.

ZYGÆNIDÆ.

Alypia octomaculata (*Fabr.*)........................ June 2.
Eudryas grata (*Fabr.*) .... July 13, Aug. 1.

---

[1] The dates assigned to the larval collections, in most instances, are those of the full maturity of the larva and preparation for pupation by the commencement of its cocoon or of its ground cell.

[2] A pair of these moths was taken *in copula* on the trunk of a sumach (*Rhus typhina*). After having been pinned, the female deposited a number of dark-brown eggs of an oval form.

Eudryas unio (*Hubn.*) [1]		July 8.
Acoloithus falsarius *Clem.*		June 25.
Scepsis fulvicollis (*Hübn.*)	Sept. 4,	Sept. 14.
Ctenucha virginica (*Charp.*)	June 4,	June 23.
Lycomorpha pholus (*Drury*) [2]	July 15,	Aug. 16.

Bombycidæ.

Euphanessa mendica (*Walk.*)	June 21,	July 10.
Callimorpha Lecontii *Boisd.*	July 23,	Aug. 2.
Arctia virgo (*Linn.*)		July 16.
Arctia Saundersii *Grote*		Aug. 20.
Arctia arge (*Drury*), larva		Feb. 28.
Pyrrharctia isabella (*Smith*) *Pack.*		June 30.
Spilosoma virginica (*Fabr.*)	May 7,	June 19.
Hyphantria cunea (*Drury*)		June 3.
Hyphantria textor *Harris*		May 19.
Ecpantheria scribonia (*Hübn*) larva		Oct. 6.
Halisidota tessellaris (*Smith*)		July 2.
Orgyia leucostigma (*Smith*)	Aug. 4,	Oct. 6.
Parorgyia cinnamomea *Gr.-Rob*		June 20.
Parorgyia parallela *Gr.-Rob.*		July 21.
Nadata gibbosa (*Smith*), larva [3]		Sept. 20.
Œdemasia concinna (*Smith*) *Pack*		May 12.
Cœlodasys unicornis (*Smith*) *Pack.*, larva		Aug. 1.
Heterocampa marthesia (*Cram.*)=*Lochmæus tessella* Pack.		July 25.
Platycerura furcilla *Pack* [4]		June 12.
Cerura borealis *Boisd.*, larva, on willow	Aug. 22,	Sept. 2.
Dryopteris rosea (*Walk.*)		June 30.
Telea Polyphemus (*Linn.*)	June 17,	July 28.
Actias Luna (*Linn.*)	May 7,	July 2.
Callosamia Promethea (*Drury*) *Pack*		June 27.
Platysamia Cecropia (*Linn.*) *Grote*	May 20,	June 27.
Hyperchiria Io (*Fabr.*)		July 9.
Anisota rubicunda (*Fabr.*), larva		Aug. 9.
Gastropacha Americana *Harr.*, larva, on birch	July 14,	Sept. 12.
Tolype laricis (*Fitch*)		Sept. 1.

---

[1] The larvæ have been collected from willow herb (*Epilobium coloratum*) but have not been observed by me on grape.

[2] Occurs frequently at Schoharie on blossoms of golden rod (Solidago). On one occasion six individuals were seen feeding together on a single plant, upon a hillside among evergreens.

[3] Feeds on maple and changes to a pupa beneath a leaf fastened by some threads to the ground. The imago is disclosed in June.

[4] The larva feeds on the pine, and attaining maturity about the middle of September, pupates in a slight cocoon among leaves on the surface of the ground.

Tolype velleda (*Stoll*) .... Sept. 10.
Tolype velleda, larva, on elm .... July 26.
Clisiocampa sylvatica *Harr* .... July 11.
Xyleutes robiniæ (*Peck*) .... June 18, July 9.
Xyleutes querciperda (*Fitch*) .... June 27.

NOCTUIDÆ.

Thyatyra abrasa *Guen* .... June 25.
Thyatyra cymataphorides *Guen* .... July 11.
Lacinia expultrix *Grote* .... July 9.
Leptina ophthalmica *Guen* .... June 12.
Bryophila palliatricula *Guen* .... July 20.
Micrococlia diphteroides *Guen* .... July 21.
Micrococlia vinnula *Grote* .... May 25.
Diphtera deridens *Guen* .... May 25.
Diphtera deridens, larva .... Sept. 10.
Acronycta dissecta *Gr.-Rob* .... July 2.
Acronycta superans *Guen* .... July 16.
Acronycta brumosa *Guen* .... July 18
Acronycta oblinita *Smith*, larva, on Polygonum .... Sept. 12.
Acronycta occidentalis *Grote*, larva, on apple .... Sept. 20.
Leucania pallens *Linn* .... June 22, July 8.
Leucania pseudargyria *Guen* .... July 12, July 30.
Leucania unipuncta *Haworth* .... Aug. 30, Oct. 1.
Gortyna nitela *Guen* .... Sept. 14.
Gortyna nebris *Guen* .... Sept. 28.
Hydrœcia sera *Gr.-Rob* .... July 22.
Nephelodes violans *Guen* .... Aug. 27.
Xylophasia apamiformis *Guen* .... June 19.
Xylophasia lignicolora *Guen* .... July 15.
Mamestra devastator (*Brace*) .... Aug. 15.
Mamestra dubitans *Walk* .... Aug. 6.
Mamestra picta *Harr.*, larva, on turnip .... Sept. 19, Oct. 12.
Mamestra adjuncta *Boisd.*, larva, on Solidago .... Oct. 9.
Apamea finitima *Grote* .... June 3.
Apamea iaspis *Guen* .... July 12.
Celæna herbimacula *Guen* .... Aug. 1.
Agrotis c-nigrum (*Linn.*) .... June 23.
Agrotis venerabilis *Walk* .... Sept. 9.
Agrotis collaris *Gr.-Rob* .... July 3.
Agrotis Cochranii *Riley* .... Sept. 20.
Noctua clandestina (*Harr.*) .... July 1.
Noctua bicarnea *Guen* .... July 27.

Noctua baja *W.-V*		Aug. 10.
Noctua plecta *Linn*		June 30.
Cirrœdia pampina *Guen*		Sept. 5.
Phlogophora Iris *Guen*		June 28.
Aplecta herbida *W.-V.*	July 18,	Aug. 11.
Aplecta nimbosa *Guen*		June 19.
Hadena xylinoides *Guen*	May 20,	Aug. 18.
Hadena distincta *Hübn*		May 7.
Hadena badistriga *Grote*		July 26.
Cucullia convexipennis *Gr.-Rob*		July 21.
Cucullia convexipennis, larva, on Solidago		Oct. 10.
Cucullia intermedia *Speyer*	May 27,	Aug. 10.
Cucullia intermedia, larva		Sept. 14.
Xylina cinerea *Riley*	April 8,	Oct. 10.
Xylina petrificata (*W.-V.*)		May 6.
Xylina Bethunei *Gr.-Rob*	March —, Sept. 12,	Oct. 7.
Rhodophora florida *Guen*		July 17.
Anthœcia bina *Guen*		June 16.
Chamyris cerintha (*Treits.*)		July 2.
Erastria synochites *Gr.-Rob*		June 12.
Erastria muscosula *Guen*	June 14,	June 20.
Erastria nigritula *Guen*	June 8,	Aug. 30.
Erastria carneola *Guen*		June 9.
Leptosia concinnimacula *Guen*		May 18.
Eriopus monetifera *Guen*		July 7.
Plusia precationis *Guen*		June 3.
Plusia balluca *Hübn*	July 14,	July 23.
Plusia æroides *Grote*	July 8,	July 28.
Plusia festucæ (*Linn.*)	July 25,	Aug. 1.
Plusia simplex *Guen*		Sept. 8.
Plusiodonta compressipalpis *Guen*		June 13.
Deva purpurigera *Walk*		July 3.
Gonoptera libatrix (*Linn.*)[1]	May 6,	Aug. 3.
Catocala piatrix *Grote*		Aug. 20.
Catocala amatrix *Hübn*		Sept. 12.
Catocala briseis *Edw*		Aug. 10.
Catocala relicta *Walk*	Aug. 9,	Sept. 14.
Catocala unijuga *Walk*		Aug. 15.

[1] The larva is of a bright velvety green color, having on each side a yellow stripe, shaded beneath with brown; longitudinally on the head is a black stripe. At maturity it measures one inch and a half in length by one-eighth of an inch in diameter. It is very sprightly in its movements. It feeds on willow, and pupates among some of the leaves drawn together by silken threads to which the pupa is attached by an anal spine. Pupation of the fall brood, from fifteen to twenty days.

Catocala parta *Guen.*, larva, on willow, ... July 7.
Erebus odora (*Linn.*) .... Nov. 2.
Ophiusa bistriaris *Hübn.* .... Aug. 15.
Drasteria erechtea (*Cram.*) .... April 22.
Euclidia cuspidea *Hübn.* .... May 16, June 17.

PHALÆNIDÆ.

Eutrapela transversata (*Drury*) .... July 25, Aug. 18.
Priocycla armataria (*Her.-Sch.*) .... June 26.
Angerona crocataria (*Fabr.*) .... July 10.
Hyperetis alienaria (*Her.-Sch.*) .... June 3.
Nematocampa filamentaria *Guen.* .... Aug. 8.
Endropia hypochraria (*Her.-Sch.*) .... June 3, June 22.
Endropia pectinaria (*W.-V.*) .... June 1.
Ellopia fiscellaria *Guen* .... Sept. 15.
Caberodes imbraria *Guen.* .... July 10.
Eurymene phlogosaria *Guen.* .... April 30, May 8.
Metanema inatomaria *Guen.* .... July 16.
Amphidasys cognataria *Guen* .... July 28.
Amphidasys cognataria, larva, on plum. .... Aug. 19.
Boarmia sublunaria *Guen.* .... June 5.
Tephrosia Canadaria *Guen.* .... May 23.
Nemoria chloroleucaria *Guen* .... June 5.
Aplodes mimosaria *Guen* .... June 5.
Acidalia enucleata *Guen.* .... Aug. 15.
Acidalia persimilata *Grote* .... Aug. 20.
Stegania pustularia *Guen.* .... Aug. 17.
Cabera intentata .... Sept. 15.
Corycia vestaliata *Guen* .... May 22.
Corycia albata *Lef.* .... May 30.
Macaria 4-signata *Walk.* .... July 25.
Abraxas ribearia *Fitch* .... July 14.
Anisopteryx vernata (*Peck*) .... April 5.
Hybernia tiliaria *Harr* .... Oct. 28.
Oporabia dilutata (*Albin*) .... Nov. 6.
Melanippe lacustrata *Guen.* .... June 6.
Coremia propugnata (*W.-V.*) .... July 19.
Cidaria gracilineata *Guen.* .... Aug. 1.
Cidaria hersiliata *Guen* .... June 23.
Heterophleps triguttaria *Her.-Sch.* .... July 10, Aug. 2.
Odezia albovittata *Guen.* .... Aug. 2.

MICROLEPIDOPTERA.

Hypena humuli *Harris*........................ May 7, July 12.
Hypena elegantalis *Fitch*............................. July 7.
Hypena scabralis (*Fabr.*)............................. Sept. 10.
Hypena erectalis *Guen*................................ Sept. 14.
Herminia morbidalis *Guen*............................. July 31.
Herminia pedipilalis *Guen*............................ June 21.
Herminia cruralis *Guen*............................... July 31.
Helia æmulalis (*Hübn*)................................ July 30.
Pyralis farinalis *Linn*............................... July 12.
Botys elealis (*Walk.*)=*B. adipaloides* Gr.-Rob........... May 22.
Pionea stramentalis (*Hübn.*).......................... Aug. 20.
Argyrolepia quercifoliana *Fitch*...................... July 1.

XI. LIST OF NORTH AMERICAN LEPIDOPTERA CONTAINED IN "SPECIES GÉNÉRAL DES LÉPIDOPTÈRES." BY A. GUENÉE.

The following list embraces above six hundred species of moths, of which, with the exception of some identical European species which have been frequently described and figured by the earlier authors, descriptions are given in the six volumes (V. to X.) of the *Species Général des Lépidoptères par M. A. Guenée*, forming a portion of the *Suites à Buffon.*

The species which are credited by Guenée to the State of New York are indicated by an asterisk (*). Elsewhere, when no *habitat* is given, it is to be understood as "North America." Species known to the compiler to occur in New York, in addition to those designated by Guenée, are marked in the list with a dagger (†).

NOCTUÉLITES.

Noctuo-Bombycidæ Boisd.

	Vol. V, pa.	No.
***Thyatyra abrasa** *Guen.*	12	2
Thyatyra pudens *Guen.*	13	5
***Thyatyra cymatophoroides** *Guen.*	13	6
***Leptina dormitans** *Guen.*	15	7
***Leptina ophthalmica** *Guen.*	15	8
†**Leptina Doubledayi** *Guen.* North. States	15	9

Bryophilidæ Guen.

***Bryophila palliatricula** *Guen.*	26	26
Bryophila corticosa *Guen.*	30	32
Grammophora hebræa *Guen.*	31	33
Grammophora cora *Hübn.* Georgia	31	34

Bombycoidæ Boisd.

***Microcœlia fragilis** *Guen.*	34	35
***Microcœlia diphteroides** *Guen.*	34	36
†**Diphtera deridens** *Guen.*	35	37
Diphtera jocosa *Guen*	37	40
Acronycta tritona *Hübn.* Georgia, Florida	42	45
Acronycta psi *Linn.* North Amer., Europe	43	47
Acronycta lobeliæ *Guen.*	44	49
†**Acronycta furcifera** *Guen.*	44	50
†**Acronycta hasta** *Guen.*	45	51
Acronycta telum *Guen.*	45	52
***Acronycta spinigera** *Guen.*	45	53
Acronycta interrupta *Boisd.* North Amer. [Georgia ?]	46	54

	Vol. V, pa.	No.
Acronycta lepusculina *Guen.* North Amer.?	46	55
Acronycta hastulifera *Abb.*	47	57
Acronycta acericola *Guen.* Georgia, Virginia	48	59
Acronycta rubricoma *Guen*	48	60
+**Acronycta oblinita** *Abb*	49	61
+**Acronycta innotata** *Guen*	50	64
+**Acronycta brumosa** *Guen*	52	68
Acronycta hamamelis *Guen.* Georgia, Virginia, etc	52	69
***Acronycta superans** *Guen*	53	71
***Acronycta clarescens** *Guen*	54	73
Acronycta longa *Guen*	54	74
Acronycta xyliniformis *Guen.* Georgia, Florida	56	77

Leucanidæ GUEN.

Leucania littera *Guen.* Florida	71	89
***Leucania pseudargyria** *Guen*	74	94
Leucania obusta *Guen*	74	95
Leucania ebriosa *Guen*	74	96
+**Leucania extranea** *Guen.* North Amer., Brazil, Java, N. Holland	77	104
Leucania videns *Guen.* Florida	78	106
Leucania extincta *Guen.* Florida	79	107
***Leucania insueta** *Guen*	81	113
Leucania linita *Guen.* Florida	81	114
Leucania juncicola *Boisd*	83	119
Leucania scirpicola *Guen.* Florida	84	120
***Leucania commoides** *Guen*	86	127
+**Leucania albilinea** *Hübn*	89	135
+**Leucania phragmitidicola** *Guen*	89	136
***Leucania pallens** *Linn.* North Amer., Europe	92	145
***Nonagria inquinata** *Guen*	104	159
Nonagria enervata *Guen.* Florida	105	163
Nonagria fodiens *Guen.* Florida	105	164
Nonagria typhæ *Naturf.* North Amer., Europe	108	172

Glottulidæ GUEN.

Glottula timais *Cram*	116	184

Apamidæ GUEN.

+**Gortyna rutila** *Guen.* Illinois	123	191
+**Gortyna marginidens** *Guen*	123	192
Gortyna limpida *Guen.* Illinois	124	193
+**Gortyna nebris** *Guen.* Illinois	124	194
***Gortyna nitela** *Guen.* Illinois	124	195
+**Hydrœcia nictitans** *Linn.* North Amer., Europe	126	196
***Hydrœcia lorea** *Guen*	126	197
***Hydrœcia immanis** *Guen*	128	201
***Hydrœcia stramentosa** *Guen*	129	202
Nephelodes minians *Guen*	130	203
***Nephelodes violans** *Guen.* Illinois	130	204
+**Scolecocampa ligni** *Guen.* Georgia	131	206
***Achatodes sandix** *Guen* [= *Gortyna zeæ* Harris]	132	207
+**Xylophasia lateritia** *Naturf.* Europe	137	215

	Vol. V, pa.	No.
***Xylophasia apamiformis** *Guen*	137	216
†**Xylophasia rurea** *Fabr.* North Amer., Europe	137	217
***Xylophasia lignicolora** *Guen.*	140	221
***Xylophasia verbascoides** *Guen.*	141	224
Xylophasia sectilis *Guen.* North Amer.?	141	225
Xylophasia mucens *Hübn.* Pennsylvania, Florida	142	226
Xylophasia confusa *Hübn.* Pennsylvania, Carolina	142	227
***Xylophasia cariosa** *Guen*	144	232
†**Dipterygia pinastri** *Linn.* North Amer., Europe	146	234
Xylomyges eridania *Cram.*=(*phytolaccæ* Abb.) North Am., [Georgia ?].	148	235
Laphygma frugiperda *Abb.* North Amer., Australia?	159	254
Prodenia commelinæ *Abb.*	162	256
Prodenia ornithogalli *Guen.* North Amer., [Georgia ?]	163	258
Prodenia eudiopta *Guen.* North Amer.?	164	261
Heliophobus fimbriaris *Guen.*	172	271
†**Mamestra arctica** *Boisd.*	193	304
***Mamestra abjecta** *Hübn.* North Amer., Europe	193	306
***Mamestra impulsa** *Guen.*	194	307
***Mamestra passer** *Guen.*	195	308
***Mamestra adjuncta** *Boisd.*	199	315
***Apamea finitima** *Guen.*	206	324
***Apamea mactata** *Guen.*	207	326
†**Apamea modica** *Guen.* North Amer.?	207	327
†**Apamea gemina** *Hübn.*	208	328
***Apamea iaspis** *Guen.*	209	330
Apamea oculea *Linn.* North Amer., Europe	210	333
Celæna festivoides *Guen.* Florida	220	348
Celæna chalcedonia *Hübn.*	221	350
Celæna arna *Guen.* Florida	222	351
Celæna exesa *Guen.* Florida	222	352
***Celæna herbimacula** *Guen.*	223	354
Perigea xanthioides *Guen.* Florida	227	361
Perigea infelix *Guen.* Florida	229	368
***Perigea vecors** *Guen.*	230	371

Caradrinidæ Boisd.

Monodes nucicolora *Guen.* Florida	241	386
Caradrina tarda *Guen.*	243	389

Noctuidæ Boisd.

Agrotis spissa *Guen*	261	415
***Agrotis jaculifera** *Guen.* Pennsylvania, Canada	262	417
†**Agrotis trifurca** *Evers.* Russia	265	423
Agrotis malefida *Guen.* Florida	267	429
Agrotis annexa *Treits.* North Amer., Antilles, Brazil	268	430
†**Agrotis suffusa**[1] *W.-V.* North Amer., Europe, East India	268	431
***Agrotis fennica** *Evers.* North Amer., Europe	270	434
Agrotis saucia *Engr.* North Amer., Europe	271	435
Agrotis incivis *Guen*	274	441
Agrotis exclamationis *Linn.* Canada, Europe	280	453

[1] Identical with *Agrotis telifera* Harris.

	Vol. V, pa.	No.
†**Agrotis nigricans** *Linn.* North Amer., Europe	286	468
Agrotis tritici *Linn.* North Amer., Europe	288	471
Agrotis obeliscoides *Guen*	293	477
Agrotis ravida[1] *W.-V.* North Amer., Europe	300	493
†**Noctua lubricans** *Guen.* Florida	323	531
†**Noctua augur** *Fabr.* North Amer., Europe	325	537
***Noctua sigmoides** *Guen*	325	539
†**Noctua plecta** *Linn.* North Amer., Europe	326	540
Noctua ochrogaster *Guen*	327	542
†**Noctua c-nigrum** *Linn.* North Amer., Europe	328	545
***Noctua bicarnea** *Guen*	329	546
***Noctua triangulum** *Hübn.* Europe	329	548
***Noctua dahlii** *Hübn*	332	554
Noctua elimata *Guen.* Georgia	333	556
†**Noctua baja** *W.-V.* Europe	335	562

Orthosidæ Guen.

Ceramica exusta *Guen*	344	574
Ceramica vindemialis *Guen.* Florida	344	576
Ceramica u-album *Guen.* Florida	345	577
†**Tæniocampa instabilis** *Rœs.* North Amer., Europe	350	586
***Tæniocampa alia** *Guen*	352	587
Tæniocampa hibisci *Guen.* North Amer. [Georgia ?]	355	591
Tæniocampa oviduca *Guen.* North Amer	357	597
Tæniocampa styracis *Guen.* North Amer. [Georgia ?]	357	598
***Orthodes infirma** *Guen*	375	626
***Orthodes cynica** *Guen*	375	627
***Orthodes nimia** *Guen*	376	628
***Orthodes candens** *Guen*	376	629
***Orthodes enervis** *Guen.=O. vecors*	376	630
Cerastis anchocelioides *Guen*	384	639
Scopelosoma sidus *Guen*	386	642
Xanthia rufago *Hübn*	392	646
Xanthia aurantiago *Guen*	394	651
***Xanthia bicolorago** *Guen*	397	655
†**Xanthia ferruginea** *W.-V.* North Amer., Europe	397	656
***Cirrœdia pampina** *Guen*	402	661
Mesogona culea *Guen.* Florida	404	665

Cosmidæ Guen.

	Vol. VI.	
Cosmia orina *Guen*	10	678

Hadenidæ Guen.

Dianthœcia capsularis *Guen.* Florida	22	693
Hecatera laudabilis *Guen.* North Amer. [Georgia ?]	30	709
Epunda onychina *Guen*	48	735
Chariptera festa *Guen.* Carolina	57	745
***Phlogophora anodonta** *Guen*	63	752
***Phlogophora iris** *Guen*	64	753

[1] Very near to *Noctua clandestina* Harris, but in all probability distinct.

	Vol. VI, pa.	No.
***Phlogophora periculosa** *Guen.*	65	755
†**Euplexia lucipara** *Linn.* North Amer., Europe	68	758
Polyphænis herbacea *Guen.*	73	764
†**Aplecta herbida** *W.-V.* North Amer., Europe	75	765
***Aplecta imbrifera** *Guen.*	76	768
†**Aplecta nimbosa** *Guen.*	77	769
†**Aplecta latex** *Guen*	78	771
***Aplecta condita** *Guen.*	78	772
***Hadena miselioides** *Guen*	89	791
†**Hadena distincta** *Hübn.*	91	795
†**Hadena chenopodii** *Albin.* North Amer., Europe	97	807
***Hadena pisi** *Linn.* North Amer., Europe	101	817
†**Hadena w-latinum** *Hufn.* North Amer., Europe	104	822
***Hadena grandis** *Boisd.* Lapland and Greenland	105	823
†**Hadena xylinoides** *Guen*	106	825

Xylinidæ Guen.

†**Cloantha ramosula** *Guen.*	114	831
†**Xylina petrificata** *W.-V.* North Amer., Europe	121	844
†**Cucullia asteroides** *Guen*	133	858
Cucullia postera *Guen*	133	859
***Cucullia florea** *Guen.*	134	860
Cucullia umbratica[1] *Linn.* North Amer., Europe	146	885
†**Crambodes talidiformis** *Guen*	152	896

Heliothidæ Boisd.

Oria sanguinea *Hübn.*	167	913
Rhodophora gauræ *Abb.* Georgia, Florida	170	917
***Rhodophora florida** *Guen*	171	918
Lepipolys perscripta *Guen.* Florida	174	921
Aspila rhexiæ *Abb.*	175	922
Aspila virescens *Fabr.* West Indies [North Amer. *Grote*]	175	923
Aspila subflexa *Guen.*	175	924
Tamila nundina *Drury*	176	925
†**Heliothis marginata** *Kléem.* Europe	178	927
†**Heliothis armigera** *Hübn.* [=*H. umbrosus* Grote.] North Amer., Eur.	181	933
†**Heliothis spinosæ** *Guen.* Canada	182	937
†**Anthœcia rivulosa** *Guen.* [=*Crambus marginatus* Haworth]	184	938
†**Anthœcia arcifera** *Guen*	184	939
Anthœcia jaguarina *Guen*	184	940
†**Anthœcia lynx** *Guen.*	185	942
Anthœcia tuberculum *Hübn*	185	943
†**Anthœcia bina** *Guen*	186	944

See Addenda.

Hæmerosidæ Guen.

***Lepidomys irrenosa** *Guen.*	202	966

Acontidæ Boisd.

Agrophila leo *Guen.*	205	968
Agrophila dama *Guen*	205	969
Agrophila onagrus *Guen*	205	970

[1] This species does not occur in North America. *C. intermedia* of Speyer has been mistaken for it.

	Vol. VI, pa.	No.
†**Acontia candefacta** *Hübn.*	216	984
†**Acontia erastrioides** *Guen.*	218	990
Acontia biplaga *Guen.*	218	991
Acontia aprica *Hübn.*	219	992

Erastridæ Guen.

†**Chamyris cerintha** *Treits*	225	1002
***Erastria carneola** *Guen.*	228	1008
†**Erastria nigritula** *Guen.* Florida	229	1009
†**Erastria muscosula** *Guen*	230	1011
Erastria albidula *Guen.*	230	1012
Bankia olivula *Guen.*	231	1013

Anthophilidæ Dupon.

†**Leptosia concinnimacula** *Guen*	238	1021
Galgula subpartita *Guen.*	239	1022
Galgula hepara *Guen*	239	1023
Xanthoptera nigrofimbria *Guen*	241	1025
Xanthoptera semiflava *Guen*	241	1026
Xanthoptera semicrocea *Guen.* [Georgia?]	241	1027

Eriopidæ Guen.

Eriopus floridensis *Guen.* Florida	293	1094
†**Eriopus mollissima** *Guen.* Florida	294	1098
†**Eriopus monetifera** *Guen.* North Amer.?	295	1099
Eriopus granitosa *Guen.*	295	1100

Eurhipidæ Guen.

Ingura delineata *Guen.* North Amer. [Georgia?]	311	1118
Ingura abrostoloides *Guen*	311	1119
Ingura cristatrix *Guen.* North Amer.?	313	1122
†**Ingura oculatrix** *Guen.* North Amer.?	313	1123

Placodidæ Guen.

***Placodes cinereola** *Guen*	316	1126
Diastema tigris *Guen.* Colombia, [North Amer. in exp. of plates]	317	1127

Plusidæ Boisd.

***Abrostola urentis** *Guen*	322	1130
***Abrostola ovalis** *Guen*	322	1131
***Plusia ærea** *Hübn.*	333	1148
†**Plusia balluca** *Hübn.*	334	1150
†**Plusia festucæ** *Linn.* North Amer., Europe	337	1157
***Plusia thyatyroides** *Guen.*	337	1158
***Plusia u-brevis** *Guen.*	341	1163
†**Plusia biloba** *Steph.* Florida	341	1164
Plusia verruca *Fabr*	342	1165
Plusia rogationis *Guen*	344	1169
†**Plusia precationis** *Guen.*	344	1170
***Plusia simplex** *Guen.*	346	1174
Plusia ou *Guen*	348	1176

	Vol. VI, pa.	No.
Plusia ni *Engr.* North Amer., Europe	349	1178
Plusia oxygramma *Hübn.*	350	1181
***Plusia mortuorum** *Guen.*	353	1187
Basilodes pepita *Guen.* Florida	358	1199
†**Plusiodonta compressipalpis** *Guen.* ——?	359	1200

Hemiceridæ GUEN.

Hemiceras cadmia *Guen.*	383	1240

Gonopteridæ GUEN.

Cosmophila erosa *Hübn.*	395	1255
Anomis fulvida *Guen.*	397	1259
†**Anomis bipunctina**[1] *Guen.*	401	1267
Anomis luridula *Guen.*	401	1268
Monogona hormos *Hübn.* Pennsylvania, Georgia	403	1270
†**Gonoptera libatrix** *Linn.* North Amer., Europe	405	1273

Amphipyridæ GUEN.

†**Amphipyra pyramidoides** *Guen.*	413	1278

Homopteridæ BOISD.

	Vol. VII, pa.	No.
Phæocyma lunifera *Hübn.*	3	1320
†**Homoptera lunata** *Drury*	12	1335
Homoptera exhausta *Guen.* North Amer.?	14	1337
†**Homoptera edusa** *Drury*	14	1338
Homoptera minerea *Guen.*	15	1339
Homoptera calycanthata *Abb*	15	1340
Homoptera obliqua *Guen.*	16	1341
Ypsia æruginosa *Guen.*	17	1342
†**Ypsia undularis** *Drury*	18	1343
Anthracia coracias *Guen.*	19	1344
Anthracia cornix *Guen.*	19	1345

Hypogrammidæ GUEN.

Campometra amella *Guen.* North Amer. [Georgia?]	25	1352
Hypogramma Andromedæ *Guen.* North Amer. [Georgia?]	36	1368
†**Allotria elonympha** *Hübn.* Georgia, Florida	37	1369

Bolinidæ GUEN.

Panula inconstans *Guen*	59	1392
Panula remigipila *Guen.* Florida	60	1393
Bolina cinis *Guen.*	62	1395
†**Syneda limbolaris** *Hübn.*	71	1416
†**Syneda graphica** *Hübn.* Georgia	71	1417

Catocalidæ BOISD.

†**Parthenos nubilis** *Hübn.*	80	1427
Catocala fraxini *Linn.* North Amer., Europe	83	1428

[1] This is said by Mr. Grote to be *Anomis Xylina*, described by Say in 1827.

	Vol. VII, pa.	No.
+**Catocala parta** *Guen.* Canada	84	1431
†**Catocala amatrix** *Hübn*	86	1434
+**Catocala cara** *Guen.* Baltimore	87	1435
†**Catocala ultronia** *Hübn*	89	1440
+**Catocala ilia** *Cram*	91	1445
Catocala uxor *Guen*	92	1446
Catocala lacrymosa *Guen*	93	1447
+**Catocala epione** *Drury.* [Long Island]	93	1448
+**Catocala insolabilis** *Guen*	94	1449
†**Catocala viduata** *Abb*	94	1450
+**Catocala desperata** *Guen.* Baltimore	95	1451
†**Catocala cerogama** *Guen*	96	1452
†**Catocala neogama** *Abb.* [Long Island]	96	1453
+**Catocala palæogama** *Guen*	97	1454
+**Catocala muliercula** *Guen.* [Georgia?]	97	1455
+**Catocala innubens** *Guen*	98	1456
+**Catocala melanympha** *Guen.* Canada	98	1457
Catocala consors *Abb.* Georgia [Alabama]	99	1458
Catocala micronympha *Guen*	102	1466
Catocala amasia *Abb.* [Georgia?]	103	1468
†**Catocala polygama** *Guen*	105	1472
Catocala connubialis *Guen.* [Georgia?]	105	1473
†**Catocala androphila** *Guen.* [Long Island]	106	1474
Catocala messalina *Guen*	107	1475

Erebidæ GUEN.

+**Erebus odora** *Linn*	167	1559

Bendidæ GUEN.

Bendis hinna *Hübn*	216	1622

Ophiusidæ GUEN.

Ophiusa Smithii *Guen.* Georgia	266	1696
Ophiusa similis *Boisd*	267	1697
+**Ophiusa bistriaris** *Hübn*	268	1699
Ophiusa consobrina *Guen*	268	1700
Agnomonia anilis *Drury*	273	1712

Euclidiæ GUEN.

Drasteria convalescens *Guen*	289	1734
+**Drasteria erechtea** *Cram*	289	1735
Drasteria erichto *Guen*	290	1736
†**Euclidia cuspidea** *Hübn*	292	1739

Poaphilidæ GUEN.

Lyssia orthosioides *Guen*	296	1745
Poaphila deleta *Guen*	300	1748
Poaphila sylvarum *Guen*	300	1749
†**Poaphila quadrifilaris** *Hübn*	300	1750

	Vol. VII, pa.	No.
Poaphila erasa *Guen.*	301	1751
Poaphila herbicola *Boisd.*	301	1752
Poaphila contempta *Boisd.*	302	1753
Poaphila flavistriaris *Hübn.*	302	1754
Poaphila perplexa *Boisd.* Savannah.	302	1755
Poaphila bistrigata *Hübn*	303	1756
Poaphila herbarum *Guen.*	303	1757
Phurys vinculum *Guen.*	304	1758
Phurys lima *Guen.* North Amer.?	305	1759
Celiptera frustulum *Guen.*	308	1767

Remigidæ GUEN.

Remigia latipes *Guen.*	314	1774
Remigia marcida *Guen.* Savannah.	317	1777
Isogona natatrix *Guen.*	323	1786
Panopoda rubricosta *Guen.*	324	1788
Panopoda roseicosta *Guen.*	325	1789
Panopoda carneicosta *Guen.* United States.	325	1790

Thermesidæ GUEN.

Thermesia gemmatalis *Hübn.* United States, Brazil.	355	1828
Marmorinia epionoides *Guen.* Georgia.	371	1853
Marmorinia geometroides *Guen.*	371	1854

ADDITIONS.

Cerastis adulta *Guen.* Georgia.	393	633 bis
Hoporina hesperidago *Guen.* Georgia.	393	644 bis

DELTOÏDES LAT.

Hypenidæ H.-S.

	Vol. VIII, pa.	No.
†**Hypena Baltimoralis** *Guen.* United States.	34	31
Hypena madefactalis *Guen.*	35	33
***Hypena scabralis** *Fabr.* Pennsylvania, Canada, etc.	40	45
†**Hypena erectalis** *Guen.* Pennsylvania.	40	46

Herminidæ DUPON.

†**Rivula propinqualis** *Guen.*	49	54
†**Herminia morbidalis** *Guen.*	56	60
†**Herminia pedipilalis** *Guen.*	57	62
†**Herminia cruralis** *Guen.*	58	65
Nodaria Hispanalis *Guen.* North Amer.?	64	77
Bleptina caradrinalis *Guen.*	67	84
Bleptina? madopalis *Guen.* North Amer.?	69	89
†**Helia phæalis** *Guen.*	76	96
†**Helia Americalis** *Guen.*	78	100
†**Helia æmulalis** *Hübn.*	78	101
Helia lituralis *Hübn.* Georgia.	79	102
†**Renia discoloralis** *Guen.*	82	107
†**Clanyma angulalis** *Hübn.*	95	128
Clanyma asopialis *Guen.*	96	130

PYRALITES.

Pyralidæ GUEN.

	Vol. VIII, pa.	No.
†**Pyralis fimbrialis** *W.-V.* [=*Asopia costalis* Fabr. sp.] Europe	118	3
†**Pyralis olinalis** *Guen.* North Amer.?	118	4
†**Pyralis farinalis** *Linn.* North Amer., Europe	119	6
Pyralis glaucinalis *Linn.* North Amer.?	122	14
†**Aglossa pinguinalis** *Linn.* Europe	127	22
Aglossa cuprealis *Hübn.* Europe, [United States, *Walker*]	127	23
Aglossa domalis *Guen.*	128	24

Ennychidæ DUPON.

Rhodaria phœnicalis *Hübn.*	173	96
Herbula subsequalis *Guen.*	177	99
†**Ennychia octomaculalis** *Linn.* Lapland, Europe	184	116

Asopidæ GUEN.

Syngama florellalis *Cram.* Central America	187	118
†**Desmia maculalis** *Westw.*	189	121
†**Samea ecclesialis** *Guen.* North Amer., Brazil, Cayenne	194	132
Samea castellalis *Guen.* North Amer., Brazil, Colombia	195	133
Samea ebulealis *Guen.* North America, Brazil	196	136
Samea huronalis *Guen.* Canada	198	141
Asopia bicoloralis *Guen.* North Amer., Brazil	205	159
Hyalea dividalis *Hübn.* United States	207	162
Agathodes monstralis *Guen.* North Amer.?	209	165
Agathodes designalis *Guen.* Brazil, North Amer.?	209	166
Spoladea perspectalis *Hübn.* North and Central America	226	192
Isopteryx aplicalis *Guen.* Georgia	229	199
Isopteryx magualis *Guen.*	230	201
Isopteryx stenialis *Guen.* Georgia	231	203

Steniadæ GUEN.?

Stenia ranalis *Guen.*	243	219
Parthenodes? xantholeucalis *Guen.* Georgia	253	241

Spilomelidæ GUEN.

Spilomela platinalis *Guen.* Missouri	282	277

Margarodidæ GUEN.

Phakellura hyalinatalis *Linn.* North Amer., Hayti, Brazil, etc.	296	302
Phakellura immaculalis *Guen.* Gaudaloupe, [North Amer., *Walker*].	297	303
Phakellura nitidalis *Cram.* North Amer., Brazil, Cayenne, Colombia,	299	311
Cliniodes opalalis *Guen.*	301	313
Margarodes quadristigmalis *Guen.*	304	319

Botydæ GUEN.

Botys ponderalis *Guen.* North Amer.?, Brazil	328	356
Botys oxydalis *Guen.* Georgia	328	357

	Vol. VIII, pa.	No.
Botys flavidalis *Guen.*	329	358
Botys extricalis *Guen*	338	382
Botys argyralis *Hübn.*	341	388
Ebulea fumalis *Guen.* Georgia	358	430
Ebulea tertialis *Guen.*	364	446
Homophysa glaphyralis *Guen.*	366	450
Homophysa sesquistrialis *Hübn.* Georgia	366	451
Pionea rimosalis *Guen.*	371	460
†**Pionea stramentalis**[1] *Hübn.* Europe	373	465
†**Pionea scripturalis** *Guen.* Brazil, [North Amer., *Walker*]	373	466
Scopula illibalis *Hübn*	395	509
Scopula rubigalis *Guen*	398	516
Nymphula similalis *Guen.*	403	524
Mecyna reversalis *Guen.*	409	531

PHALÉNITES.

Urapterydæ GUEN.

	Vol. IX, pa.	No.
Chœrodes tetragonata *Guen.* Brazil, [North Amer. in exp. of plates].	36	16
Chœrodes incurvata *Guen*	37	21
†**Chœrodes transversata** *Drury.* United States	38	22
***Chœrodes goniata** *Guen*	38	24
Eutrapela clemataria *Abb*	47	42
Crocopteryx martiata *Guen.* North Amer. ?	74	100

Ennomidæ GUEN.

	Vol. IX, pa.	No.
Apicia spinetaria *Guen.* North Amer., Brazil	85	123
Apicia juncturaria *Guen.* North Amer. ? Brazil ?	88	132
***Priocycla armataria** *Her.-Sch.*	91	136
Epione serinaria *Her.-Sch.* Cincinnati	98	149
Sicya truncataria *Guen.* Canada	104	159
***Sicya solfataria** *Guen.*	104	160
Sycya sublimaria[2] *Harris*	105	161
†**Angerona crocataria** *Fabr.*	114	175
Hyperetis nyssaria *Abb.*	118	178
Hyperetis exsinuaria *Guen.* Pennsylvania	118	179
Hyperetis amicaria *Her.-Sch.* Cincinnati	118	180
Hyperetis insinuaria *Guen*	119	181
Hyperetis persinuaria *Guen.* Near Baltimore	119	182
***Hyperetis subsinuaria** *Guen*	119	183
†**Hyperetis alienaria** *Her.-Sch.* Near Baltimore	120	184
†**Nematocampa filamentaria** *Guen*	121	185
†**Endropia pectinaria** *W.-V.*	122	186
Endropia tigrinaria *Guen.* Canada	123	187
Endropia obtusaria *Hübn.*	123	188
†**Endropia amœnaria** *Guen*	124	190
†**Endropia hypochraria** *Her.-Sch*	125	191
***Endropia refractaria** *Guen.*	125	192

[1] An Orobena, identical with the German species, but somewhat lighter in color.—SPEYER.

[2] This is *Ennomos macularia* of Harris (Lake Superior, page 392), the specific name being changed by Guenée, from having been previously used.

	Vol. IX, pa.	No.
Endropia lateritiaria *Guen.*	125	193
Metrocampa prægrandaria *Guen.*	128	195
Metrocampa perlata *Guen.*	128	197
Ellopia pultaria *Guen.*	131	201
Ellopia ? placeraria *Guen.* California	132	202
Ellopia fervidaria *Hübn.*	132	203
†**Ellopia fiscellaria** *Guen.*	133	204
***Ellopia flagitiaria** *Guen.* Canada	133	205
†**Caberodes metrocamparia** *Guen.*	137	212
Caberodes remissaria *Guen.* Pennsylvania	137	213
Caberodes imbraria *Guen.* Pennsylvania	137	214
Caberodes superaria *Guen.*	138	215
Caberodes majoraria *Guen.*	138	216
Caberodes ineffusaria *Guen.* Near Baltimore	138	217
Caberodes floridaria *Guen.* Pennsylvania	139	218
†**Caberodes confusaria** *Hübn.*	139	220
Caberodes phasianaria *Guen.*	140	221
***Caberodes interlinearia** *Guen.*	140	222
†**Tetracis crocallata** *Guen.*	141	224
***Tetracis aspilatata** *Guen.*	141	225
Tetracis ægrotata *Guen.* California	141	226
***Tetracis cachexiata**[1] *Guen.* New Holland	142	227
Tetracis truxaliata *Guen.* California	142	228
Eurymene emargataria *Guen.*	145	233
†**Eurymene phlogosaria** *Guen.* Canada	146	234
Eurymene alcoolaria *Guen.* Canada	146	235
†**Azelina Hubneraria** *Guen.*	159	249
†**Metanema inatomaria** *Guen.* Canada	171	272
Metanema forficaria *Guen.* California	172	273
Metanema quercivoraria *Abb.* [Georgia ?]	172	275
†**Ennomos magnaria** *Guen.*	174	276
†**Ennomos subsignaria** *Hübn.*	181	284

Amphidasydæ GUEN.

Ceratonyx satanaria *Guen.* Georgia	194	295
Amphidasys quernaria *Abb.* Georgia, Virginia	207	310
†**Amphidasys cognataria** *Guen.*	208	312

Boarmidæ GUEN.

Hemerophila unitaria *Her.-Sch.*	219	326
Synopsia phigaliaria *Guen.*	225	336
Boarmia pampinaria *Guen.* Baltimore	245	367
Boarmia olivinaria *Guen.* California	245	368
Boarmia frugaliara *Guen.* Georgia	246	369
Boarmia humaria *Guen.* Georgia	246	370
†**Boarmia intraria** *Guen.* Near Baltimore	246	371
Boarmia defectaria *Guen.*	247	372
Boarmia larvaria *Guen.* Canada	247	373
Boarmia momaria *Guen.* North Amer. ?	247	374
†**Boarmia sublunaria** *Guen.*	248	376

---

1 Guenée's description so accurately conforms to the New York species that it is not improbable that the locality assigned by him is an erroneous one. — SPEYER.

	Vol. IX, pa.	No.
Boarmia titearia *Cram.* Virginia	248	377
Boarmia gnopharia *Guen.*	251	383
Boarmia umbrosaria *Guen.* Georgia	251	384
Boarmia porcelaria *Abb.* [Georgia?]	252	385
†**Tephrosia cribrataria** *Guen.* Georgia	260	399
†**Tephrosia canadaria** *Guen.* Canada	263	409
Tephrosia occiduaria *Guen.*	266	413
Paraphia deplanaria *Guen.*	272	426
Paraphia subatomaria *Wood.*	272	427
Paraphia nubecularia *Guen.*	272	428
Paraphia mamurraria *Guen.* Canada	273	429
†**Bronchelia hortaria** *Fabr.*	288	462
Bronchelia dendraria *Guen.* Georgia	289	463
Stenotrachelys approximaria *Hübn.*	290	464
Exelis pyrolaria *Guen.*	324	508

Geometridæ Guen.

†**Geometra iridaria** *Guen.*	344	531
Nemoria? pistasciaria *Guen.*	348	539
†**Nemoria chloroleucaria** *Guen.*	351	546
Nemoria? faseolaria *Guen.* California	351	547
Iodis euchloraria *Abb.*	355	553
†**Dyspteris abortivaria** *Her.-Sch.* Cincinnati	363	572
Racheospila lixaria *Guen.*	374	601
Synchlora liquoraria *Guen.* California	375	603
†**Aplodes mimosaria** *Guen.* Georgia, Canada	377	605
Aplodes glaucaria *Guen.* Georgia	377	606

Ephyridæ Guen.

Ephyra culicaria *Guen.* Georgia	407	658
†**Ephyra myrtaria** *Guen.*	408	660
†**Ephyra pendulinaria** *Guen.*	414	674

Acidalidæ Guen.

Asthena lucata *Guen.* Canada	437	723
Acidalia magnetaria *Guen.* California	450	743
Acidalia sideraria *Guen.* California	451	744
Acidalia plemyraria *Guen.* Georgia	453	750
Acidalia demissaria *Hübn.*	466	781
Acidalia insulsaria *Guen.*	469	788
Acidalia placidaria *Guen.*	469	789
Acidalia pannaria *Guen.*	470	790
Acidalia hepaticaria *Guen.* Baltimore	471	793
Acidalia lævitaria *Hübn.* Georgia	471	794
Acidalia sublataria *Guen.*	474	803
Acidalia ossularia *Hübn.* Pennsylvania	475	804
Acidalia temnaria *Guen.*	476	807
Acidalia myrmidonata *Guen.*	487	838
Acidalia purata *Guen.*	488	839
Acidalia lumenaria *Hübn.*	488	840
Acidalia inductata *Guen.*	494	852
†**Acidalia nivosaria** *Guen.* Canada	499	863
†**Acidalia enucleata** *Guen.*	505	874

	Vol. X, pa.	No.
Timandra viridipennaria *Guen*	3	894

Caberidæ GUEN.

+**Stegania pustularia** *Guen*	49	978
Cabera erythemaria *Guen.* Pennsylvania, Canada	56	986
+**Cabera variolaria** *Guen.* Pennsylvania	56	987
+**Corycia hermineata** *Guen.* Canada	58	992
+**Corycia albata** *Lef.* Georgia	58	993
+**Corycia vestaliata** *Guen*	59	994

Macaridæ GUEN.

Amilapis unipunctata *Haw*	62	998
Amilapis nullaria *Hübn.* North Amer.?	63	999
Macaria distribuaria *Hübn*	76	1024
Macaria præatomata *Haw*	76	1025
Macaria bicolorata *Fabr.* Virginia	77	1026
Macaria ocellinata *Guen*	85	1051
+**Macaria granitata** *Guen.* Pennsylvania	85	1053
Macaria contemptata *Guen*	86	1054
Halia marcescaria *Guen.* California	92	1067

Fidonidæ GUEN.

Tephrina haliata *Guen.* California	97	1071
Tephrina muscariata *Guen.* California	98	1073
Tephrina neptaria *Guen.* California	99	1076
***Tephrina gnophosaria** *Guen*	99	1077
Tephrina monicaria *Guen.* California	100	1081
Tephrina unicalcararia *Guen.* California	100	1082
Tephrina lorquinaria *Guen*	101	1083
Tephrina sabularia *Guen*	105	1093
Tephrina detersata *Guen*	105	1094
Psamatodes eremiata *Guen*	109	1100
+**Numeria obfirmaria** *Hübn*	135	1139
Numeria duaria *Guen.* Canada	135	1140
Numeria hamaria *Guen*	136	1141
Numeria fritillaria *Guen.* United States	136	1142
Selidosema juturnaria *Guen.* California	147	1164
Selidosema fæminaria *Guen.* California	149	1168
Fidonia avuncularia *Guen.* California	155	1176
+**Hæmatopis grataria** *Fabr*	171	1200
Gorytodes uncanaria *Guen.* California	180	1214
+**Aspilates dissimilaria** *Hübn*	182	1216
+**Aspilates coloraria** *Fabr.* Georgia	183	1217
Aspilates sigmaria *Guen*	184	1219

Zerenidæ GUEN.

***Abraxas? ribearia** *Fitch*	208	1268
+**Zerene catenaria** *Cram*	222	1286

Ligidæ GUEN.

Doryodes acutaria *Her.-Sch.* Georgia	233	1395
Doryodes spadaria *Guen.* Florida	234	1306

Larentidæ GUEN.

	Vol. X, pa.	No.
†**Oporabia dilutata** *Alb.* Europe, [New York, *Fitch*]	262	1333
Larentia implicata *Guen.* California	284	1367
Eupithecia subapicata *Guen.* California	331	1438
Eupithecia coagulata *Guen.* Pennsylvania	339	1451
Lepiodes scolopacinaria *Guen.* United States	360	1484
Ypsipetes pluviata *Guen.*	378	1505
Melanthia albicillata *Linn.* Europe	382	1510
†**Melanthia rufiscillata** *Guen.* Canada	382	1511
†**Melanippe gothicata** *Guen.* [=*M. hastata* Linn]	388	1521
***Melanippe lacustrata** *Guen.* Canada	395	1535
***Melanippe intermediata** *Guen.* Canada, Pennsylvania	395	1536
Melanippe iduata *Guen.* Canada	403	1548
Anticlea vasiliata *Guen.* Canada	407	1557
Coremia convallaria *Guen.* California	410	1561
Coremia defensaria *Guen.* California	411	1562
†**Coremia propugnata** *W.-V.* Europe, North Amer	412	1567
†**Coremia ferrugata** *Alb.* Europe	413	1568
Coremia orthogrammaria *Led.* Georgia	417	1574
Coremia plebeculata *Guen.* California	419	1580
Camptogramma fluviata *Hübn.* Europe, North Amer	429	1601
†**Camptogramma gemmata** *Hübn.* Europe	430	1602
†**Phibalapteryx intestinata** *Guen.* Canada	432	1605
Scotosia hæsitata *Guen.* California	444	1629
†**Scotosia undulata** *Linn.* Canada	449	1640
Spargania magnoliata *Guen.* Canada	455	1653
†**Cidaria hersiliata** *Guen.* Canada	464	1672
†**Cidaria russata** *W.-V.* Europe, Canada	464	1673
Cidaria mancipata *Guen.* California	468	1674
†**Cidaria diversilineata** *Hübn.*	475	1689
†**Cidaria gracilineata** *Guen.*	476	1690

Eubolidæ GUEN.

Eubolia custodiata *Guen.* California	491	1[illegible]15

Sionidæ GUEN.

†**Heterophleps triguttaria** *Her.-Sch.* Pennsylvania	514	1751
***Odezia albovittata** *Guen.* Canada	520	1757

ADDENDA.

	Vol. V.	
Nephelodes rubeolans *Guen.* New Holland? [Halifax, N. S., *Bethune*],	130	205

	Vol. VI.	
Aspila virescens *Fabr.* West Indies, [North Amer., *Grote*]	175	923
Tamila nundina *Drury*	176	925
Anarta melanopa *Becklin.* Lapland, Alps of Switz.. [N. Amer., *Grote*]	190	950
Anarta funebris *Hübn.* Chamouni, [North Amer., *Grote*]	191	952
Anarta amissa *Lef.* Lapland, [North Amer., *Grote*]	192	953
Anarta algida *Lef.* Norway, Sweden, Lapland, Greenland	192	954
Anarta melaleuca *Becklin.* Lapland, [North Amer., *Grote*]	193	956
Anarta cordigera *Sebaldt.* Lapland, Switzerland, [N. Amer., *Grote*]..	194	957
Anomis grandipuncta *Guen.* Brazil, [Canada, *Bethune*]	400	1266
Mania typica *Linn.* Europe, [Canada, *Bethune*]	417	1286

XII. NOTES ON CUCULLIA INTERMEDIA SPEYER.

In the reference made in the following paper by Dr. Speyer* to a manuscript description received from me of the larva of *Cucullia intermedia*, an error has occurred in the translation of the description sent to him, whereby the dorsal and lateral spots, which constitute the entire colorational marking of the larva, are designated as "reddish" instead of *orange*, as originally written. The figure of the larva, to which he also refers, was a copy by a friend from a colored figure made by me. Upon reference to my original figure, the spots were found to be inaccurately colored, being represented as reddish, instead of conforming to the description. In the copy taken from it, it is possible that a still further variation from the true color may have occurred, warranting its indication by Dr. Speyer as "lilac." These errors may seem quite trivial: they would not be deemed of sufficient importance to demand a formal correction at the present time were it not that the color of the larval spots is introduced by Dr. S. as a prominent specific feature in his comparison of *C. intermedia* with *C. lucifuga*. A few omissions and other minor differences appear in the following paragraphs in the translation, as compared with my notes; the latter read as follows: "Larva shining black, covered closely with minute granulations; sides with thirteen orange spots, one on each segment except the twelfth, which has two small ones; the first four are quadrangular, and the next seven are semicircular or triangular." The representation of the larva on plate 8, fig. 7, is from my original figure, but does not faithfully portray the form of the spots.

The larva, in all probability, occurs on the common burdock (*Lappa officinalis*), as I found, several years ago (as appears from notes made in 1857), three crushed larvæ upon a sidewalk at Schoharie, by the side of which a number of burdocks were growing, with their leaves much eaten. In two or three instances in which I have taken the larvæ after their last molt and matured them on burdock, I am unable, in the absence of memoranda, to recall positively the circumstances under which they were found, but my impression is, that they occurred on the sidewalk at an early hour of the day. The habit of the larva, we may presume, is to conceal itself during the day beneath stones or other objects lying on the ground, and to come forth after dark to take its food.

As the closely allied European species, *C. umbratica*, feeds on the leaves of lettuce (*Lactuca virosa*), and on several species of sowthistle

* A. Speyer, M. D., of Rhoden, Fürstenthum Waldeck, Prussia.

(*Sonchus arvensis*, *S. oleraceus* and *S. palustris*),* it is probable that *C. intermedia* is not confined to the burdock, but may be found on others of the Compositæ.

The larvæ taken by me were fed in a box containing a few inches of earth, in which it was presumed that they would bury themselves for their transformation, but, instead of doing so, they were found to construct their cocoons upon the surface. In one instance in which the commencement of the operation was observed, the larva was seen to attach its thread to the side of the box, at a height of about an inch from the ground, and to carry it thence outwardly to the ground, at an angle of about 45°. A number of threads were thus spun backward and forward within the space of perhaps an half-inch laterally, to which other threads were attached, running in different directions beneath. When these last had been made to define somewhat of an oval form, inclosing the larva, particles of earth were taken up in its mouth, as the operation of spinning continued, and deposited on the viscid thread, until a wall was built up around it, entirely hiding it from view.

In this manner, a firm cocoon is constructed, which, to ordinary observation, appears simply as a ball of earth, but which, under a lens, shows clearly the silken threads traversing every portion of it. The cocoons rested with one end on the surface of the ground, with a side, somewhat flattened, attached to the box. One in my possession is ellipsoidal in form, and measures six-tenths by nine-tenths of an inch in diameter.

A pupa-case of *C. intermedia* in my collection (Plate 8, fig. 6), in which the abdominal rings have been contracted in the escape of the imago, so that the anal spine is opposite the tip of the extended tongue-case, measures seven-tenths of an inch in length. The free end of the tongue-case extends fifteen-hundredths of an inch beyond the wing-cases, at which point it is enlarged, and, apparently, has its apical portion recurved and folded upon itself for about one-third its length. The anal spine is smooth, short, curved, rounded at tip, and hollowed beneath. The shell is thin, translucent, of a testaceous color, with the stigmata, a mesial line on the apical portion of the tongue-case and tip of the anal spine, black.

There are two annual broods of *intermedia*. In addition to the August brood mentioned by Dr. Speyer, they have been observed by me during the latter part of May, taking their food upon the wing, at the hour of twilight, from lilac blossoms (*Syringa vulgaris*), associated with *Deilephila chamænerii*, *Amphion Nessus*, *Thyreus Abbotii* and *Sesia Thysbe*, and imitating very closely these sphinges in their method

* *Newman's Natural History of British Moths*, p. 436. 1869.

of feeding and in their flight. Fig. 5 of Plate 8 represents the female moth.

The larva of *Cucullia convexipennis* Gr.-Rob. (conspicuously marked with a dorsal stripe of brick-red on a ground of black, and with a broad lateral stripe of yellow, broken transversely into lines resembling Roman letters), which I have taken during the months of September and October, feeding first on the leaves and later on the flowers of the golden rod (*Solidago canadensis*), also constructs a cocoon of earth and silk; but, unlike that of *intermedia*, it is placed beneath the surface of the ground. Although not so firmly built as that of its congenor, it is sufficiently compact to admit of the escape of the imago through an opening made in the end, without destroying its ellipsoidal form.

We are indebted to Mr. E. L. Graef, of Brooklyn, L. I., for the able translation of the following paper of Dr. Speyer, which may justly be regarded as a model of entomological criticism. The thirty years of close study which its author has given to the Lepidoptera of Europe, have made him so thoroughly conversant with the European forms, that he is now prepared to continue their investigation with unfailing interest to himself, and greatly to the advancement of science, through their comparison with representative and allied species from other portions of the globe. The opportunity of very favorable comparison with many of the New York species has been afforded him through large collections reared from the larvæ, or carefully made in the field, and subsequently prepared in a superior manner by Mr. Meske, to whom reference has been made in these pages less frequently than his labors deserve, or my obligations to him demand.*

*Mr. Meske's field collections are made with unusual care. A gauze net is used by him, of so delicate a texture that the captured insect, in its efforts to escape, may brush against its sides without the loss of any of its cilia. As quickly as possible it is withdrawn from the net in a wide-mouthed bottle, and speedily quieted by a few drops of chloroform, poured on some cotton contained in a glass tube passing through the cork. When the insect is dead, or nearly so, it is carefully turned out on the palm of the left hand, and in that position pinned, without taking it, as is usually done, between the fingers. In this manner, even the strong-winged Hesperians may be secured without the least injury to their thoracic garniture, or to their slightly attached cilia — in an absolutely perfect condition.

I have found a lump of cyanide of potassa secured by a piece of gauze to the stopple of a bottle (a French mustard jar, with its hollow, screw stopple, forms an excellent collecting bottle), to be more convenient for use than chloroform, and nearly as prompt in its anæsthetic effects. As the larger insects soon revive after being transferred to the collecting box, unless left under the influence of the potassa for fifteen minutes or more, the field collector will find it convenient to provide himself with a duplicate bottle, for use while the occupied one is resting in an inverted position in his pocket. An insect killed in this method remains in good condition for setting, wholly free from the rigidity which often attends the use of chloroform.

One result of these comparative studies is presented in the following paper, and others appear, in part, as notes to some of the preceding pages. Reference would further be made to some recent determinations by Dr. Speyer of erroneous generic references of several of our Lepidoptera, if we were confident that, in extracting from correspondence, we would not be anticipating intended publication in European journals. Whenever published, we bespeak for them, on the part of our entomologists, the consideration to which they are entitled, as coming from one who, although his modest labors have not secured for him an extensive reputation in this country, has been pronounced by perhaps our highest American authority, "the foremost student in Lepidopterology in the world."

(E.)

[From the Stettiner Entomologische Zeitung, 31 Jahrgang, No. 10-12, 1870.]

ON CUCULLIA INTERMEDIA Nov. Spec. AND C. LUCIFUGA W.-V.

BY A. SPEYER, M. D.

Of the group of Cucullia, of which *C. umbratica* Linn. [Plate 8, fig. 4] is the most common European representative, Guenée in his well-known writings, mentions only one American species, viz., *umbratica*, which he represents (Noctuélites II, p. 147) as "commune dans toute l'Europe et l'Amérique Septentrionale." Walker also knows of but one American species of this group, not *umbratica*, but *chamomillæ* W.-V., represented in the British Museum by one specimen from Hudson's Bay and one specimen from the State of New York (*List of Spec. of Lepidop. Ins. in the Collec. of the Br. Mus.*, XI, p. 650). Through the kindness of my friend Mr. Meske, of Albany, N. Y., I have received specimens of the species which, according to his authority, is generally known in America as *umbratica* [Plate 8, fig. 5], and this species is neither *umbratica* nor *chamomillæ*, but is so nearly allied to *C. lucifuga* W.-V., that I was at first disposed to take it for a local variety of the last-named species. The receipt of a greater number of specimens from America, accompanied with a drawing and description of the larva, enable me to place the identity of the species beyond all doubt. I have named it *intermedia*, it being between *lucifuga* and *lactucæ* W.-V., having the coloring of the first-named species with the form and markings of the latter, but, in fact, allied more closely to *lucifuga*.

The question now arises, is *intermedia*, which was formerly known in America as *umbratica*, also the identical *C. umbratica* of Guenée? Guenée was too well acquainted with the differences between *umbratica* and *lucifuga* and their allied species to confound *intermedia* with the so dissimilar *umbratica*, if he really had *intermedia* before him. He does not, however, expressly say that his statement as to the occurrence of the species is founded on his own examination, for "commune" *umbratica* certainly is not, otherwise it would not have escaped my entomological friends in America. I am led to believe that Guenée's statement is only a reproduction of the error made by American collectors; he, however, cites no American authority. Neither would I take Walker's *chamomillæ* for *intermedia*. It is not probable that a species, which seems to be as common in the northern United States as *umbratica* is in Europe, should not be represented in the British Museum; and whether Walker is correct in his distinction of the species is very questionable. Of *lucifuga*, he mentions only a single European specimen in the Museum.

The following description is based on the comparison of 10 *intermedia* (4 ♂ and 6 ♀) from New York, with 7 *lucifuga* (3 ♂ and 4 ♀) from Austria, Bavaria, Switzerland, Silesia and Thüringen.

Expanse about the same; the anterior wings of the largest females of both species expand 23 mm., of the smallest males of *intermedia* 20 mm., and of *lucifuga* 21 mm. As far as I can discover, the abdominal construction is the same, as is also the sharp cut of the wings so characteristic of this genus. Anterior wings sharp, posterior margin oblique, slightly wavy and curved toward the interior angle. The anterior angle of the secondaries obtuse, rounded; posterior margin slightly undulated, irregular and somewhat wavy. Color of anteriors the same as in *lactucæ*, or a little darker and more inclined to blue, a uniform bluish-gray, with slight shadings of light mold-gray, especially in the interior margin and terminal region, the latter traversed by lighter rays, but often very indistinct and variable. The last may be said of the two zigzag lines, which are, however, formed precisely as with *lucifuga;* of these the front line is almost always distinctly visible, and the hinder one only distinct near the interior margin. Toward the anterior margin these lines become broader and macular, and here, between the two, a third line is visible, darker and stronger than either. The dorsal vein,* and those in the terminal region, appear as very fine black lines. The black ray emerging from the base is long and fine. The three rays in the terminal region — the longer and finer ray in cell no. 4†, which emerges from the outer margin of the

[*The submedian of American entomologists.]

[†Among the German entomologists, the nerves and nervules are designated by the numbers 1, 2, 3, 4, etc., counting on the posterior margin from the posterior toward the anterior angle of the wing. The first nervule of the median uniformly bears the number 2. The submedian, the internal, and whatever interior nerves may exist, are known as 1 a, 1 b, 1 c, enumerating from the internal margin. The nerve opposite the discal cell, and usually given off from the cross-vein, is no. 5, and, for the sake of uniformity, is so counted even when absent: no. 8 usually terminates just below the apex of the wing. The cells (*interspaces* of Clemens and others) are as follows: between the internal margin and nerve adjacent is 1 a; if this nerve be the internal, then the space between it and the submedian is 1 b, followed by 1 c; but when there is no internal nerve, then this latter cell becomes 1 b. Between the nervules of the median and of the subcostal, the cells bear the numbers of the nervules which precede them; thus, between the first median nervule (no. 2) and the second (no. 3), is cell no. 2; opposite the discal cell, separated by nerve 5, are cells 4 and 5, and thus, to cell 12 or 13, if the venation permit, on the basal portion of the anterior margin.]

reniform mark and runs along without quite reaching the margin; a shorter ray resting on the margin in cell no. 3; and, lastly, the short and usually strong and somewhat oblique ray, with a whitish border, in cell 1 b, near the interior angle — have the forms and positions as with *lucifuga*, but the markings are mostly fainter; in some specimens quite indistinct, but are never wanting. A row of black lines or lunettes, which, with the female, form nearly an unbroken line, are placed on the extreme terminal margin as with *lucifuga*, and the gray fringes, divided in the middle by a light line, are identical in both species. The orbicular spot is totally wanting; the outer border of the reniform spot is more or less completely marked by dark lines; most constantly its lower portion, next in frequency the outer, and lastly the upper border. These marks are not discernible in the male. On the discal cross-vein is, in some cases, an indistinct dark spot. The blackish-gray secondaries become lighter toward the base. With the male, the secondaries are always lighter colored, as is the case with all of this genus; in the female they are sometimes uniform black-gray. The white fringes are divided by a hair-like dark line, which latter is in some cases so obscure and imperfect that they seem totally white. Under side of the primaries ashy-gray, bordered along the anterior and interior margins with whitish-gray. The secondaries of the female whitish-gray, with a more or less broad hinder margin of a darker gray; in the male, almost completely dull white, with dark veins. On the discal cross-vein is a distinct, rounded, dark reniform mark, which is often connected with the base of the wing by a streak of the same color.

Color and markings of the remaining parts of the body are precisely as in *lucifuga*, with the exception that *intermedia* is more bluish-gray on the thorax, corresponding to the color of the wings. The center, between the shoulder covers of *intermedia*, is also darker, blackish-gray; on the abdomen are four distinct dark, downy tufts, as is the case with all its allies. The female is remarkable for the pointedness of its abdomen. The downy hair which adorns this part, is on the sides and on the underside of the last segment, in a greater or less degree, of a rust-yellow color. The most distinctly marked females have, on each side of the segment, a rust-yellow spot, the base of which is formed by the last incisure; and on the flat tuft, which covers the sexual organ from beneath, is a transverse spot of the same color. When I first noticed this singularity of coloring, I supposed it to be a peculiarity exclusively characteristic of *intermedia*, as I did not observe it in any others of the genus, and has, as far as I know, never been mentioned in any description. It proved, however, to be the same with *lucifuga*, and, furthermore, that this marking is not constant. Four of my female *intermedia* show the spots large and distinct, the fifth, small and less brightly colored, while with the sixth it is reduced to simply a few of the hairs of the tufts which form the gray ground, being sprinkled slightly with rust-yellow. In two females of *lucifuga* the spots are similarly conspicuous, as usually in *intermedia*, while in two other bred specimens they are apparent only in a slight sprinkling of rust-yellow upon the gray tufts.

Of the allied species which I can compare, the Russian *C. balsamitæ* (1 ♀) alone has these spots of the same shape and coloring as *intermedia*, while the much closer allied species, *lactucæ* and *campanulæ*, show no trace of them; *umbratica* ♀ is also without this mark, although some specimens have a slight sprinkling of dull rust-yellow scales in the last segment. In *chamomillæ*, and especially in its variety, *chrysanthemi*, the parts named show only an indistinct surrounding of dull yellow or rust-brown.

Invariable distinguishing marks between *intermedia* and *lucifuga* are, therefore, not to be found either in the markings or form of the imago; the coloration only is different. The color of *lucifuga* (*Hübner Noct.*, fig. 262; *Freyer N. Beitr.*, tab. 431) on the thorax and anteriors is less inclined to blue, being more ash-gray, the lighter shades more strikingly whitish, the shadings stronger and inclined to brown, while *intermedia* is devoid of all brown whatever. The yellowish-brown coloration of the reniform mark which *lucifuga* shows more or less distinctly, is also wanting in *intermedia*.

From *C. lactucæ* (*Entom. Zeit.* 1858, S. 83 fig.), with which it has a coloration in common (though somewhat darker), *intermedia* differs by the sloping cut of the anteriors, the points of which are with *lactucæ*, obtuse and rounded, and the margin more convex. The dark markings in the terminal region of this species are also much fainter, or scarcely discernible, particularly the short dark line in cell no. 3, near the margin (so distinct in *intermedia*), which in this is totally wanting. The terminal margin is also without the strong black lines or lunettes; the dark dividing line of the fringes of the primaries is, on the contrary, broader and stronger. The middle of the thorax is not materially darkened, the tufts of the abdomen are less robust and lighter, being brownish-gray. *C. campanulæ* has a similar ground color, but a very different cut of the secondaries, they not being rounded at the anterior angle, but strong, pointed and almost falcate; the hinder margin is slightly indented. The deep, black markings, especially the lines in cells nos. 4 and 1 b., are longer and stronger, and are very conspicuous on the gray ground color. Lastly, *campanulæ* has a fine comma-like line in the median region, which emerges from the middle prong of the oblique line, which, of all the other species, *umbratica* alone possesses. The last-named species also corresponds in the cut of the secondaries with *campanulæ*, and is additionally easily distinguishable by the faint, dull yellow streak on the primaries, and in the white secondaries of the male. *C. santolinæ* Ramb. and *C. chamomillæ* have indented secondaries and no black crescentic lines on the terminal margin, and are characterized by the deep black veins which run into the middle of the fringes. *Santolinæ* is also smaller than *intermedia*, and *chamomillæ* is differently colored from it. The remaining species are even less liable to be confounded with *intermedia*.

So slight as the differences between the so simply and monotonously marked imagines may appear, they are, on the contrary, very strongly marked in the brilliant and variegated colors of the larvæ. The larva of *intermedia* is no exception to this rule, although, as may be expected, in its earlier stages, it approaches nearer *lucifuga* than to the other species. The larva of *intermedia* is described by Mr. Lintner of the New York State Museum of Natural History, who is a close and reliable observer, as follows: "Sides with reddish spots, one on each segment, the first four square, the seven following semicircular, the two on the last segment small. Dorsum with two small reddish spots on the first four segments, a very indistinct spot on the fifth and one on each of the last four segments. The caterpillar is very lively in its motions and feeds generally at night." According to the figure accompanying the description, the ground color of the mature larva is dull black, head and feet the same, color of the spots, which are of about the same size as with *lucifuga*, lilac. Of its food and transformations I have not been informed. Very perfect specimens of the imago were taken by Mr. Meske in Albany, N. Y., early in August.

For the sake of comparison I give a detailed description of the larva of *lucifuga* (which is however already accurately described by Treitschke, Schmett. v. Eur. X, 2, 128) from two specimens prepared by Mr. O. Schreiner in Weimar in his masterly style. Mature larva black with three parallel rows of orange-yellow spots. The row on the dorsum consists of twenty-five small spots, two on the first segment, three each on the second and third, the first of which is rectangular, two each of a round form on the fourth to the eleventh segments, and the last two form a band instead of spots. The row on the side consists of thirteen spots, one on each segment.* Those on the second and third segments are broader, made so by the small spot in front running into the larger one behind; those on the twelfth and thirteenth are smaller and nearly touch each other; the last is a spot running

[*The author regards the body of the larva as consisting of *thirteen* segments, the twelfth segment being properly divisible, it is claimed, into two distinct portions.]

cross-wise. The rest are rounded, somewhat elongated transversely, which is also the case with those on the dorsum. The larva is entirely naked, and all except the collar thickly granulated. Head and feet black, the first granulated and wrinkled in front, with the clypeus furrowed cross-wise. In its youth the larva has a quite different appearance, it having stripes instead of spots. The young larva, (length thirty millimetres) which I have before me is velvety black with light yellow stripes, the dorsal stripe slightly mixed with reddish-yellow, and the broad side-stripes dotted with black. The skin is not granulated, but each of the middle segments is adorned by four tiny warts surmounted each by a rather long bristle. The head is smooth, black, with yellow markings in the middle and on the sides.

It is probable that the larva of *C. intermedia* changes in its appearance after its last molting in a similar manner. In this stage it has the ground color and the three rows of spots in common with *lucifuga*, but the color, number and division of the spots are different. The spots of *intermedia* are lilac, those of *lucifuga* orange; the first has thirteen dorsal spots, the other twenty-five, etc., etc. The larvæ of *lactucæ* and *campanulæ* have no resemblance whatever with that of *intermedia*. The larva of *umbratica* has more resemblance to it, but here the differences in color and markings are too striking to necessitate a comparison.

PLATE VII.

FIG. 1. *NISONIADES LUCILIUS Lintner.* ♀.
" 2. *NISONIADES LUCILIUS*, showing under surface of wings of ♂.
" 3. *NISONIADES PERSIUS Scudder.* ♂.
" 4. *NISONIADES PERSIUS*, showing under surface of wings of ♀.
" 5. *NISONIADES ICELUS Lintner.* ♂.
" 6. *NISONIADES ICELUS*, under surface of ♂.
" 7. *NISONIADES MARTIALIS Scudder.* ♂.
" 8. *NISONIADES MARTIALIS*, under surface of ♀.
" 9. *NISONIADES BRIZO Boisd. et Lec.* ♀.
" 10. *NISONIADES BRIZO*, under surface of ♀.
" 11. *NISONIADES AUSONIUS Lintner.* ♂.
" 12. *NISONIADES AUSONIUS*, under surface of ♂.

[NISONIADES]

G. B. Simpson, Del. Swinton, Lith.

PLATE VIII.

FIG. 1. *Hemileuca Maia* (*Drury*). ♂.
" 2. Pupa of *Hemileuca Maia*.
" 3. Egg-belt of *Hemileuca Maia*.
" 4. *Cucullia umbratica* (*Linn*). ♂.
" 5. *Cucullia intermedia Speyer*. ♀.
" 6. Pupa-case of *Cucullia intermedia*.
" 7. Larva of *Cucullia intermedia*.
" 8. Larva of *Ellema Harrisii Clemens*.
" 9. Pupa of *Ellema Harrisii*.
" 10. *Ellema Harrisii*, showing under surface of wings of ♂.
" 11. *Ellema Harrisii*, showing under surface of wings of ♀.
" 12. *Ellema pineum Lintner*. ♂.
" 13. *Ellema pineum Lintner*. ♀.
" 14. Chrysalis of *Melitæa Nycteis Doubleday*.

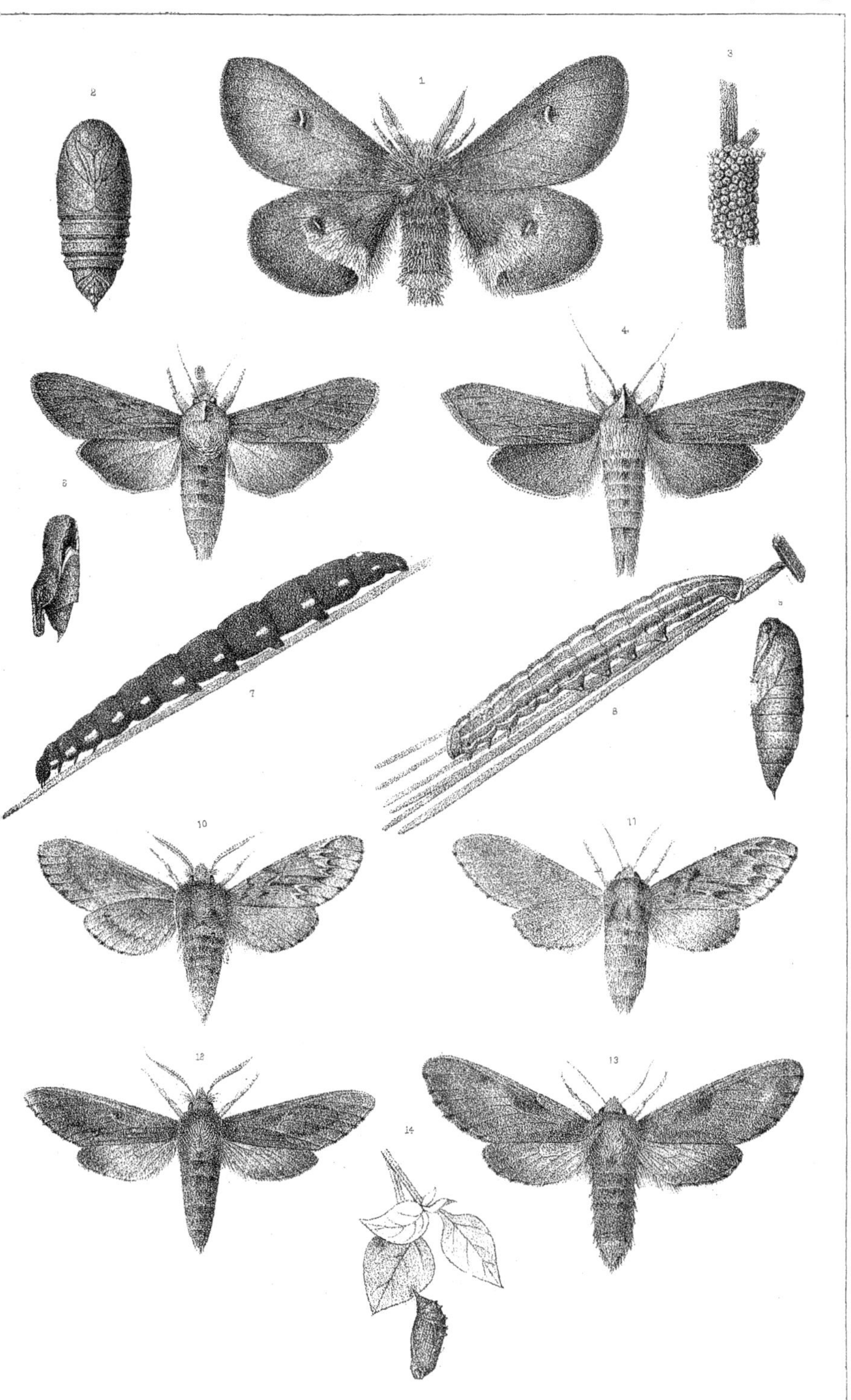

S. Simpson, Del. Swinton, Lith.

ENTOMOLOGICAL CONTRIBUTIONS—NO. II.

I. ON THE LARVA AND IMAGO OF SESIA DIFFINIS, *Harris.*

By J. A. LINTNER.

The larva of the above Sphinx, nearly mature, was taken in the vicinity of Albany, July 4th, feeding on *Diervilla trifida* (bush honeysuckle). The following description represents it at maturity:

The head is oval, with small whitish points. The body tapers moderately anteriorly, and on the last three segments; the vascular line is distinct, and of a brownish shade; the dorsal region is greyish-pink; the lateral region yellow-brown, deepening into reddish-brown below the stigmata; the ventral region is brownish-red; midway between the vascular line and the stigmata is a pale yellow line, proceeding from the posterior portion of the collar, and terminating in the horn. The granulations of the collar anteriorly are yellow; those of the body are white and double-rowed upon the first annulet * of each segment, arranged rectilinearly on the last annulet, and irregularly on the intervening ones. The stigmata are oval, black-bordered, and with a white dot at each end. The legs and prolegs are unicolored with the ventral region. The caudal horn is reddish, straight, acutely granulated, and measures two-tenths of an inch in length.

* In the Sphingidæ, and in some other families of the Lepidoptera, the segments of the larvæ are divided by impressed encircling lines into several (usually eight) subequal parts, which, by Clemens and others, have been denominated *wrinkles*. In consideration of their constant character and marked features, it would seem proper that they should be known by a name implying less of irregularity and chance, and, regarding them as subdivisions of the primary "rings" of the larva, I have, accordingly, in these notices, designated them as *annulets*.

In the Sphinges, the segments four to ten inclusive, or all the stigma-bearing segments, except the two terminal, are divided into eight of these annulets (on segment four the two anterior are not readily detected). The first three annulets encircle the body, forming complete rings; the impressed lines which define the remaining five become obsolete over the prolegs. The position of the stigma is uniformly between the third and fourth annulets, somewhat encroaching on the latter. In some of the species, the first annulet is subdivided laterally.

These annulets are a prominent feature in the ornamentation of the larva of *Ceratomia quadricornis* Harr., where each one is crowned with an elongated papilla, forming, in combination, the conspicuous dorsal row of papillæ or fleshy serrations which imparts so marked a character to that peculiar larva.

The larva measures one inch and six-tenths in repose. When observed from above in this condition, its head is not visible, being bent downward in such a position that its plane is almost parallel to the surface on which the larva rests.

The above larva fed for a day after its capture, when it fastened a leaf, by a thin irregular web of brown silk, to the side of the glass beneath which it was confined. Within this slight shelter, it transformed to a pupa on the day following, the 6th.

The imago (a female) emerged on the morning of the 19th of July, after a pupation of thirteen days. Before the wings had expanded, their entire surface was covered with scales. In fifteen minutes from the time of its escape from the pupa-case, the wings had attained their full size. They remained folded together over the back, showing their under-surface for a half-hour following, when they were brought down to their normal position. They were now seen to be overspread with light brownish scales having a golden reflection, but so thinly distributed that the usual transparent portion of the wing was plainly separable from the densely scaled marginal region.

Desirous of securing so novel a specimen in a perfect condition, I intended to leave it undisturbed for a few hours, until the danger of bleeding from too early pinning had passed. Meanwhile, the strong rays of the sun had encroached upon that portion of the breeding-cage where the moth was resting, and driven it to a shaded corner. In this change of position, its wings, after the habit of many of the moths at this stage, were probably put in vibration, but not used in flight about the cage, for the cilia and thoracic scales were intact; yet this moderate use of them sufficed to destroy the peculiar beauty of the specimen, by divesting it of the greater portion of the very slightly attached scales. Upon the anterior wings, only those remained which bordered the nervures and posterior marginal band. Over the posterior wings, they were still uniformly distributed, but not so closely as at first.

It will be interesting to those who have the opportunity of making the observation, to ascertain which of our Sesias emerge from their pupal state, with scales covering that portion of their wings which we are accustomed to see as transparent. *Sesia Buffaloensis* Gr. & Rob., is known to do so in both sexes, and to retain them, to a small extent, when taken in a comparatively fresh condition in the field. While the adherence of the scales in this species is stronger than in *S. diffinis*, in no field specimen of which have I ever seen them remaining even along the nervures, yet their attachment is very slight, for with the utmost care, I have never succeeded in preparing *S. Buffaloensis* ex larva, without an attendant loss of many of its scales. Field collections of *Sesia gracilis* Gr. & Rob., regarded as "perfect," show none of these scales.

S. diffinis appears to be eminently a day-flier, for I have never taken it at twilight when others of its congenors and of Thyreus have been abundant. In 1869, I captured it twice on lilac blossoms, but it was usually met with hovering over flowers in open spots, particularly those of *Lupinus perennis.* It was observed at Center, N. Y., during the year 1869, on May 25th, 27th, June 1st, 3d, 7th, 9th and 15th. The earliest date of its capture which I have recorded, is May 12th. Its usual time of apparition, in this State, would seem to be the last week of May. It has, as appears from the above larval observations, a spring and a summer brood; the latter, in seasons not unusually forward, may be expected about the last of July, extending into August.

The female appears to be quite rare. Among a considerable number of captured individuals of the species in my collection, not a single female occurs.

II. TRANSFORMATIONS OF SESIA BUFFALOENSIS, *Gr. and Rob.*

An egg of the above comparatively rare Sphinx was found on the snowball (*Viburnum opulus*), at Schoharie, N. Y., August —, 1868; it was nearly round, smooth, and of a pale green color. The time of its hatching was not noted.

The *young larva* was of a uniform whitish-green, with a straight, dark brown caudal horn.

Its *first molt* occurred August 24th: Length, 15-100ths of an inch; color, uniform pale green, of the shade of the midvein of the leaf. Body, under a simple lens, showing a number of delicate hairs. Caudal horn, brown, smooth and straight. When in position for the next change, its length was 32-100ths of an inch, and its diameter 4-100ths of an inch.

Second molt—August 30th: Length, 35-100ths of an inch. Head uniform pale green, with short, fine whitish hairs. Body, slightly hairy, pale green, with a lateral stripe of yellow green; segments with a yellow-green dot in front on each side of the vascular line, and a few smaller ones on the posterior half. Caudal horn straight, nearly cylindrical, light red, striped anteriorly near the base and tipped with brown, and spotted with brown intermediately; borne at an angle of about forty degrees.

At this stage the larva died from injuries received from some larger ones with which it was confined in a small box while their supply of food had become exhausted. The collection of three other larvæ of the species from the same bush, after their first molting, afforded the opportunity of continuing its history.

Third molt (six days after their preceding one), on August 24th: Length, one-half inch. Head granulated. Collar bordered anteriorly with about twelve small tubercles. Body with features as before recorded, and in addition, minutely papillose. Caudal horn light red, regularly tapering from base to tip, covered with spinules which anteriorly and posteriorly have black bases. Stigmata deep orange, with a yellow-green dot at each end; those of the proleg-bearing segments bordered before with a deep orange line. Legs at the base marked with black; prolegs on the outer side, and body beneath on the last two segments, pale red.

Fourth molt—six days later, on August 30th: Length, 65-100ths of an inch; breadth, 13-100ths. Head green, of the shade of the lower side of the leaf, its surface appearing shagreened under a lens, marked with indistinct lateral stripes, and dotted with whitish granulations, which diminish in size toward the center; mandibles yellow, black tipped; eyes on a fuscous crescent. Collar with whitish granulations, except on the anterior margin where they are orange. Body greenish-white dorsally, with a vascular line of rose-pink interrupted at the incisures; a lateral stripe of yellow-green papillæ of one to each annulet, edged above with darker green, and below with green shading deeper to the prolegs; lateral papillæ greenish-yellow; ventrally, from the fourth segment to the posterior extremity, concolorous with the vascular stripe. Caudal horn curved, rose-colored, tipped with ferruginous, with fuscous spinules anteriorly and posteriorly. Stigmata crimson, white-dotted at the extremities, and surrounded with rose. Legs and prolegs ferruginous basally, next fuscous, and rufous terminally.

As the larva approaches maturity, it becomes more white dorsally, the red of the vascular stripe changes to white, and the red surrounding the stigmata disappears. With the distention of the skin, the papillæ change to whitish ocellations.

On September 6th, the above larvæ commenced constructing their cocoons under leaves drawn against the sides of the glass in which they were confined. The maximum length of the cocoons is one inch and one-half. The silk of which it is composed is of a bronze color, and in so small a quantity as to permit the larva to be distinctly seen through its meshes. On the 11th, they changed to pupæ of a chestnut color, broadly banded with testaceous at the incisures.

The pupæ, after wintering in a cold apartment, were transferred early in April to a warm room. On the 26th of April, from the three pupæ, two males and one female were disclosed. The "vitreous space" in the wings of each is thinly covered with scales.

On September 19th of the following year, another of the larvæ was taken from the same snowball, which made its cocoon on the day following. In the month of September of the two succeeding years, leaves were noticed which had probably been eaten by this larva, indicating it to be an annual visitant of this particular shrub.

I have collected the imago also at Utica, N. Y.

III. ON THE LARVA AND PUPA OF THYREUS ABBOTII Swainson.

Male larva.—Head large, of the diameter of the first segment, subquadrate, shagreened, two broad stripes of brown on the front, behind which is a subtriangular whitish patch, occupying all the lateral portion except a small patch of brown posteriorly; the clypeus half as long as the head. Collar pale yellow, divided on each side by a narrow dark brown line, which is continued over the following segment. Body reddish-brown, with numerous longitudinal linings in darker brown; dorsally and extending half-way down the sides is a series of nine sharply defined, bright yellow spots, which commence on the second annulet of a segment and extend backward to the incisure (leaving intermediate transverse reddish-brown bands of a single annulet), and are convex on their sides; the first spot on the second segment is triangular, the second is suboval, and the others of a uniform outline; on the sides is a stigmatal series of yellow spots, one to each segment, of which the first three and the last three are confluent; the six intermediate ones are of a uniform subtriangular outline, the last four of which have their lower rounded angle reaching downward on the base of the prolegs, their anterior angle in range with the front margin of the dorsal spot and touching the inferior portion of the stigma, and their posterior angle resting in an ovoid outline on the first annulet of the following segment, or impinging in a point on the second annulet at a height of the superior portion of the stigma. On the eleventh segment is a hump, encircled at the base with a delicate black ring, and surmounted by a vitreous oval tubercle of a fuscous color inclosed by a whitish line, which line is reflected and continued in a circumscribing circle embracing a black curved spot on the front and sides of the oval tubercle. From the base of the tubercle backward to the incisure and ranging laterally with the adjacent dorsal spot is a yellow supra-anal spot. The subcordate anal shield is of a yellow less bright than the spots. The legs are pale red; the prolegs are marked exteriorly with a narrow quadrilateral black spot surrounded with a whitish shade. The ventral region is red, with two rows of pale yellow spots.

The segmental annulets of this larva differ somewhat from those of the Sphinx proper, in that the eighth is subdivided so as almost to form an additional one, and the first has a depressed line subdividing it laterally.

The larva from which the above description was drawn, was taken, July 29th, feeding on the grape-vine. It had nearly matured, measuring two and one-half inches in length.

On the 2d of August, the dorsal spots which were previously yellow had changed to brown, and the lateral ones to a sordid yellow. The larva having fully matured, endeavored to escape from the jar to seek a place for its pupal change.

August 4th, the larva formed a little cavity on the surface of the ground, and covered itself with some pieces of leaves loosely spun together, intermingled with grains of earth. It transformed to a pupa August 9th.

Pupa.—The pupa measured one inch and three-tenths long, by 35-100ths of an inch broad; color, dark brown; head-case, as seen from above, prominent, broad, rounded in front, with the eye-cases projecting; tongue-case buried, extending to the tips of the wings; antennæ-cases, reaching to the end of the middle leg-cases; dorsally, the second segment is moderately wrinkled; the third is narrowed medially by the convex margins of the second and fourth segments; caudal spine polished, short, bifid, with a rugose, flattened, triangular base of twice the length of the spine.

The imago from the above emerged on the 5th of April.

The larva of *T. Abbotii* is peculiarly interesting from the fact that its two styles of ornamentation, in marked contrast one with the other, indicate the sex of the insect, no other instance of which, among the Lepidoptera, is known to us. The dorsal and lateral series of spots, yellow as described above, but frequently and perhaps usually of a pale green color, denote the male; the female being brown, without any trace of the above spots, but with interrupted, dark, subdorsal and stigmatal bands and numerous small longitudinal patches.* The following is a more particular description of it:

Female larva.—Length, two and two-tenths inches; diameter, thirty-seven hundredths of an inch. Head semi-oval, shagreened, a medial depressed line superiorly, two broad brown stripes in front, bordered with paler brown, shading into darker brown behind the eyes. Body cylindrical, with the three anterior segments tapering; dorsally dark brown, shading to lighter on the sides; the annulets with large, subquadrangular spots of light brown; third, fourth and

*For an excellent representation of this sex see *Harris Ent. Corr.*, pl. iii, fig. 1.

fifth segments with a bluish shade dorsally; an indistinct lateral brown stripe, more conspicuous on the first three segments, running into the head stripe; the first two segments have also a brown dorsal line; below the stigmata is a brown stripe less distinct than the lateral one. Caudal tubercle moderately elevated, shining black, surrounded with dark brown. Caudal shield and plates reticulated with dark brown. Prolegs dark brown exteriorly, with a light band near the base.

In the larvæ of *Deilephila lineata* (Fabr.), variations almost as marked as the above, have been observed and figured by Mr. Riley.* It is not improbable that these differences may also prove to be sexual characteristics.

The pupation of *T. Abbotii* is not uniform, and consequently diverse statements appear respecting it. Clemens states that it "takes place in a superficial cell;" and in his generic diagnosis that "it prepares for pupation on or near the surface of the ground." Riley, who has reared the larva, also represents the pupa as "formed in a superficial cell on the ground."† Harris, in observations on some of the larvæ which he had received, writes, "the green-spotted one [male] went into the ground; the others [females] * * * chrysalis on top of ground July 14th."‡ A female brought to me on the 29th of July of the present year, buried in the ground, while the male described above transformed on the surface.

* *Third Rep. Ins. Missouri*, 1871, pp. 141, 142, figs. 61, 62.

† *Second Rep. Ins. Missouri*, 1870, p. 79.

‡ *Entomolog. Correspondence*, 1869, p. 284.

IV. ON THE LARVA OF PHILAMPELUS ACHEMON (Drury).

Egg.—Found on the under side of a grape-leaf, July —. No description was taken, but from recollection, it was of a larger size than those of other sphinges, and nearly round in form. A few days before its development, it was irregularly marked with light red. From the shell having been eaten by the larva after its escape, it could not be ascertained if the color was permanent, or if the shell was colorless and transparent, as are all the sphinx egg-shells which have come under my observation.

Young larva.—The larva emerged July 10th, measuring 11-100ths of an inch, and with a uniform light green color. Its caudal horn was very conspicuous, being one-half the length of the body, very slightly tapering, straight, of a reddish-brown color, and carried perpendicularly to the body.

First molt.—The time was not noted. No change in appearance was observed except in size. Previous to its second molting, its length was 53-100ths of an inch, with a diameter of 3-100ths of an inch. Color, light green, with white dots on the annulets, a subdorsal stripe of regular white spots, and indistinct lateral bands of similar spots directed anteriorly. Head of a uniform delicate green, without stripes or spots. Horn one-fourth of an inch long, dark reddish-brown, covered with minute cilia, and with a prominent green base, borne erect, with its tip directed anteriorly; from the base of the horn, a reddish-brown line, extending to the anal shield. Legs green. The anterior segments of the body are moderately enlarged.

Second molt—July 15th. The body marked as in the preceding stage. The horn of a lighter shade of reddish-brown, and its tip darker than before; anterior to it, on its base, of which it now occupies the posterior portion, a small triangular black spot, with its apex in front.

A day preceding its next molting, the larva measured 8-10ths of an inch in length, with a diameter at the sixth segment of 15-100ths of an inch. Its fourth, fifth and sixth segments are enlarged. The lateral stripes are well defined. The lateral bands show indistinctly, in a yellow-green spot nearly surrounding the stigma, and in another anterior to it on the second annulet; the four posterior bands are more conspicuous than the other two; horn, one-fourth of an inch long, of a fulvous color, and with its tip curving toward the head.

Third molt—July 22d. Immediately after the change it presented the following features: color light yellow-green; the annulets of the segments with whitish granulations, of which those on the lateral portions of the third and fourth segments are annulated with black; lateral bands, cream-colored, margined with black, having the characteristic outline of maturity, showing the three component sub-oval patches, of which the posterior is the largest, and incloses the stigma except at its upper part. Head smooth, of a uniform delicate green. Horn four-tenths of an inch in length, slender, bending anteriorly, ciliated, of a light reddish color, merging into a rose-color at the tip; its base occupying less than the posterior half of the tubercle; anterior to the base is a sub-triangular black wart resting on the crown of the tubercle—the development of the "small triangular black spot" of the preceding stage; the tubercle is yellow, almost surrounded near its base with a band, the extremities of which, instead of meeting, curve upward to the horn posteriorly. Anal shield and plates, legs and prolegs of a uniform green.

A short time after molting, a change of color was observed, and at the lapse of six hours it had assumed a salmon color. After a night's feeding, it was found to be of a reddish-brown color, deeper than that which it ordinarily bears at maturity, and in marked contrast with its former garb. A figure taken of it at this time represents it as one inch and one-fourth long, 12-100ths of an inch in diameter at the central segments, and with a horn, 34-100ths of an inch in length, regularly curving anteriorly from its base to the tip.

Fourth molt—July 27th. Head and body reddish-brown, but of a lighter shade than before. Caudal horn absent, its former position on the tubercle indicated by an oval spot having a central black dot. Tubercle more prominent than in the preceding stage.

The mature larva measured three and one-tenth inches in length, by one-half an inch in diameter, when at rest, at which time its small head is nearly hidden within the first segment, which is withdrawn within the second, over and in range with which projects the third segment, presenting a front almost perpendicular to the body. The posterior segment also descends almost perpendicularly from the tubercle to the terminal prolegs.

The descriptions and figures already published of this larva,* render a further account at the present unnecessary. The excellent figure given of it by Mr. Riley cannot fail of securing for it ready identification.

* Clemens: *Jour. Acad. Nat. Sci. Ph.*, 1859, p. 155. Harris: *Treat. Ins. Inj. Veg.*, 1862, p. 325, f. 150. Lintner: *Proc. Ent. Soc. Ph.*, 1864, v. iii, p. 660. Harris: *Ent. Corr.*, 1869, pl. 3, f. 11. Riley: *Sec. Rep. Ins. Mo.*, 1870, pp. 74, 75, f. 49.

V. SMERINTHUS GEMINATUS SAY, AND ITS SUPPOSED VARIETIES.

A number of eggs were deposited June 12th, by a pinned specimen of the above named Sphinx. An average of twelve of the eggs gave for their longest diameter 74-1000ths of an inch; for their shorter diameter 57-1000ths of an inch; they were slightly flattened, and of a pale green color.

On the morning of the 19th, three larvæ were found in the box, and through the transparent shells of the undeveloped eggs (appearing of a dull green), could be seen the dark caudal horn of the inclosed larva, and the black mandibles busily employed in wearing an opening through the shell. Two or three had made small openings which they were engaged in enlarging, by biting off small portions from the margins.

Young larva.—The young larva was of the length of two-tenths of an inch. Color uniform pale green. Head subglobular, one-half broader then the body, with the eyes and mandibles black. Caudal horn fuscous,.lighter at the base, slightly tapering, straight (curving forward as it emerges from the shell), and carried at an angle of about 85°.

Larvæ continued to emerge during the day, the last appearing in the evening. Upon willow leaves being given them, they commenced feeding thereon. On the third day, traces of the lateral bands and a subdorsal line in lighter green were seen. When disturbed they threw their body upward, supporting themselves on their terminal and one or two pairs of prolegs.

On the evening of the 24th, they commenced taking position for molting. At noon of the following day, one had molted, and by noon of the 26th all but one had undergone the change, presenting the following features:

First molt.—Head triangular, granulated, bordered laterally with a row of larger granulations which culminate at the apex in two tubercles. Body with whitish granulations on the annulets: lateral bands consisting of yellow-green papillæ which extend over three-eighths of one segment and the whole of the two following: subdorsal line of whitish granulations, indistinct, until before the first lateral band where it becomes a thoracic stripe of transversely elongated yellow papillæ, continuing to the collar. Caudal horn straight, pointed, spinous, yellow laterally, brown anteriorly, and with a narrow brown stripe posteriorly. Legs roseate; prolegs green.

Second molt.—June 29th, and of the last two larvæ, on the 30th (thirty in all). Length at rest 45-100ths; diameter 11-100ths of an inch. Lateral stripes of the head rectilinear from the front of the eyes to the apical tubercles. Body pale green, dorsally whitish green; in the lateral bands the first three papillæ are inconspicuous, and those on the first three annulets of the two following segments are smaller than the others and geminate; on segments one to seven, is a small red spot placed on the seventh annulet, above the subdorsal and thoracic lines and anterior to the first four lateral bands. Caudal horn slightly curved, yellow, narrowly lined with reddish-brown anteriorly.

On the morning of July 3d, several had taken their position for molting on the under surface of the leaves, suspended by their terminal, and one or at most two pairs of prolegs, with their body hanging downward. On the portion of the leaf beneath them (in most instances a part of the midvein), for about the breadth of the body and two-thirds of its length, a slight webbing of silk had been spun to serve as a foothold. The head was partially withdrawn from its case, showing a translucency at the tip, and a corresponding enlargement and extension of the first segment. The subdorsal lines had disappeared, while the thoracic lines were still conspicuous in their yellow papillæ. In the larger number of the larvæ, the row of small red subdorsal spots was quite distinct, especially on segments five to eight inclusive, where they occupy the seventh annulet and encroach on the eighth. Length at rest, one inch; diameter 12-100ths of an inch.

Third molt.—July 4th probably, as when again observed on the morning of the 6th, all but two had molted, and most of them had increased materially in size, the largest measuring one inch and six-tenths in length. Immediately after the molting, they were one inch and one-fourth long, by 14-100ths of an inch in diameter. When their colors had developed, they presented the following appearance:

Lateral stripes of the head, yellow. Lateral abdominal bands and thoracic stripe, pale yellow with whitish papillæ, the bands occupying of three segments, three-eighths, the whole, and six-eighths respectively. Caudal horn straight, acutely granulated, pale violet. Anal shield with white granulations as the body. Legs roseate. Stigmata elliptical, dark red: above each stigma (except at the extremities) at about three-fourths its length, is a short delicate hair, proceeding from a minute granulation on the fourth annulet, and directed downward; beneath the stigma, at a distance of about its length, are two similar hairs directed backward, proceeding from the summit of a slightly larger granulation.

On the morning of the 12th of July, quite unexpectedly, it was found that six of the larvæ had left the willows, had undergone a change of color to a sordid apple-green, and were endeavoring to escape from confinement, thereby indicating their having attained maturity, and a readiness for their pupal change. It was the first instance which had come under my observation of pupation in the sphingidæ preceded by only three moltings. That the three above noted, were all that these larvæ had undergone, was beyond all doubt, not alone from the careful observations made, but each cast head-case had been carefully preserved, and of these there were but the three sets.

Mature larva.—Length at rest, one inch and nine-tenths to two inches. Color pale green, whitish dorsally. Head triangular, the apex not rising above the first segment, granulated in pale green anteriorly and in white laterally; the lateral stripes yellow, having within them a row of larger granulations increasing in size to the apex, where the two superior ones are papilliform and of an orange color. Body moderately tapering in the anterior segments: the seven lateral bands pale yellow, except the posterior one which is bright yellow, the anterior one obsolete; their extent, two-eighths to three-eighths, the whole, and from five-eighths to seven-eighths, respectively, of three segments: thoracic stripe with pearl white papillæ larger than those in the bands, commencing on the superior portion of the first segment, and merged into the first lateral band on the fourth segment at its seventh annulet, midway between the stigmata and the vascular line. Anal shield and plates granulated, of a darker green concolorous with the ventral region. Caudal horn straight, sometimes slightly curved, granulated, violet, with fuscous acute granulations at its tip occasionally. Stigmata elliptical, red, except the first which is orange. Legs roseate interiorly, fuscous exteriorly; prolegs green.

Pupation.—On the evening of the 12th, the above larvæ were placed, for their pupation, in a 9×15 box, containing five inches of earth mixed with one-third part of hard wood sawdust, and moistened to a packing consistency. Showing a disposition to travel over the surface of the ground, holes were made with the finger in which they were dropped head downward; they readily availed themselves of the assistance, and in a few minutes all had buried themselves out of sight. On the morning of the 13th, twenty-one additional larvæ were ready for pupation, and were placed in the same box with the above, in which they soon buried themselves; none reappeared above the ground, as the Smerinthini in most of my experiments in rearing them heretofore, have shown a tendency to do, probably from the omission of such a preparation of the ground as was made for the present colony. On the 14th and 15th, the remaining nine larvæ entered the ground and remained therein, in a small box (6 × 9 × 4), divided into compartments by card partitions, as a preventative against interference in the construction of their cells.

Perfect insect.—On the 30th of July, a male imago was found in the morning to have emerged from pupa, eighteen days after the first larva entered the ground; on the 31st, five emerged; August 1st, nine (five males and four females); on the 2d, two; on the 3d, eight (six males and two females); on the 4th, four; on the 10th, two, and the last. From thirty-six larvæ, thirty-one imagines were obtained.

Metamorphoses.—The length of time required for the several changes above recorded is as follows:

From deposit of egg to disclosure of larva	7	days.
disclosure of larva to first molt	6	
first molt to second molt	4	
second molt to third molt	5	
third molt to earthing	8	
earthing to the pupa, probably *	5	
the pupa to the imago	13	
Development of the ovum	7	
Development of the larva	28	
Development of the pupa	13	
Development of the insect	48	

Double brood.—These observations establish the fact not previously recorded of two annual broods for this species, occurring in the months of June and August; of these, captures have been made by me as early as June 9th, and as late as August 16th. It is probable that the larvæ from which are produced the first brood of moths, will be found to undergo four moltings.

Variety.—Among the above imagines was a female, having but a single blue pupil on the black ocellated spot of the secondaries. The occurrence of this variety is peculiarly interesting from the fact, that from specimens differing from the type of *S. geminatus* mainly in

* As observed in *Ceratomia quadricornis*, Proc. Ent. Soc. Ph., vol. i. p. 291, *Sesia Buffaloensis* and *Thyreus Abbotii*, pp. 113, 115, of this report.

having but a single pupil, two other species seem to be based, viz.: *Sphinx ocellatus Jamaicensis* of Drury, and *Smerinthus Cerisyi* of Kirby.

Supposed varieties.—Drury's *Jamaicensis* is cited by Clemens as a synonym of *S. geminatus*, although its habitat is given by Drury as "Jamaica." Grote and Robinson include it in their catalogue of North American Sphingidæ, as a distinct species,* remarking of it "it seems to us, judging from Drury's figure and description, quite distinct from the northern species from the Atlantic district."†

A careful comparison of Drury's figure with our variety, leaves scarcely a doubt of their identity. In the shape of the wings they correspond closely, the principal difference being in the less rounded anterior angle of the secondaries of the figure, which difference, however, is less than that presented in the apices of the primaries of the figure, showing the representation to be not strictly reliable. It is probably accidental, that in the markings of the primaries, the figure conforms more closely to our variety, in the near approach just below the first median nervule of the two bands crossing the middle of the wing, than it does to any other specimen of *S. geminatus* which we have seen. The secondaries of the two correspond in their general color, margins, central red shade, and black spot with its hook like process running from it to the anal angle, quite as well as could be expected, from the indifferent execution of the figure: in the location of the pupil on the black spot, a strict agreement between the two, is hardly possible, for while in our variety, the suboval pupil occupies the superior half of the spot, in the figure the right hand oval pupil is placed centrally on the spot, and the round left hand one occupies the inferior portion. The description ‡ is faithful to our variety in every particular.

Additional evidence of Drury's *Jamaicensis* being a simple variety of *S. geminatus*, and probably erroneously assigned to Jamaica, may be

* *Proc. Ent. Soc. Ph.*, 1865, vol. v., p. 160.

† Loc. cit., p. 185.

‡ *Upper side.*—Antennæ pectinated and brown. Head and thorax soft dun-colored, but dark brown above. Abdomen dun. Anterior wings delicate fine grayish, light brown next the shoulders and tips; the remaining parts being clouded with dark olive brown colors. Posterior wings red in the middle, but along the external edges dun-colored; having a large black spot placed near the abdominal corners, the middle of which is blue, and imperfectly resembling an eye. All the wings are angulated.

Under side.—Breast and abdomen dun. Anterior wings red in the middle, but along the anterior edges ash-colored, which runs to the tips where it forms a crescent, the inner part being dark olive brown; the external edges are olive brown, but lighter than the crescent. Posterior wings clouded with olive brown and ash-color; having a double ash-colored bar crossing them, which rises at the anterior edges of the anterior wings, and running circularly, ends at the abdominal edges of the posterior.—*Westw. Drury's Illus. Exot. Ent.*, 1837, v. ii., p. 47.

found in the fact that Mr. Grote has not met with the species in the "very large entomological material" received by him from Cuba, constituting the Poey collections, and embracing fifty-four species of Sphingidæ, nor indeed with a single member of the tribe of Smerinthini.* The species could hardly fail of representation in these collections if it occurred in the neighboring island of Jamaica, a locality, it may be remarked, still more remote than Cuba from the "Atlantic district" (Leconte), to which our American Smerinthini would seem almost to be confined.

S. Cerisyi of Kirby, the description of which is appended for comparison,† is, in all probability, a simple variety of *S. geminatus*, in which the superior of the two blue markings has retained its normal crescentic form, and the inferior one instead of its usual suboval shape, has also become crescentic—the tips of the crescents approximating, with their concavities directed toward one another, thus presenting "a black pupil, *nearly but not quite* surrounded by a blue iris." In some of my specimens, quite an approach to this form is shown.

Kirby's figure better represents our species than Drury's, the primaries being very well portrayed, except in the addition of a moderate excavation of the external margin, between the second and third median nervules. The black spot of the secondaries is less extended toward the base than usual in *S. geminatus*. In this latter particular and in the general shape of the spot, the figure approaches the European *S. ocellatus*, though differing materially from that species in the excavated apex of the primaries (acute in *ocellatus*), and in a more conspicuous excavation at the posterior angle of the same wings.

* List of the Sphingidæ, Ægeridæ, Zygænidæ and Bombycidæ of Cuba.—*Trans. Amer. Ent. Soc.*, 1870, v. iii., p. 183.

† Body ash-colored: thorax with a large trapezoidal brown spot dilated next the abdomen: primaries angulated ash-colored, with a transverse series of brown submarginal crescents in a paler band, between which and the posterior margin is another obsolete paler one; above the crescents is a straight whitish band, and a linear angular forked one, under the internal sinuses of which the wings are clouded with dark brown; underneath, the above markings of the wings are very indistinct: the secondaries are rose-color, paler at the côstal and posterior margins; underneath they are dusky cinereous, with a whitish band coinciding with that of the primaries, a transverse series of crescents and a dentated brownish band, all rather indistinct: but the most conspicuous character of the secondaries is a large eyelet situated at the anal angle, consisting of a black pupil, nearly but not quite surrounded by a blue iris, and situated in a black triangular spot or atmosphere, which extends to the anal angle, and is surmounted by some blue scales: the abdomen above is dusky ash colored.

This insect appears to be the American representative of *S. ocellatus*, from which, however, it differs considerably. It comes very near *S. geminatus* (Say, *Am. Ent.* i. t xii,) but in that the eyelet has two blue pupils. Taken in North America, locality not stated.—*Faun. Boreali-Americana*, 1837, vol. iv., p. 301.

The omission from the figure of an angle in the margin of the posterior wings at the submedian nervure, must, we think, be an error in representation, as also the termination of the anal process of the black spot, not in the anal angle, but wholly within the internal margin. Errors so obvious and other probable ones, must necessarily afford a poor basis on which to sustain a valid species.

Mr. Grote, in his valuable papers on American Sphingidæ, has advocated the specific distinctness of *Cerisyi*.* In one of them he remarks: "The fact that *Cerisii* Kirby, is certainly distinct from *S. geminatus* Say, an opinion I have entertained since studying Kirby's description and figure, has been recently ascertained by the discovery of specimens, as I am informed by Mr. S. Calverley." At the present I have no means of determining the character of the specimens referred to, but I cannot believe that they will prove to be different from the exceptional form obtained by me from the deposit of *S. geminatus* eggs above recorded.

Mr. W. H. Edwards informs me that he has regarded *S. Cerisyi* as a distinct form. He has, in his collection, a specimen taken far north, by Kennicott, believed to be the only one in the country.

Kirby's type is probably in the collections of the British Museum, where, it is stated, the insects described in *Fauna Boreali-Americana* were deposited.

From the very brief description of *S. opthalmicus* given by Boisduval,† it was thought by Clemens to be possibly a variety of *S. geminatus*, having but a single eye in the ocellated spot.‡ Grote and Robinson in their catalogue of N. A. Sphingidæ, have recorded it as a distinct species.§

Through the kindness of Mr. James Angus of West Farms, N. Y., I have had the privilege of examining a beautiful specimen of the species, received by him from California. It is structurally distinct from *S. geminatus*, and is closely allied to *S. ocellatus* of Europe, from which however it differs materially.

As near as I could determine without dissection, the antennæ consist of about forty joints having longitudinally on them a single series of thin, nearly square laminæ, each equal in length to the joint

* Notes of Cuban Sphingidæ.—*Proc. Ent. Soc. Ph.*, 1865, vol. v., p. 40.

† Le *S. opthalmica* assez rapproche de notre *ocellatus*, plus voisin de *Gemina* de Say, mais l'oeil n'est pas double et il differe de toutes les especes du meme groupe par sa large bande brune, anguleuse, qui traverse le milieu des ailes superieures.—*Ann. Soc. Ent. France*, t. iii., 3me ser. xxxii.

‡ *Jour. Acad. Nat. Soc. Ph.*, 1859, p. 184.

§ *Proc. Ent. Soc. Ph.*, 1865, vol. v., p. 160.

upon which it is placed; the laminæ bear on their two sides two rows of fine cilia, extending from a common point on the middle of their base to the two outer angles, regularly increasing in length as they recede from the originating and diverging point, the rows slightly curving toward one another and uniting at their tips. As seen from above, the connivent cilia fringing the antennal stem, are alone visible, in their greatest length nearly equaling the pectinations of *S. geminatus*. In this latter species, from the middle of each antennal joint, are given out two broadly diverging, slender, curved, cylindrical, apically rounded pectinations which are margined with short and fine cilia; the pectinations are thirty-nine or forty in number.

The apex of the primaries is acute in *S. opthalmicus* as in *S. ocellatus* (excavated in *S. geminatus*), but less curved apically on the anterior margin; it is without the white-bordered semioval brown patch which is a feature in *geminatus*. The excavations of the hind margin approach nearer to *ocellatus* than to *geminatus*. The posterior wings are less developed costally than in either of the above two species. The ocellated spot is quite small, having a diameter between the crescents of about one-half that of the thoracic spot (in the other two the diameter exceeds that of the thoracic spot); it rests anteriorly on the second median nervule, centers on the first, extends to midway between the latter and the submedian nervure, and is removed one-half its longest diameter from the outer margin. It consists of a black spot and two slender, subequal crescents almost uniting at their tips, of which the anterior one is placed just within the anterior margin of the spot which is lost beneath the long, rose-colored basal hairs, and the posterior one forms its posterior margin — the whole presenting a well defined ellipse, having its transverse diameter on the submedian nervure. From opposite the center of the spot interiorly, disconnected from it by a brownish line, a short black dash points toward the anal angle, but is merged in an ochreous-brown shade running to the angle, and thence acutely reflected toward the base. The thoracic spot is ochraceous-brown, straight in front, covering all of the thorax except a white bordering to the gray tegulæ. The colors of the abdomen and wings differ materially from those of the two species with which it is compared, for while they are characterized by shades of deep brown, in this, the colors are fawn or pale ochraceous-brown. Its expanse is two and six-tenths inches; length of body one inch.

Synonymy.—The following table of reference and synonymy is presented, in the belief that it will prove to be correct.

Sphinx ocellatus Jamaicensis Drury. Illus. Nat. Hist., 1773, v. ii, p. 43, pl. 25, figs. 2, 3.
Smerinthus geminatus Say. Amer. Ent., 1824, v. i, p. 25, pl. 12.
S. geminata Harr. in Cat. An. & Pl. Mass., 1835, p. 71.
S. Jamaicensis Westw.-Drur. Ill. Ex. Ent., 1837, v. ii, p. 47, pl. 25, figs. 2, 3.
S. Cerisyi Kirby. Faun. Bor.-Amer., 1837, v. iv, p. 301, pl. 4, figs. 4, 5.
S. geminata Harr. in Amer. Jour. Sci.-Ar., 1839, v. xxxvi, p. 291.
S. geminatus Walk. Cat. Br. Mus., Lep., 1856, pt. 8, p. 246.
S. geminatus Clem. in Jour. Acad. Nat. Sci. Ph., 1859, p. 183.
S. geminatus Morr. Syn. Lep. N. Amer., 1862, p. 210.
*S. excæcatus** Lint., in Proc. Ent. Soc. Ph., 1864, v. iii, p. 665 (larva).
S. geminatus Gr. & Rob., in Proc. Ent. Soc. Ph., 1865, vol. v, p. 160.
S. cerisii Gr. & Rob. ibid., et List Lep. N. A., 1868, p. 4.
S. jamaicensis Gr., & Rob., in Proc. Ent. Soc. Ph., 1865, vol. v, p. 160.
S. geminatus Pack. Guide Stud. Ins., 1869, p. 275 (vena. post. wing).

* An erroneous determination, the larva described on page 666 of the Proceedings being that of *S. excæcatus*.

VI. TRANSFORMATIONS OF DAREMMA UNDULOSA WALKER.

A moth of this species, with broken and denuded wings and contracted abdomen, which was taken on the 8th of July, deposited eight eggs the following day in the box in which it was confined, and died three or four days thereafter. Upon examining the box on the 14th, there were found the transparent shells of six of the eggs from which the larvæ had emerged, the other two proving infertile. Three of the larvæ had escaped, and the remaining three were quite feeble from their compulsory fast of probably a day or two, but very soon commenced feeding on some tender leaves of ash (Fraxinus) which were given them. To insure them a continual supply of fresh food, they were placed on a leaf of a growing ash and inclosed in a gauze net for their protection.*

Young larva.—The larva is of a very pale green color throughout, showing no stripes nor bands. Its length is 18-100ths of an inch. Its caudal horn is cylindrical, straight, of a light green color, except its tip, which is brown, and measures 8-100ths of an inch.

* The rearing of larvæ upon growing plants, as above referred to, is becoming a favorite method with lepidopterists, for several reasons: It affords a constant supply of suitable food, free from the partial decomposition which commences almost immediately upon the plucking of the leaf, or from the rapid change, commonly known as "souring," which a twig or stem undergoes when placed in water. The great sensitiveness of young larvæ to improper food, is well known to those who have reared from the eggs, by the usual method of plucked food, broods of our sphinges, in which a mortality of one-half has often been encountered before the first molting. The larvæ require but little attention; during their infancy, the few leaves upon which they are at first placed, may suffice for a week or two; as they attain a size which demands a larger supply of food, as often as the inclosed leaves are consumed, it is only necessary to open the net, turn out the excrementitious matter, clip off the defoliated portion of the twig, and tie again farther down the stem. It prevents the injuries which so frequently prove fatal to young larvæ, when the slightest degree of force is employed in removing them to fresh leaves. In addition to natural food, it also gives a natural exposure, an important consequent of which is, that a description of the larva need not be imperfect from abnormal coloration, which so frequently is the result of in-door rearing.

There are, however, some risks to be incurred by this method. The larva sometimes deserts the leaf for its enveloping net, when of so small a size, as not to be able to protect itself against being seized and destroyed by ants or other insects; and this traveling propensity is often the first indication of approaching molting, which is always accompanied by diminished powers of defense. As I know of no means by which to prevent such occurrences, which are very annoying when the larva happens to be rare, I would, in such cases, defer placing them out of doors until their second change had given them a degree of safety in the size attained. As they approach maturity, the protection of the net does not wholly exempt them from the attack of their natural enemies, the Ichneumonidæ, which, readily drawn thither when several larvæ are associated, may often be seen prospecting over the net for a position whence they may reach with their ovipositor the body of their prey within. Having thus lost a number of rare larvæ, I am now usually successful in preventing its recurrence, by constructing the net of a large size, and extending it with two or more wire rings, so as to place the inclosed larvæ while upon the leaves, which they rarely leave after their second molting, out of ovipositor reach.

When nearly ready for their pupal change, they should be removed from the tree, and fed to maturity in a box or wired breeding cage, to guard against their liability to escape at this period, by forcing an opening in the netting.

First molt.—On the 21st of July their first molting occurred. On the 26th, when about to undergo their next change, they had attained the length of seven-tenths of an inch, and presented the following features: Head light green, with a yellow lateral stripe. Body light green, marked conspicuously with a subdorsal yellow stripe, of the breadth of two of the annulets, which commences on the first segment and terminates in the last lateral band. The lateral bands are seven in number, of a yellow color, but are less conspicuous, except the posterior one, than the subdorsal stripe; each band commences at the anterior margin of a segment in range with the stigmata; it enters the subdorsal stripe at the fifth annulet, and emerges therefrom at the incisure, whence, after rising slightly above the stripe, it is recurved and touches it again, nearly reaching the following incisure. The caudal horn is straight, 9-100ths of an inch long, and of a red color. The stigmata are not visible with an ordinary magnifying glass, but their position is indicated by a short horizontal yellowish line, sub-centrally on the segment. The legs and prolegs are green.

Second molt.—This change occurred during the night of July 26th and 27th. When in readiness for the next molting on the 31st, their length was 85-100ths of an inch. The head shows granulations and has broad lateral stripes of yellow which nearly meet at the apex. The subdorsal stripe is obsolescent, being more distinct on the thoracic segments; lateral bands quite distinct, of a bright yellow; the three anterior and the seventh with a shade of red margining them in front. Caudal horn straight, with short spines which are brown on the front of the horn and behind. The legs are light red, and the prolegs green.

Third molt—August 1st–2d. On the 6th of August, a cessation from feeding, a fixed position, and a partial withdrawal of the head from its case, indicated the near approach of another molting. Length of the larva at this time, one inch and three-tenths, with a diameter of eighteen-hundredths of an inch. The head is light green, tuberculated superiorly, with the lateral stripes broad and of a whitish shade. The body is pale green; the subdorsal stripe obsolete; the lateral bands are whitish-green, bordered anteriorly with darker green, except the first and seventh, which are yellow anterior to the stigma; the bands commencing at the incisure, cross the second annulet in range with the lower part of the stigma, are contracted over the stigma, extend thence in a straight line to the next incisure, and are continued somewhat deflected over from one to three annulets of the following segment (over three on the seventh and eighth segments); this posterior portion of the band is not edged with darker green. The caudal horn is very slightly curved, rose-colored anteriorly at the base, tipped with yellow, and is covered with spinules which, except the lateral ones, are black. The anal shield and plates have black granulations. The stigmata have a white dot at each extremity, and are bordered with orange. The legs are rose-colored and the prolegs green.

Fourth molt.—The last molting during the larval state occurred during the night of August 7th–8th. On the following morning they were found feeding on the dry leaf upon which they had been resting motionless for the twenty-four hours preceding their change. Their position in each molting has been on the midvein of the leaf. The withdrawal of the head from its case at the commencement of the molting appears to be accomplished in a very brief time, if not at a single effort. In *Ceratomia quadricornis* Harris, the operation is so gradual, that its progress can be followed for a day or more. In this species, when a careful inspection has shown no indication of the separation of the case, an hour thereafter, it has been found wholly withdrawn within the skin of the anterior segment, through which the lateral bands of the head could be very distinctly seen. In *Darapsa Myron* (Cramer), the corresponding operation appears to be as quickly accomplished.

All the above moltings of *Daremma undulosa* have taken place during the night.

Food plant.—The larva, according to information given to Dr. Clemens as stated in his description of *Ceratomia repentinus* (determined by an examination by Grote and Robinson of the typical specimen of Walker to be identical with this species*), has been taken on the ash, upon which the individuals above described were reared. Mr. Grote states that he has observed it numerously on the lilac (*Syringa vulgaris*) on Long Island. It will probably also be found on the privet (*Ligustrum vulgare*),—a larva which must have been either this species or *Sphinx cinerea* (*chersis* of Hübner), having been reported to me as occurring on this food-plant.

Pupation.—Before entering the earth, it undergoes a marked change in color. One taken from a fence August 27th, presented a soiled white appearance, in which only a trace of its original green was visible; the position of its lateral bands could with difficulty be

* *Trans. Amer. Ent. Soc.*, 1868, vol. ii, p. 76.

traced. It buried in the ground the following day, and constructed its cell of the usual ovoid form of the cells of the sphinges, at a depth of four inches. The imago emerged the following June, enabling me to determine the species, which the altered appearance of the larva did not permit of doing.

Pupa.—Length one inch and three-fourths; diameter one-half inch; color dark brown; head-case depressed, shagreened; eye-case slightly prominent, with a smooth, impressed line inferiorly, and a central one on the crescent, which is wrinkled transversely. Pronotum shagreened, quite depressed anteriorly, with a medial line; stigma fusiform. Mesanotum minutely shagreened, with an inconspicuous medial line. Metanotum with a transverse line anteriorly, posterior to which it is minutely wrinkled longitudinally. Abdominal segments punctulated, and each divided superiorly in about four parts by depressed transverse lines. Eleventh segment with a dorsal, transverse, oval depression. Tongue-case buried, reaching just below the tips of the middle leg-cases, having anteriorly a few transverse plaits near its medial line, and a few longitudinal ones near the antenna-case. Antennæ-cases in the female extend to half-way between the tips of the anterior and the middle leg-cases, showing the joints distinctly, with a granulation on each. Anterior leg-cases broad, prominent and rugose over the femur. Wing-cases somewhat granulated at their basal region, and smooth elsewhere. Spine rugose, subtriangular, constricted at the base.

As will be seen from the above description of the larva, it has no structural affinity with that of *Ceratomia quadricornis*.* The species has therefore very properly been removed from the genus in which Dr. Clemens had been led to locate it, from representations made to him by one who claimed to have reared its larva repeatedly, and described it as strongly resembling that of *C. quadricornis*. To Mr. Grote belongs the credit of discovering structural differences in the imagines of *C. quadricornis* and *D. undulosa*,† which differences are fully sustained in their earlier stages.

* *Proc. Ent. Soc. Phil.*, 1862, vol. i, p. 290.
† *Proc. Ent. Soc. Phil.*, 1865, vol. v, p. 190.

VII. NOTES ON PLATARCTIA PARTHENOS (HARR.) PACK.

Some eggs deposited by a captured moth, disclosed their larvæ on July 20th. The young larvæ were one-tenth of an inch in length, of a fulvous color, with black tubercles and long fuscous hairs.

The first molting occurred July 26th and 27th. The larvæ were now one-fourth of an inch long, with two prominent black subdorsal tubercles on the fourth and tenth segments; the hairs were one-tenth of an inch long.

The second molting commenced July 30th and terminated on August 1st.

The third molting commenced August 6th.

The fourth molting commenced August 11th, after the larvæ had maintained a fixed position for twenty-four hours.

The fifth molting commenced August 16th.

The sixth molting extended from the 23d to the 25th of August, inclusive.*

The seventh molting: two of the larvæ which had taken positions in an angle of the box on the 28th, and had spun over them a thin web covering, molted on the 31st. On the 13th of September, two others molted; the remainder did not undergo this change. After the middle of September they ate very sparingly, many of them resting for days in one position. A few of the brood having died, about the middle of October, the remaining ones (eighteen in number) were transferred to winter quarters within a box containing chips and sawdust, and inverted on the ground beneath a bedding of leaves and earth. The few larvæ which had undergone their seventh molting, had at this time attained a length when in motion of two and one-fourth inches.

On the 1st of April, eleven of the larvæ were found to have survived the winter. These were provided with growing plants, beneath glass, for food, but manifested an indisposition to eat, seeming in a feeble condition. On the 27th of April, two of the number were observed to be feeding nicely. The others died without partaking of any food.

The two larvæ without again molting, or materially increasing in size, spun cocoons of a dark colored silk interwoven with

* A description of the larva at this stage, is given by Mr. W. Saunders, in the *Canadian Entomologist*, 1871, vol. iii, p. 225.

their hairs, of a texture permitting the inclosed pupa to be seen through the threads. One only disclosed its imago (a male). Of its period of pupation, no note was made.

The moth conforms very closely to the description given by Packard.* In the female from which the eggs were obtained, the median band on the secondaries which in the male consists of approximate orange spots, becomes a continuous orange band from its enlargement at the costal margin to near the internal margin (one-third of the distance between it and the submedian), constricted opposite the cell and on the first median nervule ; the nervules which intersect it are dotted with black scales. On the fold between the median and the submedian nervures, is an orange vitta, attenuated anteriorly for two-thirds its length, enlarged and rounded posteriorly, and extending nearly one-half the distance across the wing. † The sides of the abdomen, terminal segment and anal tuft are orange, concolorous with the ground of the posterior wings ; the black of the dorsum extends medially in a point over the terminal segment. Expanse of wings 2.80 inches ; in the male 2.40 inches.

* *Proc. Ent. Soc. Phil.*, 1864, vol. iii., p. 110.

† No trace of this feature appears in the figure given in *Agassiz' Lake Superior* (pl. vii., fig. 4,) of a male taken on the northern shore of the lake, nor in a male of my collection.

VIII. NOTES ON EUPREPIA AMERICANA (HARRIS.)

A moth of this species, captured in a room late in the evening* where it had been drawn by a brilliant light, deposited, after having been pinned, a number of eggs, of which, about one-half (seventy by count) were given me to rear.

Most of the eggs were deposited in an irregular mass, and a few were lying loosely in the box. They were white, of an obovate form, very slightly compressed, with a length of 8-100ths of an inch, and a breadth of 6-100ths of an inch. The larvæ emerged from the larger end, leaving a shell firm and opaque, of which only an aperture was eaten. Less than one-half of the eggs disclosed larvæ.† The date of their appearance was not noted.

The first molting was on the 13th of August. Previous to the following molt they were three-eighths of an inch long, of a reddish brown color, with intermingled white and black hairs at the extremities, and with a lateral lead-colored stripe. The head and tubercles of the body were black.

The second molt extended from August 22d to the 25th.‡ Length, one-half inch; anterior segments fulvous, with hairs of the same color. The third molt commenced August 30th; the fourth, September 9th; the fifth, and last recorded, September 20th.

The moltings were not at all uniform throughout the brood, some of the number being at this time an entire molting in advance of others.

Several days after the last change above noted, the larvæ varied in length from six-tenths of an inch, to one inch and a tenth. Having ceased feeding (and a few having died), they were placed in winter quarters with the *P. parthenos* larvæ of the preceding paper.

When uncovered in the spring, but one of the colony was found alive. It fed for about two weeks, increasing in size during that

* In two other instances of the capture of this moth within-doors, in the same locality, during the same season, it was observed that its appearance was between the hours of *ten* and *eleven* P. M.

† The very large number of eggs borne by this moth—stated by Dr. Fitch to be seven hundred and forty-four in an instance observed by him — might account for so many (the last deposited) having failed of fertilization.

‡ It is possible, in consideration of the long interval between this and the first molt, as compared with the corresponding interval in *P. parthenos*, that an intermediate molting may have escaped observation.

time from nine-tenths of an inch to one inch and a fourth, but died before attaining maturity.

Dr. Fitch, from a comparison of colors and markings, regards this species as identical with the European *E. caja* (Linn).* Dr. Packard finds indisputable distinctive features in the stouter body, shorter wings and prominent antennal pectinations of our species. Specimens, however, have been taken in Labrador, which Packard has determined to be *E. caja*, and which he represents as giving evidence of introduction. He states that "the coloration and the markings are the same, and it can scarcely lay claim to be considered as a climatal variety. The patagia are white in the Labrador specimen, and brown in the English; this is the principal distinction."

Of its larva, Packard states: "It occurred at Gore Island, in Southern Labrador, wandering over the herbage. At Caribou Island, they were found in July, in various stages, feeding on *Potentilla anserina*. The larva was also found, full fed, crawling over herbage, on June 15th, at Little Mecatina Island, and it had no doubt hybernated in this state. The body was black, with large white papillæ, from which, on the thoracic rings, rise short yellow hairs, like those on the sides of the body. Above, the white papillæ are large and conspicuous, and from them arise long, thin, mostly irregular fascicles of pale gray hairs, with shorter and fewer black hairs, the longer ones equaling in length the breadth of the body. It is of the usual size, and its tricolored hairs and white papillæ give a striking appearance to this handsome larvæ. It began to spin a cocoon June 26th, and the moth appeared July 27th."†

I know of no description of the mature larva of *E. Americana* to compare with the above. That of Dr. Fitch (loc. cit.) does not pertain to our native species, (although associated with an excellent description of the imago), but is obviously taken from European sources, and refers to *Euprepia caja*.

E. Americana has been taken in several places in Canada and in New York, in Massachusetts and on Lake Superior.

* *Noxious Insects of New York*, Reports, 6–9, p. 234.
† *Proc. Bost. Soc. N. H.*, 1868, vol. xi, p. 34.

IX. NOTES ON EUCHÆTES EGLE (Drury).

A colony of the young larvæ was taken on the milkweed (*Asclepias cornuti*) July 20th, collected in an irregular cluster on one of the leaves. Their appearance at this stage was so unlike that presented when more advanced, that the species would not have been suspected, but from the food plant on which they occurred.

On the 21st, having consumed the leaf on which they were feeding, they moved to the top of the jar beneath which they had been placed, where they collected in a body. They were now three-tenths of an inch in length. The head was subquadrate, glossy black. Collar fuscous. Body deeply incised, obscure green, with round fuscous tubercles dorsally, and oval ones laterally, from which radiate white hairs of unequal length, those on the four anterior and two posterior segments being longer than elsewhere, and intermingled with dusky hairs. Legs spotted with fuscous, and prolegs with a fuscous spot outwardly.

On the 23d they were found to have undergone a molting, and were traveling with a very rapid motion in every direction about the jar. Later in the day, they had again collected in a cluster, when fresh food was supplied them, upon which they arranged themselves with some degree of regularity, but not with the striking parallelism which characterizes their feeding when met with in the field.

The larva now appears with twelve rows of tubercles, disposed in two ranges on each segment, the tubercles alternating on the anterior and posterior portions of the segment.

The first segment has some short white hairs projecting over the head. On the second segment are four black pencils, each with a single hair projecting beyond the others; the two inner pencils are the longer, and the single hair extends to nearly twice the length of the pencil. The third segment has its interior pencils like those of the second, but its exterior ones are without the long hair. The fourth segment has a double pencil issuing from the two anterior dorsal tubercles, with four white pencils from the four tubercles next below. Segments five to nine inclusive have each four orange pencils curving inward and forward, of which the two anterior proceed from the two superior tubercles, and the two posterior from those next below; beneath these is a row of black pencils above a blue line which incloses the stigmata. On segment ten is a similar arrangement of pencils, but the two posterior ones are white instead of orange. On segment eleven a double, dorsal black pencil from the two anterior tubercles, two white ones posterior to these, and two black ones exterior. On segment twelve are four black pencils directed backward. The two lower rows of tubercles on segments one and three, and the three rows on four and five, have fascicular bunches of short white hairs radiating from them, as have also the corresponding tubercles on the three posterior segments. The lower row of tubercles on segments six to nine inclusive, have a few similar hairs proceeding from their lower portion. Body of the larva, pale brown. Legs shining black; prolegs fuscous exteriorly.

On August 1st, another molting occurred. On the 3d, from one of the larvæ, several parasites emerged, and enclosed themselves in small white cocoons of a slight texture, enveloped in loose wool-like threads: from these seven hymenopterous imagines of an undetermined species (all females) were disclosed on the 18th.

On the 4th, one of the larvæ had spun its cocoon, through the walls of which some parasitic larvæ emerged. By the 9th, several had made their cocoons, and all had ceased feeding. The last of the colony made its cocoon on the 12th.

On the 26th of August, an *Egle* imago emerged from one of the cocoons, which proved to be the only one obtained from the entire colony. Late in the ensuing spring the cocoons were examined, when about half of the number were found to contain untransformed shriveled larvæ, and the remainder dead pupæ.

Other attempts to rear broods of this larvæ, have been attended with about the same success, showing it to be a very difficult species to carry through to the imago state, under treatment which proves successful with many others.

On the 24th of August,—a month later than the date of collection above noticed—another colony of this larva which had apparently very recently undergone the first molting was observed feeding on Asclepias.

X. TRANSFORMATIONS OF LAGOA CRISPATA PACKARD.

A cluster of eggs subsequently ascertained to be the deposit of this moth, was found July 7th, 1869, at Center, N. Y., on the under side of a leaf of *Quercus ilicifolia*, arranged somewhat in a segment of a circle, three-eighths of an inch in length, and covered with a yellow-white down.

The eggs are pale green, of an elongate-oval form, measuring 2-100ths by 45-1000ths of an inch in diameter. They were attached to the leaf by their sides in two rows, with the enveloping down extending beyond them on the leaf for a space somewhat exceeding the longest diameter of the egg.

Young larva.—The larvæ were disclosed July 13th, with a length of 62-1000ths of an inch. They were of a pale yellow-green color, and were thickly covered with long, soft, white hairs, many of which were twice the length of the body. The larvæ fed on the upper surface of the leaf.

First molt—July 21st to 23d: larva limacodiform in appearance, white, oval, flattened superiorly between the subdorsal rows of short, cylindrical, fleshy, white tubercles, apparently nine in number exclusive of the terminal ones; counting these last in the stigmatal row, it consists of twelve similar tubercles; from each of these tubercles, long, white hairs of unequal length radiate, the longest of which measure one-fourth of an inch. The head is not visible from above, and the extremity to which it belongs is with difficulty distinguishable from the posterior when the larva is at rest. Its prolegs can only be seen from beneath when moving on a transparent surface.

Second molt—July 28th: length of larva, 27-100ths of an inch, diameter 12-100ths; the hairs are three-tenths of an inch long, usually uncinate, more numerous than before and nearly concealing the body. The larvæ still eat only the upper surface of the leaf within the veinlets.

Third molt—Commenced August 3d: length of larva, 31-100ths of an inch, diameter, 15-100ths. Three rows of tubercles are visible on each side of the body, of which the substigmatal ones are round, the lateral and subdorsal ones elliptical; from these tubercles proceed the long hairs, which, diverging and interlacing, cover the body.

The larva no longer confines itself to the surface of the leaf, but commencing at the margin, now eats the entire body except the veins. Its method of eating is peculiar. Extended on the surface of the leaf (usually the lower surface) at a right angle with its margin, it bends its head over the edge in a position to grasp it with its mandibles. Its collar, which is quite extensible, is thrown forward with its sides appressed closely to the two surfaces of the leaf, entirely enveloping the head. From its position, one-half of the collar is hidden beneath the body; the other half is seen on the opposite side of the leaf as a triangular fleshy piece, having its anterior edge in range with the line of the body. On one occasion while feeding, when a small bit of leaf had been detached and was held by the anterior pair of legs, the favorable position of the larva for its observation, displayed both of the lateral edges of the collar folded on the piece, and holding it between them, while within, as disclosed by the regular motion of the body and the gradual disappearance of the leaf, the mouth was in active operation, wholly concealed from even a direct front view, except when a slight elevation of a portion of the collar chanced to disclose a section of a black mandible.

When in readiness for its fourth molting, the larva at rest, measures 7-10ths of an inch long, and 28-100ths of an inch broad. From having been up to this period entirely white, the body now shows patches of coloring. Two or three days after it has affixed itself for its molt, the colors become defined in shade and outline.* There now are seen blackish bands on the terminal segments, a line of elongate black spots above the legs, a white stigmatal band with a row of blackish spots above, and an obscure salmon shade suffusing the entire dorsal region. In these colors are revealed the new clothing nearly matured which the larva is about to assume, closely folded to the body in separate pencils of hairs, and partially seen through the translucent skin.

Fourth molt—August 20th, and several following days: length 75-100ths, breadth 3-10ths of an inch. The withdrawal of the larva from its old integument occupies about five minutes of time. As the segments successively emerge, the hairs appear as wet pencils appressed to the body, which rise up as they are released from their confinement, and very soon becoming dry, diverge and entirely cover

*A larva taken at Center, after its third molt, took its position for its fourth change during the night of August 27-8; the bands were visible on the 31st, and the molting occurred on the afternoon of September 1st, nearly four days having been required to complete the change.

the body without any outward indication of originating in a few isolated points. The larva now presents a remarkable contrast to its former appearance, in features as follows:

The head is round and white; the eyes and mandibles are black. The body is white throughout. The collar has four patches of short slate-colored hairs — the inferior patches round and the superior elongated. On the first segment are eight tufts of slate-colored hairs; on the second segment are six tufts, of which the four inferior are slate and the two superior, slate mingled with ochreous; on the third and following segments, exclusive of the terminal, the lower tuft is slate with black basally, the lateral is slate and ochreous, and the dorsal is ochreous; on the terminal segment the position of the tufts was not ascertained further than that the lower is slate-colored. The slate-colored hairs of the thoracic segments superiorly, and the ochreous ones of the terminal segments, are long and projected over the extremities, and are so fine as to be readily moved by the breath; the abdominal hairs are shorter, somewhat coarser, appressed, and meet over the dorsum in a ridge. The stigmata are white, hemispherical, with an elongated subcylindrical papilla beneath the anterior one and behind each of the others. Legs five-jointed, bearing upon them several short bristles. The five pairs of prolegs on segments 6–9 and 12, are well developed and have their plantæ armed with series of black hooklets; on their base exteriorly is a small pencil of short hairs, and three still smaller contiguous pencils anteriorly, visible with a lens: the two pairs of conical prolegs on segments 5 and 10 (completing the seven pairs ascribed to this larva*), are rudimentary, without plantæ, bearing apically a few short hairs.

Sting of larva.—On several occasions subsequent to the third molting, while transferring the larvæ to fresh leaves, a slight pricking sensation was felt, which was not, however, sufficiently decided to arrest attention; a larva seemingly so inoffensive in its downy dress, and so timid as to roll itself up in a ball at any rude touch, was not to be suspected of the possession of a method of defense bestowed, in like degree, upon only two other associate Bombycideans. Later, with their increased growth, the pricking became more acute, and its source could no longer be doubted. If a larva was pressed with the back of a finger, or permitted to drop upon it from a moderate height, a sharp stinging would be felt, soon becoming more acute, and followed in a few minutes with a redness of the skin, and elevated white spots

* *Proc. Ent. Soc., Ph.*, 1864, vol. iii, p. 336.

similar to the "nettle-rash." As would naturally be expected from the comparative size of the larvæ, the sting is not so severe as that of *Hemileuca Maia* (Drury), or *Hyperchiria Io* (Fabr.)*

A critical examination of the larva, by a partial removal of its hairs, revealed the existence of clusters of short, slender, acute, white bristles, directed upward from the several tubercles of the lateral and subdorsal rows, the presence of which had previously been unnoticed, under their covering of the long hairs surrounding and effectually concealing them. Upon touching the bristles with the hand, they were found to be the source of the sting experienced.

Fifth molt.—The fifth molting through which the larvæ were observed to pass, and their subsequent changes, were not recorded. Additional collections of larvæ made at Center were thoughtlessly added to the colony, which, with other circumstances, prevented the completion of their history. Of the above collections, one of the number spun its cocoon on the 29th of August, attached to the bottom edge of the bell-jar confining it.

An alcoholic matured specimen in my possession, which had attained more than ordinary size, measures one inch and three-tenths in length, by one-third of an inch broad. Its head, drawn within the first segment is black, as are also the tips of four or five of the anterior stigmatal papillæ.

A larva taken at New Baltimore, N. Y., feeding on plum leaves, and brought to me August 31st, was more elongate than usual; its three anterior segments were clothed with hairs of a brownish shade, and the hairs of its posterior extremity were prolonged so as to form a short tail. It made its cocoon September 1st, but did not develop the imago.

Cocoon.—The cocoon is of a tawny-brown color, oval in form, often modified by the surface to which it may be attached, and occasionally contracted near its apical end; its average size, taken from fifteen specimens, is three-fourths of an inch in length, by one-third of an inch in diameter. It is of a firm, parchment-like texture, capable of sustaining considerable pressure from the edge of the nail before yielding to it. Its exterior is irregularly covered with a thin web of rather coarse threads, in which, under the microscope, a few of the plumose hairs of the caterpillar may be observed; underneath this, the outer coating of the cocoon is thinly extended over its apical end, to the extent of about one-tenth of an inch. Upon stripping

* *Twenty-third Report on the N. Y. State Cabinet*, 1872, p.

away this portion, a flat surface is seen beneath, covered with a matting of intermingled hairs and silk; if this matting be removed, a moderate pressure of the fingers upon the sides of the cocoon, will cause the flattened apex to detach itself in part from the body, disclosing a perfectly fitting lid, of which its connection with the cocoon for about one-third of its circumference, serves as a hinge. The margin of the lid is slightly recurved; that of the cocoon to a greater degree, so as, in some examples, to render this portion as broad as the central diameter.

The lid is woven by the caterpillar separately from the rest of the cocoon, and is not a section cut from it after its completion. This interesting fact, at variance with the generally received opinion upon the subject, was clearly shown, when, upon springing open the lid, there was found to be resting upon and partly overlapping its margin, and quite distinct from the matting inclosed, a packing of stout, parallel threads and hairs, not forming an entire circle, but interrupted by returns at each side of the hinge, over which its extension would not be needed; underneath this, and more closely united to the margin was a more slender silken cord arranged in like manner. The construction of the packing and its curious disposition, prior to the formation of the lid, as from their relative position it must necessarily be, is ample evidence that, in continuation of the plan adopted, the lid is subsequently woven to its shape, with the packing serving for its base and guide.

The lids are usually of an ovate form, but in some specimens they assume a subtriangular shape, or present one or more projecting angles; in such instances, these irregularities are met by complementary variations in the margin of the cocoon, by which a perfectly fitting lid is insured.

A tranverse section of the cocoon displays an ingeniously contrived structure for the firm support of the margin of the lid, without which, notwithstanding the partial support provided for it in the inflated margin of the cocoon, it might too easily yield to outward pressure from some inquisitive enemy. While as before stated, the main body of the cocoon consists of a single wall, its superior third is seen to divide in several laminæ (seven were counted in one specimen), slightly separated and carried to the proper height to meet and sustain the lid; the interior ones are thin, while that forming the inner wall is firm and smoothly coated by a gummy secretion probably from the mouth of the larva, uniform with the remaining portion of the cocoon. With the same economy of labor and of material shown in the construction of the interrupted "packing," these laminæ do not encircle the cocoon, but disappear beneath the hinge.

The separate construction of the lid, as shown in the remarkable evidences of design above recorded, is also confirmed by microscopic observation. Under a high magnifying power, the parallelism of the threads composing its margin is distinctly seen, in marked contrast with the ragged projecting ends of an excised portion.

Pupation.—The larvæ made their cocoons between the leaves on which they had fed, or those lying on the surface beneath. In several instances, a half dozen or more were found associated between a couple of leaves, and so firmly attached to one another that they could with difficulty be separated. Of perhaps eighty cocoons obtained, eight only developed the moth during the last of October and early part of November, after a pupation of about two months. None were disclosed in the spring, at the regular time for its apparition during the month of June, as we may infer from the collection of its eggs on the full grown leaves of oak. In a number of the cocoons subsequently opened for examination, were found the shrunken remains of untransformed larvæ, and in others, the apparently fully matured pupa, seen through the thin case, perfect in all but the extension of its wings.

It is worthy of remark that this species, *Hyperchiria Io* and *Hemileuca Maia* have each, in my experiments in rearing them, produced a portion of their brood in the early fall, at a time when their exclusion could not be the result of an indoor temperature, which at that period did not exceed that of their natural exposure.

When in readiness for its final metamorphosis, the pupa forces upward the lid of the cocoon, and withdraws itself through the opening, until only its terminal segments are held by the pressure of the lid and enveloping convergent threads. As the pupa is wholly destitute of the dentiform processes which encircle the pupal segments of those of our moths (Ægeriadæ, Cossidæ et al.), which are known to extrude themselves partially from their cocoons while still in their pupal state, and which apparently are dependent on aid afforded by these processes for their release, some other provision is required by *crispata* to serve in its work of extrication. This is found in the motion permitted its encased limbs while yet a pupa. As a general rule among the Lepidoptera, when the pupa first divests itself of its larval covering, its antennæ-, leg-, and wing-cases are readily separable from the body-case on which they lie, but in a brief time are firmly cemented to it by the drying and hardening of the viscid coating which overspreads it. This species, however, is an exception to the rule, and the first which has been observed by me. * Upon opening its cocoons, the above mentioned organs are found disconnected (except basally) from the pupal body. In its extruded pupa-case may usually be seen the antennæ-cases extended in the form of the antique lyre quite in advance of the other members, the leg-cases brought up from beneath the wing-cases, the latter quite separated at their apices from the abdominal region, and giving indication of having rendered efficient service in the escape from the cocoon.

Pupa.—Its color is essentially that of the contained imago showing through the translucent shell, being ochreous on the thorax and attached members, and lutescent on the abdomen. It is of an oval shape, slightly contracted at the base of the abdomen, its extremities rounded, the terminal segment blunt and without processes. The head-case projects moderately beyond the prothorax; the eye-cases are prominent with a shining mamilla intermediately. The antennæ-cases showing distinctly at their mesial carination the curved tips of the pectinations, extend in the male to the tips of the wing-cases, and in the female to those of the anterior leg-cases. The posterior leg-cases protrude from beneath the wing-cases, nearly across the eighth segment. The wing-cases are rounded at their inner angle, and extend half-way over the seventh segment; under a lens, they show distinctly the crinkled black hairs of the disc of the wing. The thoracic divisions are distinct, not being cemented together; the pronotum is thrice as broad as long, excavated in front, convex behind, depressed medially, its posterior angles subquadrate, with a protuberance near its anterior margin on each side; the mesanotum is one-half longer than broad, its sides subparallel, and its hinder margin rounding over the metanotum to nearly its posterior margin; the metanotum is a little longer than the pronotum, and corrugated longitudinally on each side. The abdominal segments, under a lens, have fine longitudinal wrinkles anteriorly; the incisures are rather deep. The eight abdominal stigmata are visible; the seven anterior ones are broadly oval, with prominent margins, and have a small tubercle behind each; the last one is linear, without a raised margin

* In some of the Tineidæ the limbs are partially free.—*Packard, in The American Naturalist*, 1871, vol. v., p. 712.

or accompanying tubercle. Length from five-tenths to six-tenths of an inch; diameter from two-tenths to one-fourth of an inch.

The larvæ were found very abundantly at Center on the 6th and 20th of August, 1869, feeding on the different species of Quercus, on Vaccinium, on *Pteris aquilina*, and on other plants; they were all, at this time, in their white coats. On the 27th of August, at one locality at Center, on a gently sloping hill-side, a thousand individuals could have been taken by a collector in an hour's time: at this date, a few had assumed the brown coat indicative of their fourth molt, and by the 8th of September, nearly all had undergone this change. At a locality frequently visited, in Bethlehem, near Albany, but one individual was observed during the season, on September 14th.

Notwithstanding the remarkable abundance of the larvæ at Center, the imago has not been observed by me, either in that locality or elsewhere.

During the last of August, 1870, the larvæ were again observed in large numbers at Center, but not so abundantly as in the previous year. Of about twenty collected, nearly all, when in their third and fourth stages, gave out a parasitic larva, which transformed into pupæ, apparently of some species of Tachina, but of which I did not succeed in obtaining an imago. None of my collections of the preceding year were thus affected. Mr. C. V. Riley informs me that his collections of the larvæ, made in 1870, in the vicinity of St. Louis, were also destroyed by probably the same parasite, which he was equally unsuccessful in carrying to maturity.

XI. TRANSFORMATIONS OF HYPERCHIRIA IO (Fabr.).

Of the above species, a small company of sixteen larvæ was found at Center, July 15th, arranged side by side in perfect parallelism on a leaf of *Populus tremuloides*. They had evidently, at the time of their collection, undergone their first molt.

Second molt—July 18th; the leaf on which the larvæ were taken having become dry, they abandoned it and passed to the side of the jar occupied by them, in regular procession and in an unbroken line, moving in single file, unlike *Hemileuca Maia*, whose processions are in files of two's or three's. Later they had arranged themselves in the form of an **S** on the table on which the jar rested, still maintaining their line of march, with the head of one in contact with the terminal legs of the one in advance. A twig of *Populus balsamifera* was given them, which they refused, and ate in preference a fragment of a dried leaf of *P. tremuloides*.

Their cast head-case has none of the spines of the first segment adhering to it, as has that of *H. Maia*.* The exuviæ are eaten by the larvæ.

On the 20th, they were one-half inch in length. The head at this stage, is pale red, with the clypeus fuscous, bordered with pale red; the eyes are on a black patch, surrounded with light red. The body is rufescent, with eight lighter lines: there are six rows of spines, or eight if the inferior row, interrupted on the proleg-bearing segments be included, which have black trunks with white branches; the lateral spines have their branches terminating in a black bristle; in the dorsal rows, except on the terminal segments, the branches are without the bristle, and some are black tipped; those of the substigmatal row are without the branches, having only bristles instead. The legs are marked with fuscous outwardly, and the prolegs are rufescent.

Third molt—From July 25th to 27th. Head fuscous anteriorly, dull green superiorly, as also above the eyes and margining the clypeus. Body white-dotted, and marked with a conspicuous stigmatal orange-red band bordered below with white, a pale rufescent vascular stripe, and two subdorsal and two lateral ones in which are

* *Twenty-third Report on the N. Y. State Cabinet*, 1872, p. 143.

the spines; spines pale green, with the tips or branches black. Legs fuscous; prolegs with a red patch exteriorly.

Fourth molt—Date not noted. Head pale green, black beneath, thence a black line extending upwards, dividing to inclose the eyes; the clypeus marked with black inferiorly and with an abbreviated black line external to it. Body pale green, with a pale yellow subdorsal and lateral line, an orange-red stigmatal line bearing the stigmata centrally, bordering which below is a narrow white stripe; the stripes commence on the third segment. Ventral region ocellated, more conspicuously in two rows of brown dots which range with the prolegs.

Fifth molt—August 10th. The appearance of the larva immediately succeeding this molt was not noted.

Mature larva.—Its length is two inches, and its diameter four-tenths of an inch. Its head is smooth, round, pale green, with a few short white hairs. The body tapers from the seventh segment moderately toward the head, and more considerably posteriorly: color of the body, white dorsally, pale green ventrally, with a yellow green lateral stripe and a quadrangular patch resting thereon on the posterior half of each abdominal segment; beneath this, commencing on the fourth segment, a narrow, sanguineous, stigmatal stripe inclosing the stigmata, having upon it some whitish piliferous dots, and bordered beneath with a narrow white stripe which it overlaps, except on the crown of each segment: ventrally ranging with the bases of the prolegs and on the anterior half of the segment, two rows of triangular sanguineous spots dotted as the stripes; the caudal plates are also sanguineous in continuation of the stigmatal stripes, and the prolegs are marked exteriorly with a similar colored spot. The number of spines on the several segments are, $\frac{1-5}{8}$, $\frac{6-9}{6}$, $\frac{10}{8}$, $\frac{11}{5}$, $\frac{12}{7}$: their trunks are green, of a conical form, with cylindrical green branches which are black tipped; those of the two superior rows are of the same length with the lateral ones, and have their branches contracted suddenly to an acute tip, except on the first segment where the upper branches (black on their superior half or three-fourths) are cylindrical throughout, and have implanted in their summits a bristle nearly or quite as long as the branch; on the second segment, the spines have a few of the bristle-branches, and two or three are also to be seen on the three superior spines of the twelfth segment; the lateral spines have each three or four of these bristle-branches, and the remainder like those of the dorsal rows; in the stigmatal row, the branches are all bristle-pointed, as also in the interrupted substigmatal row: the caudal plates have at their posterior angle, a rudimentary spine.

Pupation.—On the 21st of August, one of the larvæ made its slight cocoon between a couple of leaves. On the 17th of the following month, a male imago emerged from pupa; on the 21st, a second male was disclosed, and some other of the moths emerged during the fall.

The cocoons were kept in a warm room, and some time during the month of January, a crippled imago was found in the box. On the 7th of February, a second crippled specimen was obtained, and on the 10th, a perfect one, small and unusually dark colored.

In the irregularity of its disclosure, and in its extension over the fall and spring months (extending to the latter when not prematurely developed by warmth) this species resembles *Hemileuca Maia*. Its shortest period of pupation as above observed, was less than half that of *Maia*, being but twenty-seven days, and in *Maia* fifty-eight days.*

As an addition to the history of this moth, the following extracts are taken from notes made several years since:

Eggs were deposited July 10th. They are elliptical, somewhat flattened, five-hundredths of an inch in diameter, with a small black spot on each end and a larger orange one on the side. The caterpillars emerged July 22d. They are one-eighth of an inch long, of a reddish color, and have the body covered with long bristles.

On the 27th of July, occurred the first molting, when they measured one-fourth of an inch in length. The head was black, body rufescent, with black branching spines, and several stripes.

The second molt was on August 3d: length one-half inch. The larvæ are still associated in groups while feeding.

At the third molt on August 9th, they had attained a length of seven-eighths of an inch. The black spines have a few of their upper branches black, the others being white as before.

Fourth molt, August 17th: length of larva one inch and one-fourth. The fifth molt, pupation, et cet., were not recorded.

I have taken the larva feeding on locust (*Robinia pseudacacia*), on choke-cherry (*Cerasus virginiana*) on willows and other plants. A colony found on a willow, the leaves of which had become partially

* *Twenty-third Report on the N. Y. State Cabinet*, 1872, p. 148.

dried while being brought to me, deserted the twig for one of choke-cherry standing near it, on which they continued to feed.

In the *Entomological Correspondence* of Harris, clover, elm, oak, and balm of Gilead are given as food-plants of the larva.

I retain for this moth the specific name by which it has long been known, instead of adopting the one proposed for it by Walker (*varia*) and adopted by Packard in his "Synopsis of the Bombycidæ of the United States," in which he remarks that "our species has been confounded by authors with Cramer's species *Io:* judging by Cramer's plate his 'Io' from South America, belongs to a different genus." Dr. Speyer, the eminent German lepidopterist, has critically examined a number of specimens of the moth sent to him, and has found that it was correctly described under the name of *Io*, by Fabricius, in *Syst. Ent.* 1775, p. 560, and its habitat given as North America.

XII. TRANSFORMATIONS OF EACLES IMPERIALIS (Drury).

A pair of these beautiful and rare moths was taken in Greenbush, *in coitu*, and remained in that state while being brought across the river to Albany. In the box with them were some twigs and leaves of chestnut (*Castanea vesca*), with a number of eggs already deposited on them, from which circumstance, in the absence of any accompanying statement, it is to be presumed that they were captured upon that tree. A large number of eggs were subsequently deposited by the moth, of which, through the kindness of Mr. Louis Sautter, eighty-five were brought to me, which were said to have been laid on the 25th of June.

Eggs.—The eggs are flattened ellipsoids, having their diameters respectively 12-1000ths, 11-1000ths, and 8-1000ths of an inch. When examined under a high magnifying power the shell presents the appearance of having its surface studded with numerous short, capitellate setæ, somewhat curved at the base, and arranged in a degree of regularity at a little more than their length from one another; but as no setæ are seen in relief when looking across the surface of the shell, the forms observed undoubtedly pertain to its structure, and as, from the focal adjustment which their examination requires, they evidently connect the inner and outer surfaces, they can scarcely be anything else than pores traversing the shell. When the eggs were received by me, on the 30th, they all presented a circular depression on their flattened surface, which, in the eggs of many of our moths, indicates a stage in their development. They were of a light honey-yellow, with some reddish spots or clouds maculating their circumference. By the 2d of July, the larvæ could be plainly seen in frequent motion in a few of the eggs, through the transparent shell. On the following day, the larval bands were quite visible.

Young larvæ.—Four of the larvæ were disclosed July 4th, and twelve additional during the five following days; of these the last ones to emerge were quite feeble, four of them dying without partaking of food. None other of the eggs developed, probably from failure in fertilization, resulting from a disturbed coition. The newly emerged larva measures one-fourth of an inch in length. The head is red, round and smooth. Body of a dull red color, armed, except on the last two segments, with six rows of bristle-tipped spines: the subdorsal spines on the second and third segments are nearly one-third the length of the body, black, rugose, bifurcated, each prong tipped with a white acute bristle; on the top of the eleventh segment is a similar spine resting on a red, conical tubercle. The segments are annulated with three fuscous bands terminating laterally at the stigmatal flexure, of which one precedes, and two follow the spines: the terminal segment declines considerably from the plane of the others. Legs, black; prolegs, red.

The larvæ feed only at long intervals, passing most of their time in wandering over the leaves or resting on their petioles.

First molt.—Of one individual on July 11th; on the 12th, of two others, and on the night of the 14th, of four. Length of the larvæ, one-half inch. Head glossy, ferruginous, fuscous at the clypeus and about the eyes. Collar and terminal segment, ferruginous. The segments are testaceous centrally, shading into an obscure red at the incisures, the transverse bands which previously marked them having disappeared. The spines are glossy black with branches tipped with white acute bristles: the two long spines of the second and third segments each and the medial one of the eleventh, which are about one-fifth the length of the body, are directed slightly forward; their two forks are of unequal size; the last mentioned spine is in addition to the six of the preceding segments, and ranges with the four substigmatal and lateral spines, the two subdorsal being placed farther back on the segment: the terminal segment has thirteen spines, viz., six occupying the usual position, a seventh medial one behind the range of the preceding, four on the anal shield, of which the two anterior are the larger (four others are indicated by acute granulations on the posterior margin), and a small one on each terminal leg exteriorly. The stigmata are broadly elliptical, fuscous, and situated on a distinct, elliptical, testaceous spot. Legs and prolegs testaceous, marked outwardly with fuscous.

On the 16th two larvæ were in position for molting, indicating progress in the change by their translucent, vacant head-cases and heads covered by the skin of the first segment.

Second molt—July 17th. Length, six-tenths of an inch. Immediately succeeding the molt the head is pale red, and the long spines before noticed, now appearing as horns, are pearl white.

Three days thereafter, the larva measures eight-tenths of an inch in length. The head is dull ferruginous, with fuscous centrally and laterally. Body of an umber-brown, lighter at the incisures, gray dorsally with a dark vascular line; segments with a few white hairs, the longest of which surround the subdorsal spines; horns of second, third and eleventh segments curved, glossy black, with base luteous; spines dull black. Anal shield marked with a cordiform, glossy black spot, having central and marginal rufescent granulations; anal plates with a subtriangular, granulated, fuscous impression. Stigmata surrounded with a dark brown ring. Legs shining black; prolegs with a black spot exteriorly, and with fuscous near the plantæ.

Third molt—July 30th and August 3d, of the two larvæ surviving this change. Length, one inch. The head and color of the body are as before. A marked feature at this stage is the presence of long white hairs given out from the central portion of the segments, of which the superior ones are nearly twice the length of the thoracic horns, and the lateral ones shorter; similar hairs of medium length project laterally over the proleg-bases. The horns are 18-100ths of an inch long, of a honey-yellow color, and are studded with conical projections (of which the two apical are fuscous), bearing a short, acute, fuscous spinule. The spines of the two subdorsal rows are 5-100ths of an inch long, of the color of the head, and (except the two exterior to the horns) have two fuscous, spinule-tipped projections. The lateral row consists of tubercles, of which those on the interior segments are simple, and on the terminal ones branched, of a darker shade of color than the subdorsal spines. The substigmatal row is composed of still smaller simple tubercles. Anal shield brown with whitish granulations, bordered with tubercles, of which two are branched; anal plates fuscous centrally. Legs ferruginous; prolegs fuscous on the outer side.

Fourth molt of the sole survivor, August 15th. Length, one inch and three-tenths. A marked change occurs in the horns at this molting. From being heretofore cylindrical they are now conical, are armed with stout spinules, and have become shorter; the length of the thoracic ones is 12-100ths of an inch, of the posterior one, one-tenth of an inch. The anal plates are conspicuously marked with whitish granulations. The stigmata are brown, with a central line and border of white, surrounded with fuscous on a subquadrangular testaceous patch.

On the 18th of August the larva died of diarrhœa attended with an extraordinary retroversion and protrusion of the intestinal canal, resulting probably from its having been fed for so long a time on a food-plant unnatural to it. The chestnut leaves which were at first given to the young larvæ were refused. It not being convenient to provide them with buttonwood, on which Harris represents them as occurring, oak, mentioned by Abbot as one of their food-plants at the South, was procured for them, upon which they fed, but at no time in a very earnest manner. An attempt was afterward made to transfer them to pine, on which Dr. Fitch states that they are almost invariably found in the northern States,* but they were unwilling to make the change.

Although the larvæ above described were undoubtedly dwarfed by their spare diet, the small dimensions after the fourth molt, as compared with their mature size (three inches in length), would denote at least one additional molting prior to pupation. This would appear to be established by observations made on larvæ subsequently collected.

During the following month (September, 1869), from the 7th to the 16th, fourteen individuals were taken by me, and as many more by Mr. Meske, of Albany, from the lower branches of a number of pines (*Pinus strobus*) bordering a road in the Forbes manor, at Bath. Their presence on a tree was in most instances readily revealed by the large pellets of their excrement lying upon the smooth graveled road beneath, when, from the robust form of the larva in marked contrast with the slender leaves surrounding it, its resting-place was not difficult to detect. On the 7th, one was taken which had just completed its last molting; on the 9th one was observed in the process of molting, which, from some irregularity attending it, had fallen to the ground; and on the same day one which had already assumed the brown or tawny hue indicative of its full maturity was taken while moving down the trunk of a tree to seek its place for pupation. The most advanced one of the others collected, matured on the 11th, and transformed to a pupa on the surface of the ground on the 16th. Most of the remainder entered the ground, where they constructed cells of moderate dimensions for their pupal transformation.

The pupæ were kept in a cold room during the winter. About the 1st of March they were removed to a warm apartment. April 28th, May 3d and 7th, male imagines emerged, after which females were disclosed until near the end of the month.

* *Third, Fourth and Fifth Reports on the Insects of New York*, 1859, Section 271.

In the fall of 1870 diligent search was made for the larva in the locality at Bath, where it had been abundant the preceding year, as above recorded, without finding a single individual. Its non-occurrence indicates a marked periodicity in the appearance of the species or, possibly, an exhausted locality from the collections made.

A single specimen of the closely related species *Citheronia regalis* Hübner, has been taken near Albany, by Mr. Sautter, and I am informed by Dr. M. Cooke, of Utica, N. Y., that its larva has been found, on one occasion, in the vicinity of that city. I have not met with it in my field collections.

XIII. LARVAL NOTES ON ANISOTA SENATORIA (Smith).

Moths were observed at Center, N. Y., July 7th, depositing their eggs on the under surface of leaves of oak, in regular distribution in a single layer, and in contact with one another. A leaf of *Quercus prinoides* of ordinary size was collected, having one-half of its surface covered with the eggs. From a count of a portion of the deposit, the whole number was estimated at five hundred; still larger patches have been observed. From the number usually occurring in these deposits it may be presumed that the moth places all her eggs on a single leaf unless disturbed during the operation.

The eggs hatched July 11th. The head of the young larva is oval and glossy black. The body is pale yellow-green, with a few short hairs; on the second segment are two smooth, straight, subcylindrical, black horns, arising from a green base, and with a slight enlargement at the apex, where they give out two black diverging setæ of the length of two-thirds that of the horn.

The young larvæ feed in company, and occupy both surfaces of the leaf, the entire substance of which they consume, except the veins and veinlets, leaving frequently a very good skeleton of the leaf.

The *first molting* occurred on the 18th and 19th of July. At this stage the body is obscure green with seven fuscous lines, of which the dorsal and stigmatal ones are narrow; the subdorsal and lateral ones broader, having in them a row of short spines. Collar centrally and anal segment, shining black. Legs, black; prolegs, with a black spot outwardly.

Second molt—July 28th and 29th. Length of larva, 37-100ths of an inch. Head and collar, glossy black. Horns, slightly spinose, enlarged at the tip, and usually with apical spines. The abdominal stripes are black, with yellow-brown intermediately, showing a broad stigmatal stripe. The terminal segment is spinose, and of a glossy black.

Third molt—August 4th and 5th. Length, six-tenths of an inch. The larva is glossy black, with eight yellow stripes, of which the lower one is geminated by a crescent on the central portion of each segment inclosing a spinule; ventrally from the fifth segment is a yellow-green interrupted stripe. The horns are slightly tapering, clubbed at their tips, and two-tenths of an inch long. The legs and prolegs are black.

Fourth molt.—Extending from August 14th to 16th. Immediately following the molting, the head, collar, horns, anal shield, anal plates and legs are flavescent; in a few hours they become shining black. The horns are but slightly enlarged at the tip, being less so than previous to this molt. The body is covered with numerous, minute, shining, elevated points of the color of the ground upon which they are placed.

The mature larva is so fully and accurately described by Dr. Fitch* as not to need redescription here.

Subsequent collections of larvæ were made and inadvertently added to the above, preventing the observation of the date of pupation of the brood. The pupation of the last occurred about the 15th of September. Larvæ were still observed in the field on the 30th of September.

Dr. Eights, of Albany, has informed me that a number of years ago he observed on the line of the New York Central railroad, between Albany and Schenectady, a species of caterpillar so exceedingly abundant on and about the railroad track that the numbers crushed on the rails by the passage of the trains caused the slipping of the wheels of the engines to the extent of proving a serious inconvenience in ascending grades. A notice of the interesting incident was communicated by him to one of the journals of the day, in which some account of the caterpillar was given. Although from the long time which has elapsed since the event he is not able to indicate positively the species, he believes it to have been *A. senatoria*, and the locality of its occurrence in the vicinity of Center.

This larva is found annually at Center in great abundance. In the more favorable years for its multiplication, it abounds so excessively that the smaller oaks, although very numerous there, are almost as effectually defoliated as if a fire had swept over them. I have no information of its occurrence in equal numbers at any other locality.

The congeners of this species, *pellucida*, *stigma*, and *rubicunda*, are rarely taken in the neighborhood of Albany. Of the latter species the larva has not been observed, but a wing of the imago has been found, by Mr. Meske, at Center.

* *Third, Fourth, and Fifth Reports on the Insects of N. Y.*, 1859, section 322.

XIV. CALENDAR OF BUTTERFLIES FOR THE YEAR 1870.

In the following table is contained a record of seventy-three species of Rhopalocera observed at six localities in the State of New York on thirty-five days during the spring and summer of 1870, commencing with the 26th of April, the date of the first observed apparition of *Thecla Irus*, and ending on September 22d.

The figures immediately below the several months give the day of collection or observation. Underneath these the locality is indicated in roman characters, I representing Schoharie; IV, Center; V, Bethlehem; VII, Bath; Sharon Springs designated by the letter A; and Glen, in Warren county, by B. In a few instances where the apparition of fresh individuals of a new brood was noted, the date is indicated by the insertion of a larger star (*).

The greatest number of species observed in one day was twenty-nine, at Center, on the 16th of June. The time of observation was usually between the hours of 10 A. M. and 2 P. M.

The last column but partially represents the comparative abundance of the several species, its more direct import being the continuance of the brood or a succession of broods. Thus, while *C. Philodice* is recorded on twenty-eight occasions, *L. comyntas* on twenty-two, *C. Americana* on nineteen, *P. Troilus* on sixteen, *L. misippus* on fifteen, *M. tharos* on thirteen (each of these being double or triple brooded), of none of them were as many individuals seen as of *Thecla Irus*, which was observed on but eight occasions. If from this comparison *C. Philodice* be omitted, the number of *T. Irus* observed was at least three times as great as of any other of the species.

The single observations recorded of several of the species, viz., *A. Atlantis*, *A. Idalia*, *M. Harrisii*, *G. Dryas*, *P. cardui*, *P. Atalanta*, *H. Sassacus* and *H. Leonardus*, faithfully indicate their rarity during this year at least, for of each of these but a single individual was collected or recognized.

The observations at Sharon Springs are by Mr. O. Meske, as are also many of those at Bath, Bethlehem and Center.

	April.	May.									June.					July.											August.			September.						Times observed.
	26	3	4	6	14	16	21	22	25	31	5	7	16	22	29	2	4	6	9	10	13	16	21	24	28	30	7	26	31	2	6	9	14	19	22	
	IV	IV	VII	IV	IV	V	IV	VII	IV	IV	V	IV	IV	V	IV	IV	A	V	IV	A	IV	IV	IV	I	V	IV	B	V	IV	VII	VII	V	V	IV	VII	
Papilio Turnus *Linn*	.	.	.	.	.	.	.	.	*	*	.	*	*	*	*	*	*	*	.	.	.	.	.	.	.	.	.	.	.	.	.	.	.	.	.	9
Papilio Troilus *Linn*	.	.	.	.	.	.	.	.	.	*	.	*	*	*	*	*	.	*	*	.	.	*	*	.	*	*	.	*	*	.	.	*	*	.	.	16
Papilio Asterias *Fabr*	.	.	.	.	.	.	.	.	.	.	.	.	.	.	.	.	.	.	.	.	.	.	.	*	*	.	.	.	.	.	.	*	*	.	.	4
Pieris oleracea (*Harris*)	.	*	.	*	.	*	.	.	.	.	.	.	.	*	*	*	.	.	.	.	.	.	.	*	*	*	.	.	.	.	.	.	.	.	.	9
Pieris rapæ (*Linn.*)	.	.	.	.	.	.	.	.	.	.	.	.	.	.	.	.	.	.	.	.	.	.	.	.	.	.	*	*	*	*	*	*	*	*	*	9
Colias Philodice *Godt*	.	.	*	.	.	*	.	.	*	*	.	*	*	*	*	*	*	*	*	*	*	*	*	*	*	*	*	*	*	*	*	*	*	*	*	28
Danais Plexippus (*Linn.*)	.	.	.	.	.	.	.	.	.	*	.	*	.	.	.	.	.	*	.	.	.	.	.	.	*	*	*	*	*	*	*	*	*	.	.	12
Argynnis Aphrodite *Fabr*	.	.	.	.	.	.	.	.	.	.	.	.	.	*	.		*	*	*	.	.	*	*	*	*	*	.	*	.	.	.	*	*	.	.	12
Argynnis Cybele *Fabr*	.	.	.	.	.	.	.	.	.	.	.	.	.	*	*	*	.	*	*	.	*	*	*	.	*	*	.	.	.	.	.	*	*	.	.	12
Argynnis Atlantis *Edw*	.	.	.	.	.	.	.	.	.	.	.	.	.	.	.	.	.	.	.	.	.	.	.	.	.	.	*	.	.	.	.	.	.	.	.	1
Argynnis Idalia *Fabr*	.	.	.	.	.	.	.	.	.	.	.	.	.	.	.	.	.	.	.	.	.	.	.	.	.	.	.	.	.	.	.	.	*	.	.	1
Argynnis Bellona *Fabr*	.	.	*	.	.	*	.	*	.	.	.	.	.	*	.	*	.	*	.	.	.	.	.	.	*	.	.	*	.	.	.	*	*	.	.	10
Argynnis Myrina *Fabr*	.	.	.	.	.	.	.	.	.	.	.	.	*	*	.	*	.	*	.	.	.	.	.	.	*	*	.	*	*	.	.	*	*	.	.	10
Melitæa tharos *Boisd. et Lec*	.	.	.	.	.	.	.	*	.	*	.	.	*	.	.	*	.	.	.	.	.	.	.	.	*	*	.	*	*	*	*	*	*	.	*	13
Melitæa Marcia *Edw*	.	.	.	.	.	.	.	.	.	.	.	.	.	.	*	*	*	.	*	.	*	*	*	.	.	.	.	.	.	.	.	*	.	.	.	8
Melitæa Harrisii *Scudd*	.	.	.	.	.	.	.	.	.	.	.	.	*	.	.	.	.	.	.	.	.	.	.	.	.	.	.	.	.	.	.	.	.	.	.	1
Melitæa Batesii *Reak*	.	.	.	.	.	.	.	.	.	*	*	*	*	.	.	.	.	.	.	.	.	.	.	.	.	.	.	.	.	.	.	.	.	.	.	4
Melitæa Nycteis *Doubl*	.	.	.	.	.	.	.	.	.	.	.	.	*	.	*	*	.	.	*	.	*	*	*	.	.	.	.	.	.	.	.	.	.	.	.	7
Melitæa Phaeton *Fabr*	.	.	.	.	.	.	.	.	.	.	.	.	*	.	*	*	.	.	.	.	.	.	.	.	.	.	.	.	.	.	.	.	.	.	.	3
Grapta comma (*Harris*)	.	.	.	.	.	.	.	.	.	.	.	.	.	.	.	.	.	.	.	.	.	.	.	*	.	.	.	.	.	.	.	.	.	.	*	2

Grapta Dryas *Edw*	.	.	.	.	.	.	.	.	.	.	.	.	.	.	.	.	.	.	.	.	.	.	.	*	.	.	.	.	.	.	.	.	.	.	.	1
Grapta Progne (*Cram.*)	.	.	.	.	.	.	.	.	.	.	.	.	.	*	.	.	.	.	.	.	.	.	.	*	.	.	.	.	.	.	.	.	.	.	.	2
Grapta J-album (*Boisd. et Lec.*)	.	.	.	.	.	.	.	.	.	.	.	.	.	.	*	.	.	.	.	*	.	.	.	.	.	.	.	.	.	.	.	.	.	.	.	2
Vanessa Antiopa (*Linn.*)	.	.	.	.	.	*	.	.	.	*	.	.	.	.	*	*	.	*	*	.	*	*	.	.	.	.	.	.	*	*	*	.	.	.	.	11
Pyrameis huntera (*Fabr.*)	.	.	.	.	.	.	.	.	.	.	.	.	.	.	.	.	.	.	.	.	.	.	.	.	.	.	.	.	.	.	.	*	*	.	.	2
Pyrameis cardui (*Linn.*)	.	.	.	.	.	.	.	.	.	.	.	.	.	.	.	.	.	*	.	.	.	.	.	.	.	.	.	.	.	.	.	.	.	.	.	1
Pyrameis Atalanta (*Linn.*)	.	.	.	.	.	.	.	.	.	.	.	.	.	.	.	.	.	.	.	.	.	.	.	.	.	.	.	.	.	.	.	*	.	.	.	1
Limenitis misippus (*Fabr.*)	.	.	.	.	.	.	.	.	.	*	.	.	*	.	*	*	.	.	.	.	*	*	*	*	*	*	.	*	*	.	*	.	*	*	.	15
Limenitis arthemis (*Drury*)	.	.	.	.	.	.	.	.	.	.	.	.	*	.	.	.	.	*	.	.	.	.	.	.	.	.	.	.	.	.	.	.	.	.	.	2
Limenitis Ursula (*Fabr.*)	.	.	.	.	.	.	.	.	.	.	.	.	.	.	.	.	.	*	*	.	.	.	.	.	.	.	.	*	.	.	.	*	.	.	.	4
Satyrus Alope (*Fabr.*)	.	.	.	.	.	.	.	.	.	.	.	.	.	.	.	.	.	*	.	*	*	.	.	.	*	*	.	.	.	.	.	.	.	.	.	5
Satyrus Nephele (*Kirby*)	.	.	.	.	.	.	.	.	.	.	.	.	.	.	.	.	*	.	.	*	.	.	.	*	*	.	.	.	.	.	.	.	.	.	.	4
Neonympha Eurytus (*Fabr.*)[1]	.	.	.	.	.	.	.	.	.	.	*	*	.	.	.	.	.	.	.	.	.	.	.	.	.	.	.	.	.	.	.	.	.	.	.	2
Neonympha Canthus (*Linn.*)[1]	.	.	.	.	.	.	.	.	.	.	.	.	.	.	.	.	.	.	.	.	*	*	.	.	.	.	.	.	.	.	.	.	.	.	.	2
Thecla Irus (*Godt.*)[2]	*	*	.	*	*	.	*	.	*	*	.	*	.	.	.	.	.	.	.	.	.	.	.	.	.	.	.	.	.	.	.	.	.	.	.	8
Thecla Augustus *Kirby*	.	*	.	*	.	.	.	.	.	.	.	.	.	.	.	.	.	.	.	.	.	.	.	.	.	.	.	.	.	.	.	.	.	.	.	2
Thecla niphon *Hubn*	.	*	.	*	.	*	.	.	.	.	.	.	.	.	.	.	.	.	.	.	.	.	.	.	.	.	.	.	.	.	.	.	.	.	.	3
Thecla Melinus *Westw.*	.	*	.	*	.	.	.	.	.	*	.	.	.	.	*	*	.	.	*	.	*	*	.	.	.	*	.	.	*	.	.	.	*	.	.	11
Thecla Calanus (*Hubn.*)	.	.	.	.	.	.	.	.	.	.	.	.	.	.	.	.	*	.	.	*	.	.	.	.	.	.	.	.	.	.	.	.	.	.	.	2
Thecla liparops *Boisd. et Lec.*[3]	.	.	.	.	.	.	.	.	.	.	.	.	.	.	.	*	.	*	.	.	.	.	.	.	*	.	.	.	.	.	.	.	.	.	.	3
Thecla Edwardsii *Saund*	.	.	.	.	.	.	.	.	.	.	.	.	.	*	*	*	.	*	*	.	*	*	*	.	.	.	.	.	.	.	.	.	.	.	.	8
Thecla Acadica *Edw*	.	.	.	.	.	.	.	.	.	.	.	.	.	.	*	*	.	.	.	.	*	.	.	.	.	.	.	.	.	.	.	.	.	.	.	3
Thecla Mopsus (*Hubn.*)	.	.	.	.	.	.	.	.	.	.	.	.	.	*	.	.	.	.	*	.	*	*	*	.	.	*	.	.	.	.	.	.	.	.	.	6
Lycæna violacea *Edw*	.	*	*	.	*	.	.	.	.	.	.	.	.	.	.	.	.	.	.	.	.	.	.	.	.	.	.	.	.	.	.	.	.	.	.	3
Lycæna neglecta *Edw*	.	.	.	.	*	*	*	.	*	*	.	*	*	.	.	.	.	.	.	.	.	.	.	.	.	.	.	.	.	.	.	.	.	.	.	7
Lycæna comyntas (*Godt.*)	.	.	.	*	*	.	*	*	*	.	.	.	.	*	*	*	.	*	*	.	*	*	*	.	*	*	*	*	*	.	*	*	*	*	.	22

	April.	May.									June.					July.											August.			September.						Times observed.
	26	3	4	6	14	16	21	22	25	31	5	7	16	22	29	2	4	6	9	10	13	16	21	24	28	30	7	26	31	2	6	9	14	19	22	
	IV	IV	VII	IV	IV	V	IV	VII	IV	IV	V	IV	IV	V	IV	IV	A	V	IV	A	IV	IV	IV	I	V	IV	B	V	IV	VII	VII	V	V	IV	VII	
Lycæna Scudderii *Edw.*	.	.	.	.	.	.	.	.	*	*	.	*	*	.	*	*	.	.	*	.	*	*	*	.	.	*	.	.	.	.	.	.	.	.	.	11
Chrysophanus Americana (*Harris*)	.	.	.	.	.	.	*	.	.	*	.	*	*	.	.	.	.	*	*	.	*	*	*	.	*	*	*	*	*	*	.	*	*	*	*	19
Chrysophanus Hyllus (*Cram.*)[4]	.	.	.	.	.	.	.	.	.	.	.	.	*	*	.	*	.	*	.	.	.	.	.	.	.	.	.	.	.	.	.	.	*	.	.	5
Eudamus Pylades *Scudd.*	.	.	.	.	.	.	.	.	.	.	.	*	*	*	*	*	.	*	*	.	*	.	*	.	.	.	.	.	.	.	.	.	*	.	.	10
Eudamus Tityrus (*Fabr.*)	.	.	.	.	.	.	.	.	.	*	.	*	*	*	*	*	.	*	*	.	.	.	.	.	*	.	.	.	.	.	.	.	.	.	.	9
Nisoniades Juvenalis (*Fabr.*)	.	.	.	*	*	.	*	*	*	*	.	*	*	.	.	.	.	.	.	.	.	.	.	.	.	.	.	.	.	.	.	.	.	.	.	8
Nisoniades Persius *Scudd.*	.	*	.	*	*	.	*	*	*	*	.	*	.	.	.	.	.	.	.	.	.	.	.	.	.	.	.	.	.	.	.	.	.	.	.	8
Nisoniades Martialis *Scudd*	.	.	.	.	*	.	*	.	*	*	.	*	*	*	.	.	.	.	*	.	*	*	.	.	.	.	.	.	.	.	.	.	.	.	.	10
Nisoniades Lucilius *Lintn.*	.	.	.	.	.	*	*	.	*	.	.	.	.	.	.	.	.	*	.	.	.	.	.	.	*	.	.	*	.	.	.	*	*	.	.	8
Nisoniades Icelus *Lintn.*	.	.	.	.	.	.	.	.	*	*	.	*	*	.	.	.	.	.	.	.	.	.	.	.	.	.	.	.	.	.	.	.	.	.	.	4
Hesperia Metea *Scudd.*	.	.	.	.	*	.	*	.	*	*	.	.	*	.	.	*	.	.	.	.	.	.	.	.	.	.	.	.	.	.	.	.	.	.	.	6
Hesperia vialis *Edw.*	.	.	.	.	.	.	*	.	*	*	.	*	*	.	.	.	.	.	.	.	.	.	.	.	.	.	.	.	.	.	.	.	.	.	.	5
Hesperia Zabulon (*Boisd. et Lec.*)[5]	.	.	.	.	.	.	.	*	*	*	*	*	*	.	.	.	.	.	.	.	.	.	.	.	.	.	.	.	.	.	.	.	.	.	.	6
Hesperia Hianna *Scudd.*	.	.	.	.	.	.	.	.	*	*	.	*	*	.	.	.	.	.	.	.	.	.	.	.	.	.	.	.	.	.	.	.	.	.	.	4
Hesperia Sassacus *Harris*	.	.	.	.	.	.	.	.	.	.	*	.	.	.	.	.	.	.	.	.	.	.	.	.	.	.	.	.	.	.	.	.	.	.	.	1
Hesperia Peckius *Kirby*[6]	.	.	.	.	.	.	.	.	.	.	*	.	*	*	.	*	*	*	.	.	.	.	.	.	*	*	.	*	.	.	.	*	*	.	*	12
Hesperia Mystic *Edw.*	.	.	.	.	.	.	.	.	.	.	*	.	*	.	.	.	.	*	.	.	.	.	.	.	.	.	.	.	.	.	*	*	.	.	.	5
Hesperia Taumas (*Fabr.*)[7]	.	.	.	.	.	.	.	.	.	.	*	.	*	*	.	.	.	*	.	.	.	.	.	.	.	.	.	*	.	.	.	*	*	*	.	8
Hesperia bimacula *Grote-Rob.*[8]	.	.	.	.	.	.	.	.	.	.	.	.	*	*	.	.	.	.	.	.	.	.	.	.	.	.	.	.	.	.	.	.	.	.	.	2
Hesperia Metacomet *Harris*	.	.	.	.	.	.	.	.	.	.	.	.	.	*	.	.	.	*	.	.	.	.	.	.	*	.	.	.	.	.	.	.	.	.	.	3

Hesperia Ætna *Boisd.*[9]	·	·	·	·	·	·	·	·	·	·	·	·	·	*	·	*	·	*	*	*	·	·	·	·	*	·	·	·	·	·	·	·	·	·	·	6
Hesperia verna (*Edw.*)	·	·	·	·	·	·	·	·	·	·	·	·	·	*	·	·	·	*	*	*	·	·	·	·	·	·	·	·	·	·	·	·	·	·	·	4
Hesperia Manataaqua *Scudd*	·	·	·	·	·	·	·	·	·	·	·	·	·	*	*	·	·	·	*	·	·	·	·	·	·	·	·	·	·	·	·	*	·	·	·	4
Hesperia Logan *Edw.*[10]	·	·	·	·	·	·	·	·	·	·	·	·	·	·	*	*	·	·	*	·	*	*	·	·	·	·	·	·	·	·	·	·	·	·	·	5
Hesperia Massasoit *Scudd.*	·	·	·	·	·	·	·	·	·	·	·	·	·	·	·	·	·	·	*	·	·	*	·	·	·	·	·	·	·	·	·	·	·	·	·	2
Hesperia Leonardus *Harris*	·	·	·	·	·	·	·	·	·	·	·		·	·	·	·	·	·	·	·	·	·	·	·	·	·	·	·	*	·	·	·	·	·	·	1
Thymelicus Numitor (*Fabr.*)[11]	·	·	·	·	·	·	·	·	·	·	*	·	*	*	·	·	·	·	·	·	·	·	·	·	*	·	·	*	*	·	·	*	*	·	·	8
Number of species observed	1	7	3	8	8	7	10	6	15	22	8	19	29	24	20	26	7	27	22	7	18	18	13	9	22	16	6	16	13	6	8	22	22	6	6	

1 *Hipparchia Boisduvalii* Harris.
2 *Thecla Arsace* Boisd. et Lec.
3 *Thecla strigosa* Harris.
4 *Polyommatus Thoe* Boisd. et Lec.
5 *Hesperia Hobomok* Harris = *H. Pocahontas* Scudd., *H. Quadaquina* Scudd.
6 *Hesperia Wamsutta* Harris.
7 *Hesperia Ahaton* Harris.
8 *Hesperia Acanootus* Scudd.
9 *Hesperia Egeremet* Scudd.
10 *Hesperia Delaware* Edw.
11 *Heteropterus marginatus* Harris.

The following are some notes made during the year 1870 on the abundance, condition, time of appearance of sexes, successive broods, larvæ, et cet., of some of the species recorded in the preceding list:

May 3d.—*Thecla Irus* of each sex abundant at Center. Of *T. Augustus*, eight individuals were collected; of *T. niphon*, five, and of *T. Melinus*, one. Of the Nisoniades, a few of *Persius* only were observed. *Pieris oleracea* was taken by me, for the first time in Albany county.

May 14th.—*Thecla Irus*, with each sex in good condition, still abroad at Center; the only Thecla observed. A few females of *N. Persius* were taken for the first time this season, and one of *N. Juvenalis;* of *N. Martialis*, two males. Of the earliest Hesperian, *Metea*, two males were obtained.

May 16th.—Five males of *Nisoniades Lucilius* were captured while hovering over blossoms of *Aquilegia Canadensis*, and as many more were observed. *Argynnis Bellona* was abundant in wet meadows. Five of *Thecla niphon* were taken, all of which were females; the larger number of the captures of this species prove to be of this sex. Bethlehem.

May 19th.—*Pieris oleracea* about gardens, in woods and its margins in meadows. *N. Lucilius* taken. Schoharie.

May 21st.—*N. Persius* abundant, and many quite fresh, with a few only of *N. Juvenalis* and *Martialis*. A female *N. Lucilius* which had just emerged from the chrysalis was taken while sitting on a twig. Among numerous *Lycæna neglecta* no females were seen, and of *L. comyntas* but a single female. The abundant brood of *T. Irus* was represented by only a few worn specimens. Two males of *Hesperia vialis* were collected and several of *H. Metea*. A female *Chrysophanus Americana* was taken, indicating the species to have been abroad for several days. Center.

May 25th.—Of the Nisoniades, *Juvenalis* and *Icelus* were abundant, *Martialis* and *Lucilius* quite few in number. Several of each sex of *Hesperia Hianna* were obtained, and one *H. Zabulon*. Of *Lycæna Scudderii*—its first observation for the season—a single one only was seen. One female of *L. neglecta* occurred. Center.

May 28th.—At Schoharie, a few *P. oleracea* were seen, the first brood having nearly disappeared. Its eggs and some young larvæ were found on horse-radish. On June 12th none of the butterflies could be seen.

May 31st.—*L. neglecta* abounded in flocks; *L. Scudderii* was not rare. Of *Hesperia Hianna* twenty captures were made, and more could have been secured. Center.

June 7th.—Collected twelve *Melitæa Batesii*, one of which was a female. Females of *L. Scudderii* have appeared.

June 16th.—*Melitæa Batesii* were abundant; of *M. Nycteis*, only a few were observed. Of *Hesperia bimacula*, seven males and one female were collected; of *Chrysophanus Hyllus*, three, and of *Melitæa Phaeton*, six.

June 29th.—Melitæas very abundant at Center; of *M. Nycteis*, only males occurred. No Nisoniades were seen.

July 2d.—Females of *M. Nycteis* were abundant.

July 6th.—At Bethlehem, males of *Argynnis Aphrodite* and of *A. Cybele* abundant. In a number of captures of these species, no females occurred. *Satyrus Alope* abundant.

July 9th.—At Center, the second brood of *Lycæna Scudderii* very abundant in the roads, on flowers and on leaves; threw the net over fifteen individuals at once. On a small patch of damp ground in the road a large number had assembled, estimated at two hundred. Several Theclas were taken while resting on the flowers of *Ceanothus Americanus* (Jersey tea). Six of *Hesperia Logan* were captured, all of which were males.

July 13th.—*L. Scudderii* quite abundant, and among them many females. *Limenitis misippus* was more numerous than ever before observed by me. Individuals of a second brood of *Nisoniades Martialis* were taken which show some difference of color from those of the first brood; they were at first believed to be a distinct species. Center.

July 17th.—The second brood of *Pieris oleracea* which has been numerous for a time past, has disappeared, only one individual having been observed. Schoharie.

July 20th.—A larva of *Danais Plexippus* changed to a chrysalis, from which it emerged on the 28th.

July 21st.—Larva of *Papilio Troilus* found on sassafras, and two of *P. Turnus* on wild-cherry, resting each on a web spun over the upper surface of a leaf, drawing the sides somewhat together and depressing the midvein a little distance beneath the web. *Limenitis misippus* abundant; *Lycæna Scudderii* diminishing; a few good *Thecla Mopsus* still abroad; *T. Edwardsii* quite worn; all the other species of diurnals observed were worn except *C. Americana*, of which there are successive broods throughout the season.

July 24th.—*Pieris rapæ* was recognized, for the first time, in Albany. A few were seen flying about piles of cabbages exposed upon the sidewalk at some vegetable stands in the south part of the city, and in one instance alighting upon a cabbage as if to deposit an egg. On the 27th many were seen and several were captured in a vegetable garden at the extreme southern part of the city, near "the Island," upon which cabbages are extensively cultivated. Upon visiting the island, they were found to be so abundant that several could be observed at any moment hovering over or alighting on the plants. Many were attracted by the blossoms of *Lappa officinalis* (burdock) growing abundantly upon the bank of Island creek; on one plant ten were counted, intent on taking their food from the flowers. Of the thirty individuals collected, two-thirds were males which were nearly all in good condition, while the females were worn. This butterfly is more difficult to capture on the wing than *oleracea*, for while the flight is not more rapid than of that species, it is undoubtedly more erratic, for less than one-third of my attempts to inclose them in the net were successful.

July 25th.—No *P. oleracea* were observed at Schoharie, but some full-grown larvæ were collected from horse-radish, and a few of its chrysalides were found.

July 28th.—A large number of larvæ of *Nisoniades Lucilius* were found resting concealed on the under surface of leaves of *Aquilegia Canadensis*, growing abundantly in an elevated rocky locality in Bethlehem. Their shelter, as observed in numerous specimens collected at this time and in larvæ subsequently taken, is constructed in a very ingenious manner. Shortly after the larva leaves its shell, and with its first feeding it commences to cut a narrow channel in the leaf from the margin inwardly a short distance; this completed, from another point on the margin not far removed from the first, a second channel is cut, curving toward the former, the two not uniting, but frequently running parallel for a short space. The portion thus nearly separated retains its connexion with the leaf by only a pedicel-like attachment. Its own weight carries it downward to nearly the position which it is to assume, when a very slight effort by the young larva serves to bring it to its desired place, almost in contact with the lower side of the leaf, to which it is then fastened by threads passing between the two surfaces at several points. Sometimes, as if with the object of economizing time or labor, the lobe of a leaf is selected of which to construct this shelter, when but a moderate amount of cutting at its base gives the requisite size and desired form.

Resting upon the inside of this recurved portion, the larva may always be found, except during the brief time that it leaves its concealment to take its food from some neighboring leaf. Its rapid feeding soon satisfies its appetite, when it moves quickly back and resumes its position. In localities where the larva occurs, these hiding-places may be readily found by bending over the stems of the Aquilegia, when these little bits of the bright green upper surface of the leaf, in marked contrast with the grayish-green of the lower side upon which they rest, at once disclose their presence. Should one of them be found deserted, its former occupant may perhaps be discovered on a leaf near by, within a larger retreat of similar construction. From the gradation of sizes observed, it is probable that following each molting a new shelter is constructed, of a size sufficient to cover the larva during that stage of growth, until at the last larval molting, when an entire leaf is simply folded over, or two or more leaves have to be brought together in order to afford the necessary concealment.

Some of the larvæ taken at this time had undergone their third molt, many their second, and the larger number their first. About one hundred were collected, and two eggs near their development.

Fresh broods of *Argynnis Myrina* and *A. Bellona* were observed on mint blossoms; also many *Hesperia Peckius*, *Limenitis misippus* and *Thymelicus Numitor*. Bethlehem.

July 30th.—*Limenitis misippus* abundant and easily captured. A few *Hesperia Peckius* seen. Some *Lycæna comyntas* in good condition, but *L. Scudderii* quite worn. A few *Thecla Mopsus*, a little worn, on blossoms of Jersey tea. Some *P. oleracea*, but no *P. rapæ* observed. Center.

July 31st.—The third brood of *P. oleracea* quite abundant at Schoharie.

August 6th.—At Saratoga Springs, saw a number of *Pieris rapæ* in company with *P. oleracea*, flying about gardens and blossoms of burdock in vacant lots.

August 8th.—At the Glen, Warren county, *Pieris rapæ* was numerous, and its larvæ were found on garden cabbages, usually feeding on the tender central leaves of the plant, unlike *oleracea*, which more frequently occur on the older outer leaves. On a small central leaf of a plant commencing to form a head, three larvæ were feeding, the ravages of which, in their progress to maturity, in all probability would have prevented the heading of the plant. Of several of the plants the central leaves had been nearly consumed, and an amount of excrementitious matter was distributed about their stalks by larvæ which had matured and probably transformed to the butterflies which were then flitting about the garden. A rapæ chrysalis was observed, attached to the midrib of one of the larger leaves. Between Saratoga and the Glen several of *Danais Plexippus* were seen in graceful flight. At the Glen, *Colias Philodice* occurred in companies, upon damp patches of ground. *Argynnis Cybele* and *A. Atlantis* were captured on the flowers of Canada thistle. *Chrysophanus Americana* was very abundant. A single *Lycæna comyntas* was seen, and a Grapta in flight, of which the species could not be determined.

August 15th.—At Schoharie, *P. oleracea* of the third brood, more abundant than at any time previously this year.

Two colonies of *Vanessa Milbertii* larvæ were taken on nettle (*Urtica dioica*), the one apparently after the first molting, and the other after the third. The larvæ of the former were feeding in company near the tip of a stem, and on one of the terminal leaves was a cluster of the egg-shells from which they had emerged.

August 18th.—Fourteen larvæ of *Pieris rapæ* were collected, all of which had transformed to chrysalides by the 21st; the last imago from these emerged on the 30th.

August 22d.—Larvæ of *Pyrameis Atalanta* abundant within folded leaves of the nettle; some of the larvæ were nearly mature, and others about half-grown.

August 26th.—A few young larvæ of *Limenitis misippus* were found on poplar and willow. One on the willow had, at this early period, commenced the construction of its hybernaculum, and another had built its peculiar structure of bits of leaf on the mid-vein.

Collected ten *Nisoniades Lucilius* of the third (?) brood, most of which were males. A few of the larvæ, nearly full-grown were found within their leaf shelters on the Aquilegia, but none of the chrysalides or their cases could be discovered, although careful search was made for them.

Hesperia Peckius, *Argynnis Bellona* and *A. Myrina*, were abundant on mint. *Danais Plexippus* was more numerous than I had ever seen it.

August 29th.—The third brood of *P. oleracea* is nearly gone. As near as can be determined from visits regularly made to Schoharie, at intervals of a week, the three broods of the above species have had, during the present year, a continuance of about a month each, the last brood probably continuing for a few days longer. Approximate periods of their duration would be, of the first brood, from May 5th to June 5th; second brood, from June 20th to July 20th; third brood, from July 28th to September 1st.

Nisoniades Lucilius emerged after thirteen days in the chrysalis state. The larva had been reared in confinement, and was transformed to a chrysalis among the leaves of its food-plant.

August 31st.—Eight males of the hitherto rare *Hesperia Leonardus* were collected on the summit of a hill at Center, indicating it to be partial to elevated ground. Several *L. comyntas* of a new brood were taken, and also fresh specimens of *Vanessa Antiopa.*

September 9th.—At Bethlehem collected several males of *Nisoniades Lucilius*, with some of its larvæ just from the egg, and others half-grown. *Argynnis Myrina* were observed *in copula.* Of *Hesperia Taumas*, usually rare, six males and one female were collected.

September 14th.—*Argynnis Idalia* was taken for the first time in the vicinity of Albany. Young larvæ of *L. misippus* and *N. Lucilius* were observed. Bethlehem.

October 21st.—On "the Island" collected numbers of the chrysalides of *P. rapæ* which were attached to the under surface of the stems of the coarser weeds lying on the ground. Several were taken from the lower side of prostrate pieces of timber. None were found upon the trunks of trees or standing weeds, nor on fences, except at one place where tomato vines had grown over the lower board, affording a partial protection. A few larvæ were found suspended, and nearly ready for transformation, and a few were still remaining on the plants.

From a portion of the above collection of chrysalides placed in a box, there were obtained during the winter ten male and thirteen female imagines.

XV. DATES OF COLLECTION OF SOME HETEROCERA FOR 1870.

Sphingidæ.

Sesia diffinis *Harris*	May 21.
Sesia gracilis *Gr.-Rob.*	May 25.
Sesia Buffaloensis *Gr.-Rob.*, larva	Aug. 22.
Philampelus Pandorus (*Hübner*)=*P. Satellitia* Harris	June 21.
Philampelus achemon (*Drury*)	June 23.
Smerinthus geminatus *Say*	June 4.
Ceratomia Amyntor (*Hübner*)=*C. quadricornis* Harris	June 19.
Macrosila quinquemaculata (*Haworth*)	June 24, Sept. 26.
Sphinx chersis (*Hübner*)=*S. cinerea* Harris	June 5, June 24.
Sphinx drupiferarum *Sm.-Abb*	June 11, June 24.
Sphinx kalmiæ *Sm.-Abb*	June 24.
Sphinx gordius *Cramer*	May 19, June 24.

Bombycidæ.

Eudryas grata (*Fabr.*)	July 6.
Eudryas unio (*Hübn.*)	July 6.
Scepsis fulvicollis (*Hübn.*)	Sept. 9.
Ctenucha virginica (*Charp.*)	June 6.
Spilosoma virginica (*Fabr.*)	April 27, May 21.
Hyphantria textor *Harris*	May 25.
Ecpantheria scribonia (*Hübner*), larva	Aug. 15.
Ecpantheria scribonia, ex larva	Nov. 9.
Lagoa crispata *Packard*	June 16.
Empretia stimulea *Clemens*, larva	Aug. 30.
Limacodes scapha (*Harris*), larva	Sept. 16.
Perophora Melsheimerii *Harris*	June 7.
Cerura borealis *Harris*, larva	Aug. 26.
Hemileuca Maia (*Drury*)	Sept. 19.
Eacles imperialis (*Drury*), larva	Sept. 2.
Anisota stigma (*Sm.-Abb.*), larva	Sept. 9.
Anisota senatoria (*Sm.-Abb.*)	June 16.
Clisiocampa decipiens *Walk*	July 17.
Lithædia bellicula	May 16.

NOCTUIDÆ.

Leucania unipuncta *Haw.*	Sept. 9.
Gortyna nitela *Guen.*	Sept. 5.
Hydrœcia sera *Gr.-Rob.*	July 22.
Achatodes sandix *Guen.*	July 25.
Prodenia autumnalis *Riley*	Sept. 28.
Cirrœdia pampina *Guen.*	Sept. 14.
Phlogophora anodonta *Guen.*	July 18.
Euplexia lucipira (*Linn*)	June 8.
Aplecta latex *Guen*	May 15.
Hadena subjuncta *Gr.-Rob.*	June 1.
Hadena chenopodii (*Albin*)	July 22.
Xylina Bethunei *Gr.-Rob.*	Sept. 18.
Cucullia florea *Guen.*	July 8.
Cucullia intermedia *Speyer*	May 28.
Rhodophora florida *Guen.*	July 27.
Abrostola urentis *Guen.*	July 12.
Plusia simplex *Guen.*	Sept. 14.
Amphipyra pyramidoides (*Linn.*)	July 25.
Catocala Clintonii *Grote*	July 10.
Catocala cerogama *Guen.*	July 22.
Catocala relicta *Walk.*	July 25.
Catocala cara *Guen.*	Aug. 22.
Catocala parta *Guen.*	Sept. 6.
Catocala amatrix *Hübn.*	Sept. 22.
Drasteria erechtea (*Cram.*)	May 16, Sept. 9.
Drasteria erechtea (*Cram.*), larva	Oct. 21.
Euclidea cuspidea *Hübn.*	May 21.
Poaphila quadrifilaris (*Hübn.*)	May 21, June 7.

PHALÆNIDÆ.

Eutrapela transversata (*Drury*)	July 27.
Angerona crocataria (*Fabr.*)	June 9.
Endropia ferruginaria *Pack.* MS	June 7.
Tetracis lorata *Grote*	May 31.
Cleora pulchraria *Minot*	Sept. 6.
Boarmia sublunaria *Guen.*	May 15.
Corycia semiclarata ? *Walk.*	May 31.
Lozogramma petraria *Hübn.*	June 15.

Numeria obfirmaria *Hübn*............................ May 21.
Fidonia bicoloraria *Minot*............................ May 25.
Hæmatopis grataria (*Fabr.*)............................ May 25.
Aspilates dissimilaria *Hübn*........ .... June 29.
Crochiphora accessaria *Hübn*......................... May 25.
Zerene catenaria (*Cram.*)............................ Sept. 19.

ENTOMOLOGICAL CONTRIBUTIONS—NO. III.

By J. A. LINTNER.

I. ON THE LARVA OF EUDRYAS UNIO HUBN. AND ALLIED FORMS.

On the 9th of September the larvæ of this moth were found feeding on *Epilobium coloratum* growing in a swampy portion of a pasture. About thirty individuals were collected during a few minutes search, two or three of the larvæ, in some instances, occurring on the same plant. They had nearly attained their maturity; some of their number, a day or two after their collection, buried themselves in the moist sand in which were inserted the plants upon which they were fed, and on the 16th a pupa was observed, partially extruded from the sand. Only four of the larvæ were carried through to their pupal change, it having been inconvenient to supply them with suitable food. The pupæ were kept during the winter in a moderately warm apartment, and on the 8th of April the first disclosed its imago.

The larva bears a strong resemblance to those of *E. grata* (Fabr.) and *Alypia octomaculata* (Fabr.) in shape, markings and colors. Its prominent features are its bands on each segment of white, black and orange (a single orange one occurring on the center of the segment), and a hump on the eleventh segment. A detailed description is as follows:

Head rounded, its diameter somewhat exceeding one-half that of the body, orange with black spots, of which there is an oblong one near the base of the clypeus, two semi-ellipsoidal ones surmounting its apex and a small quadrangular one on each side; a perpendicular row of five spots on each side of the clypeus of which the second superior one is the largest, a spot above the ocelli, and a row of three behind them. *Body* tapering regularly toward the head, from the eleventh segment, which is elevated in a hump. First segment white, with two transverse bands of black spots, and with two black bands only seen when extended. The abdominal segments have each three white and three black bands on each side of a central orange band. The orange band is the broadest; it is marked dorsally on its anterior margin by two transversely elongated black spots resting on the black line margining it, and laterally by two geminate similar ones, of which the upper is the larger and the lower embraces the stigma; behind the lower margin of the stigmatic spot, centrally on the band is a small rounded black tubercle bearing a short hair; on the posterior margin of the band, resting on the bordering black line, are two subdorsal semi-elliptical black spots, forming with the two anterior spots a "trapezoid"; between these subdorsal spots are two or four black points, of which the two interior sometimes assume the form of a "dove-tail" medial process of the black band; the orange band extends downward to the black bases of the prolegs, midway between which and the stigmata, on or in range with the third black band, is an elongated hair-bearing black spot, and posteriorly another similar one, lower and running into the black bordering the prolegs. The white band preceding the orange is interrupted or greatly contracted on the medial line by an enlargement of the black band anterior to it, and is marked with a small piliferous black dot in front of the stigma. On the second and third segments the orange band is marked with a row of eight spots, of which the six superior are located in the middle of the band, and the two inferior coalesce with the black band margining it behind. On the eleventh segment the trapezoidal fuscous spots are of a well-defined oval form; above the stigma is another similar spot. On the twelfth segment the corresponding spots are round, and the trapezoid has its broadest side in front. The anal shield bears two spots centrally and five marginal ones, of which the medial one is elongated. On the sides of the larva a yellowish shade rests on the incisures. Ventrally, white and black interrupted bandings are observable on the abdominal segments when extended; the thoracic region is almost wholly white; on segments four and five the orange band is continued beneath, inclosing on the former four and on the latter six rounded black spots. The legs are dull yellow, tipped or edged on the two joints with black, and dotted with black interiorly. The prolegs are dull yellow, with a velvety black base, and with two lateral lines and three black spots (one small); the terminal pair have a black line outwardly and a cluster of black spots behind, which, as well as all of the black spots noticed in the above description, are piliferous, having the hair somewhat longer and stouter than in *grata*.

Length of the mature larva one inch and one-eighth; diameter three-sixteenths of an inch.

The larva has not, that I am aware of, been previously described, nor can I find any positive record of its observation. It seems to have occurred at Otisco, N. Y., for, in reply to some inquiries directed from that place to Mr. C. V. Riley, the answer is returned that "the Eudryas larva which feeds on *Epilobium coloratum*, or Purple-veined Willow-herb, is in all probability *E. unio* Hübner, although we cannot determine positively unless specimens are sent."* Harris, Fitch and Riley describe the moth, but were doubtless unacquainted with its larva; for Harris states his ignorance of it; and, although Dr. Packard asserts that Fitch has raised both *grata* and *unio* from the grape, † there is reason to believe that Dr. Fitch had assigned to *unio*, without any knowledge of its habits, the food-plant of which it was natural to suppose it would partake in common with its congener; ‡ and Mr. Riley also probably includes it among his "blue caterpillars of the vine," § without personal observation, but from a reliance on the usual accuracy of the statements of Dr. Fitch.

At present we have no information of its having been found on any other plant than *Epilobium coloratum*. It is quite remarkable that two species, so closely allied, should have such dissimilar food-plants. The fact suggests an interesting inquiry, whether *unio* be confined to Epilobium, or if it occurs on other of the Onagraceæ, or even ranges to some other order. As *grata* is known to feed on Ampelopsis as readily as on the grape, it is not improbable that a careful examination, during the month of September, of the common evening primrose (*Œnothera biennis*), may be rewarded by a discovery of *unio* upon it.

Two other larvæ occur in New York, viz., *Alypia octomaculata* and *Psychomorpha epimenis* (Drury), which bear so strong a resemblance to the Eudryades, that the four are liable to be confounded, not only by the casual observer, but by the entomologist who may not have acquainted himself with their characteristic features.

Of *A. octomaculata*, Harris remarks, ‖ "It resembles the larva of *Eudryas grata* in its colorings and markings so much, that, before I

* *The American Entomologist*, 1870, vol. ii, p. 59.

† *Proc. Essex Institute*, 1864, vol. iv, p. 27.

‡ *Third Report on the Insects of New York*, p. 81.

§ *Second Report on the Insects of Missouri*, 1870, p. 83.

‖ *Entomological Correspondence of Thaddeus William Harris, M. D.*, Boston, 1867, p. 116.

was acquainted with its manners, I have frequently taken the one for the other;" and again, when writing of *E. grata* (loc. cit. p. 138), he says: "The position of the larva in repose, with its head depressed, and the third and fourth segments arched upwards, give it a hunch-backed appearance; the attitude, disposition of the colors and the habitat, are similar to those of the larva of *Alypia octomaculata.*"

Several of the larvæ of *P. epimenis*, sent to Mr. Riley by correspondents and also collected by himself from grape-vines, were referred by him, although with some doubt, to *A. octomaculata*,* and were figured in association with the imago in one of his plates. Subsequently he was able to rear *octomaculata* from its larva, which he figures and describes, correcting the first erroneous reference.† But in continuation of the confusion, the *epimenis* larva is now made to stand (with a reservation) for *E. unio*, the larval state of which was then unknown; and only in a following report does it find its true name and proper place beside the beautiful imago which it produces. I mention the above, not to reflect, in the slightest degree, upon Mr. Riley, whose able reports are conceded to be very valuable acquisitions to science, but as an illustration of the close resemblances existing among these larvæ. If they are capable of thus puzzling so accurate an observer, there certainly is need of faithful description, or at least a statement of prominent features and differences, that their identification, whenever met with, may not be a matter of doubt.

I regret that I have no memoranda, or material at hand, to enable me to institute a full comparison between the most nearly allied of these larvæ, viz., *octomaculata*, *grata* and *unio*. I have only at command two alcoholic examples of *unio*, three immature forms of *grata*, and one collected several years ago and labeled *grata* but which I believe to be *octomaculata.*

The comparative length of the hairs will, in all probability, prove a sufficient distinction between the last two. Harris (*Ent. Corr.*, p. 286,) describes the mature larva of *octomaculata*, taken July 16th, as transversely banded with orange and dotted with black, the dots being in two alternate rows, and all of them emitting distinct, long, whitish hairs. In a young larva found by him July 2d, between one-fourth and one-third of an inch long, the hairs were very distinct. Of *grata*, occurring abundantly on the grape-vine, August 10th, he

* *First Report on the Insects of Missouri*, 1869, p. 136.

† *Second Report on the Insects of Missouri*, 1870, p. 80.

writes (loc. cit., p. 306), "larva entirely naked;" and, on page 307, he institutes a comparison between the caterpillars of the Agaristiadæ, "which are sparingly covered with hairs," and those of Eudryas, "in which the caterpillar is not at all hairy." The two figures of *grata* given in the *Treatise on Insects Injurious to Vegetation*, represent the larva as hairless. Riley (*2d Rep. Ins. Mo.*, p. 80) says of *octomaculata*, "each spot or tubercle gives rise to a white hair," and of *grata* (l. c., p. 83), that it differs from the preceding by the hairs being less conspicuous. Of the latter species Mr. W. Saunders * states that "the bands are dotted with round black dots, from each of which arises a single short brown hair."

In the examples of the larvæ (about half-grown) of *grata* before me, the hairs do not exceed in length the breadth of the central band, and are noticeable only on close observation. In *octomaculata* they are quite long, equaling in length the diameter of the body, if we may refer to this species the description by Dr. Packard † of some larvæ collected by Mr. Putnam on the grape-vine, and deposited as *grata* larvæ in the Museum of Comparative Zoölogy at Cambridge. The description of the *grata* larva, given in the *Guide to the Study of Insects*, pp. 281–2, with its hump on the eighth ring, and each segment having across it a row of tubercles which give rise to three fascicles of hairs, evidently refers to some other form.

The following may be noticed as distinguishing features of these closely allied forms, which should serve to remove all occasion for confounding the two first mentioned with one another or with the Eudryades:

The larva of *Psychomorpha epimenis* (also a grape-vine feeder) has on each segment four white and four black bands (four-banded on a white ground), and is without the orange band which exists in the other three. The spots which conspicuously mark the others are obsolete in this. In Fig. 1 the larva is represented at *a*; *b* is an enlarged representation of one of the segments, and in *c* is given the marking of the hump on the eleventh segment. The male imago is shown in Fig. 2.

Fig. 1.

Fig. 2.

* *First Ann. Rep. on the Noxious Insects of the Province of Ontario*, 1871, p. 35.

† *Notes on the Family Zygænidæ*, in Proc. Ess. Ins., vol. iv., p. 28.

The larva of *Alypia octomaculata* is marked on each segment with eight black bands (counting the two which border the broad central orange band), as shown in *a* of Fig. 3, and more distinctly in the enlarged view of one of the segments at *b*; from the black dots long white hairs are given out (represented too short in the figure), and below the stigmata, on segments four to nine, is a row of white spots, with a large white spot extending over the incisure of the tenth and eleventh segments. At *c*, a view is given of the imago of this species.

FIG. 3.

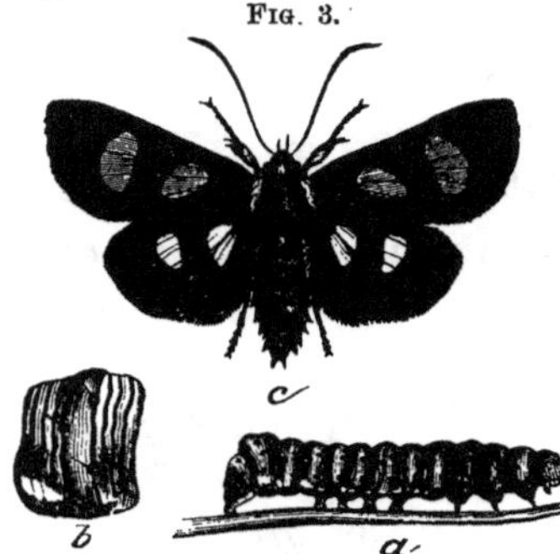

The larvæ of *Eudryas grata* and of *E. unio* have six black bands and a central orange one on each of the principal segments; they are without the white lateral patches, and have a prominent hump on the eleventh segment. A representation of the imago of *E. grata* is given in Fig. 4, for comparison with *octomaculata* and *epimenis*, and to illustrate the fact well known to those who have studied the larvæ of our Lepidoptera, that very dissimilar moths are frequently produced from larvæ closely resembling one another.*

FIG. 4.

I regret my inability to give, at the present, the differential features of *grata* and *unio*. At maturity they differ in size, *unio* being invariably the smaller. In markings they resemble one another so closely, that Mr. Riley, after a critical comparison of examples of *unio* which I had sent to him, with some examples of *grata* in his possession (probably alcoholic), writes me, that he finds the two species absolutely indistinguishable.

While specific differences may not be detected in alcoholic individuals, there is scarcely any doubt but that sufficient characters for their identification could be obtained from a comparison of the colors of the living forms. Having lately seen a large number of *grata* larvæ (at Portland, Me., in August, 1873), I may safely assert that the two species differ materially in their general aspect. While unable to state from recollection what these differentia are, I believe that *unio* will prove to be characterized by more prominently marked

* The figures illustrating this article are from the excellent drawings of Mr. C. V. Riley.

black bands laterally and ventrally, and by the yellowish shade resting on the incisures laterally.

It is possible that the minute description which Mr. Riley proposes shortly to give of *grata*, drawn from a large number of living larvæ, when compared with the detailed description which I have endeavored to give of *unio*, may develop such points of difference as will permit of the ready recognition of these two forms, peculiarly interesting from the close resemblance existing between them.

Since the above has been put in type, I have been able to compare my two examples of *E. unio* larvæ with six alcoholic examples of *E. grata*, and, as the result of such comparison, I am compelled to disagree with Mr. Riley as above quoted, for while the two are very similar in their ornamentation, yet I find such differences that (provided the features to be referred to prove constant in larger numbers) I would have no difficulty in selecting a single mature individual of either species from among a thousand of the other. Through the kindness of Mr. Riley, I am able to accompany this paper with an excellent figure from drawings made by himself of the larva of *E. grata*, and also (for the first time) a representation of the beautiful egg of this species. At *a* the larva is shown of its natural size; at *b*, one of the segments (the fifth) enlarged; at *c*, the ordinary ornamentation of the collar, differing in some examples by the addition of several (to the number of eight) central dots; at *d*, the usual marking of the hump on the eleventh segment; *e* represents the egg as seen from above, and *f* is a side-view of the same (natural size shown with the enlarged figures).

Fig. 5.

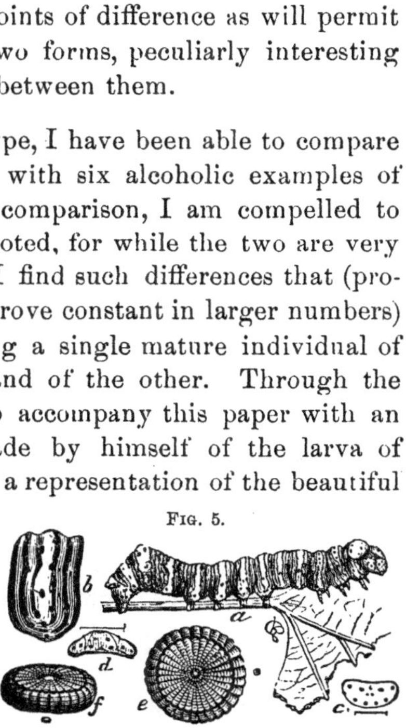

The following are the principal differences that I find in the two species:

Contracted by their preservation in alcohol, the two *unio* larvæ average in length 1.05 in.; the six *grata* 1.29 in. They differ in form, in that the latter presents much the more prominent hump on the penultimate segment, and is angulated at that point to a degree that were it a vertebrate, it would suggest the idea of its terminal portion dragging from having been broken at that point; in *unio* the hump is moderate and the peculiar angulated form, well represented in the figure, is not seen.

Unio is the more heavily marked with black, both in its bands and spots. In none of the examples of *grata* are the black bands broader than one-half the width of the intervening white ones, while in one *unio* their average width is nearly double that of the white. The spots on the head are the same in position in both species, but are smaller in *grata.* In that species there are usually two distinct piliferous spots on the base of the clypeus; in two of my examples these are confluent, running together by slender projections in a broad V-form: in *unio* the two are united as a band across the clypeus. In *unio* a black spot, broadly rounded beneath, following the curved line of the ocelli, and tapering to a point above, incloses the four superior ocelli: this is not present in *grata*, but in two examples some of the ocelli are indistinctly annulated with black.

At *c*, in Fig. 5, the spots on the collar of *grata* are faithfully represented in position, but their size might have been slightly enlarged. In *unio*, the four spots of the anterior row are separate, but those of the posterior row, from their greater size as compared with *grata*, are confluent, except the two medial ones: in *grata*, these spots are separated by spaces varying from one diameter of a spot to two and one-half diameters.

The spots on the caudal hump of *grata*, shown at *d*, in the figure, are isolated, while in *unio* those in each row are connected by the black band to which they are united.

Similar comparisons might be instituted of all the other spots of the two species, but the above may indicate their differences. The feature which should serve better than any other to distinguish *unio* is the blackish coloring (its outline not permitting its designation as a stripe), above the prolegs and continued on the two following segments, the three piliferous spots above the prolegs being connected with it; this is entirely wanting in *grata.* It results, apparently, from the increasing breadth and coalescence of the black bands as they descend to the ventral region. In one of the examples, the ventral region of the proleg segments is essentially blackish, which feature was also observed in a number of the living larvæ, according to my recollection and that of Mr. Meske, who also collected the larvæ and bred from them several imagines.

The differential features above indicated are not entitled to the reliability that would attach to them, were they drawn from living examples; but should they prove to be sustained by future observations, there need be no necessity of failing in the determination of these species, when either may chance to be collected.

II. TRANSFORMATIONS OF SOME BOMBYCIDÆ.

Platysamia Cecropia (*Linn.*).

Two larvæ, measuring .45 in. in length (after, probably, their first molting), were found, July 13th, feeding on leaves of mountain ash (*Pyrus Americana*). The body was dull orange, bearing six rows of spines, four to six-branched at the tip. In the two dorsal rows the spines were black, except in front where they were orange, concolorous with the body ; spines of the remaining rows wholly of a shining black. The eleventh segment, with but two spines, the tenth with four, the ninth with five, and the others with six each. On each segment two black dots between the spines, making them centers of squares of four dots. Head and legs, when the larvæ were first taken, of a dark red, subsequently changing to black.

After the second molting, they measured one inch in length and were of a yellow-green color. On the second and third segments superiorly, each, two globular-headed red tubercles with seven black bristles; on the seven following segments two dorsal rows of yellow tubercles, swollen apically, of which the two on the fourth segment are larger than the others and bear seven bristles each, while the following twelve have but five. On the eleventh segment, medially, is a single yellow tubercle with eight bristles; on each side of these yellow tubercles are oblong black spots. The two lateral rows of tubercles light blue and setiferous, and beneath these, on the three anterior segments, a black pointed tubercle. Head with two converging black lines. Legs with black tarsi, and prolegs each with a black spot exteriorly.

The subsequent moltings were not observed.

A captured *Cecropia* deposited over two hundred eggs, and from her body, after death, nearly a hundred were taken, most of which were of full size; the entire number was three hundred and five. In their longest diameter they measured one-tenth of an inch; in their shortest diameter .083 of an inch.

Callosamia Promethea (*Drury*).

A deposit of eggs hatched July 6th, laid nineteen days before. On the 14th the first molting occurred. Length of larva 35-100ths of an inch. Body pale green, with yellow bands bordered by black; rows of tubercles are apparent.

From having previously fed in companies of from twenty to thirty, there are now seldom more than six collected on a single leaf. A larva usually commenced eating into the leaf at a point in its margin, where it would be joined by others, cutting into the body of the leaf, until often the entire interior was consumed, leaving an unbroken margin (except at the entering point) of a breadth barely sufficient to serve as a support for the larvæ.

At the second molting, on the 20th of July, their length was six-tenths of an inch. Body light yellow-green, with black transverse interrupted markings; on the second and third segments each, two clubbed yellow tubercles and one on the eleventh; six rows of smaller black tubercles. Legs yellow, with a white spot.

August 1st, the larvæ molted for the third time. The subsequent molting was not noted. On August 9th, some of the colony commenced the construction of their cocoons, fifty-two days from oviposition, and thirty-three days from the disclosure of the larvæ.

A measurement taken of some eggs of *Promethea* gave for the diameters .077 and .063 of an inch. They are of a white color, with an ochreous-yellow spot on the upper side.

Actias Luna (*Linn.*).

From an oviposition of one hundred and eighty eggs, larvæ were developed on July 25th. Body pale green, with a brown lateral stripe and a dorsal one on the anterior segments, and with rows of tubercles bearing bristles. Head crossed with a brown stripe.

On July 30th, larvæ molted for the first time. Length three-eighths of an inch. Color pale green. On first segment four red-tipped tubercles; on the second and third two similar ones, and a medial one on the eleventh; the tubercles elsewhere on the body are yellow-tipped. The head is marked on front with four black spots forming nearly a square.

Second molting on August 3d. Length one-half an inch. The tubercles of the first, second and eleventh segments, above mentioned, are tipped with deep red, and have several hairs branching from them. The tubercles of the lower lateral row are also red with hairs; those of the upper row green and without hairs. The yellow stripe of the side and the yellow bands marking the incisures now appear. The two superior of the four spots of the head are no longer seen. The larva, after its molting, consumes its exuvia.

Third molting on August 9th. Larva seven-eighths of an inch in length, with no material change in appearance from that presented in the preceding stage.

The fourth molting was on August 17th, developing all the features of the mature larva. Length one inch and one-fourth. Color a pale apple-green, shading darker below the stigmata; incisures yellow, and a yellow line on the upper margin of the substigmatal fold. Six rows of small pink warts, each with one or more black hairs. Scattered over the body are a few white hairs, some of which are of a clavate form. Anal shield brown, triangular, yellow bordered; anal plates brown, bordered anteriorly with yellow.

By the 31st of the month all of the larvæ had inclosed themselves in cocoons.

For several days prior to the disclosure of the moth, the pupa (which is fastened by its terminal hooks to some threads in the end of its cocoon) may frequently be heard in motion, as if rotating from side to side. When the time for its transformation has arrived, the pupal-shell is broken by the muscular force of the inclosed limbs and a wet spot appears on the end of the cocoon, indicating the point at which the moth is to emerge. A sound like gnawing is now heard, which is probably produced by the friction of the base of the fore-wings against the cocoon in the effort to force an opening. After these periods of activity, in which the motion is often sufficient to produce a considerable movement of the cocoon, intervals of quiet follow. The wet spot increases in size until its diameter about equals that of the body of the moth. At length the end yields, and the head of the moth is seen through the still connected threads. It partially withdraws itself, and then again resumes its effort to escape. After one or two more rests, the antennæ are protruded, shortly followed by the first pair of legs, when the moth rapidly disengages itself from the cocoon, usually emerging with its back downward, and quickly seeks some position where it can attach itself, with its small wet wings hanging downward over its back. The anterior wings are the first to expand; next the body of the posterior wings and last the tails. In about three hours' time the wings are fully expanded, and, by a quick, muscular action, are folded over to their natural position against the surface on which it rests (if an extended one), and the insect has attained its full maturity.

In a search made for *Luna* cocoons beneath a number of hickories (*Carya alba*) at Schoharie, on May 9th, nine were found in a space of eight feet square. Of eight others collected at this time not more than one was found under a tree. The first imago from these cocoons emerged May 18th, a male, followed by three other males, after which females and other males appeared.

From the following record in my note-book, it would seem that the season of 1857 was very prolific in Luna moths at Schoharie:

"June 27th. Fine specimens of *Attacus Luna* are brought to me almost daily, most of which have been taken when the moths had but recently emerged and were resting on trunks of hickories. In three instances where seemingly fresh examples were pinned out of doors in the evening, males were found in the morning copulating with them."

"July 2d. In a walk of two hours, four females of *Attacus Luna* were found resting on trunks of hickories, at about two feet from the surface of the ground."

III. DESCRIPTIONS OF THE LARVÆ OF SOME BOMBYCIDÆ.

Parorgyia parallela *Gr.-Rob.*

The larva was taken at Schoharie during the month of June, 1859, feeding on the plum. It was tufted similar to that of *Orgyia leucostigma*, with mouse-colored feathered hairs; the pencils (from memory) were black. It made a thin cocoon July 5th, in which its hairs were loosely woven. The moth emerged July 21st. At rest, its wings slope like the roof of a house, and its front legs are extended, giving it an attitude like that of *Eudryas grata.*

On the 25th of July, of the same year, a female moth of this species was taken, which, after having been pinned, deposited a number of eggs from which ten larvæ were obtained.

The tufts and pencils of hairs marking the larvæ were developed at the second molting (date not noted).

In preparing for their third molting, they spun on the side of the box in which they were confined a thin web, somewhat larger in extent than their body, upon which they took position; their molting occurred two days thereafter. They continued resting in the same position for another day, when they commenced to travel slowly about the box, but refused to eat of any of the tender leaves which were placed in their path.

Two or three days later (October 1st) it was noticed that they had again resumed a fixed position on newly spun webs. As, without feeding and growth, another molting could not be impending, there was scarcely any doubt but that they were now commencing their period of hybernation, in accordance with their habit at this stage of their growth. This was evident a month later (November 5th) when they were found still maintaining their fixed position, but showing equal sensitiveness upon being touched to that manifested at the commencement of their rest. They were accordingly set aside in a cool room for their winter's repose.

With their heads closely appressed to the surface on which they rested, they presented the following features:

Length .18 of an inch. Body densely covered with light brown or fawn-colored hairs, short and even on the back and upper portion of the sides, and lower down with a margin of longer and unequal ones, projecting also behind. On the anterior portion of the body dorsally, a semicircular dark brown brush-like tuft, convex in front, and slightly elevated above the surrounding hairs; on the posterior portion of the body (11th segment ?) a similar round tuft of longer hairs. Two slender pencils of dark brown feathered hairs project in front of the head. Legs and prolegs light fawn-color. Ventral region black.

The attempt to carry the above larvæ through their hybernation met with the ill success that, in nearly every instance, attends similar experiments. In the spring they were found still fastened to their webs, but dead.

Apatelodes Angelica (*Grote*).

Head subrotund, dark brown, the clypeus and two lines on the front lighter brown. *Body* with the thoracic segments tapering; terminal segments tapering and flattened posteriorly; ventral region flattened; the anal legs projecting behind. Color of the body gray; numerous fine black linings, among which may be traced two forming a vascular stripe, and two similar lateral stripes on each side. On segment one, anteriorly, are four dorsal white lines, posteriorly black; segment two is black anteriorly, behind which are irregular black linings; segment three as the preceding one; on segments five to ten the dorsal black linings assume a V shape, the apex resting on the suture and inclosing centrally two yellow-green subelliptical spots, with a similar spot exterior to each within the superior lateral stripe.

From the first segment, long, whitish-brown hairs project over the head, nearly concealing it; from the middle of the second and third segments whitish hairs project forward, of which those on the latter segment are shorter and arranged somewhat in tufts, beneath which, when extended, some short stiff red hairs are seen; laterally below the stigmata are two rows of fascicles of white hairs of unequal length, mingled with a few longer brown ones, extended rectangularly with the body until to its middle, whence the remainder are directed backward; from the terminal segment white and brown hairs, of greater length than elsewhere on the body, project horizontally, brush-like, backward; short, whitish hairs are scattered sparsely over the body. (The larva escaped before its description could be completed, and the remainder is from memory). On the vascular line, on each segment, is a tuft of black hairs about .06 in. long, the ends of which converge to a point. The prolegs project laterally, almost hidden by the hairs. Ventrally is a broad fuscous stripe.

Eight or ten of the larvæ were collected at Bath (near Albany) during the early part of September, feeding on ash (*Fraxinus*); also by Mr. Meske, at Sharon Springs, on lilac (*Syringa vulgaris*). When not eating, they usually occurred resting on and closely appressed to a twig. The first transformation to a pupa was on September 14th. The larva has a marked gastropachean aspect. It is now for the first time described.

Cœlodasys unicornis (*Sm.-Abb.*) *Pack.*

Larva taken August 3d, feeding on hazel (*Corylus Americana*).

Head large, ovate, green, with delicate red markings, and with two black stripes on its front, as shown in Fig. 6 at **B**. *Body* with the thoracic segments apple-green, with a double brown dorsal stripe extending from the head to a long, fleshy, red-tipped spine on the fourth segment, broadly forked at the tip and bearing two hairs as at **C**.. Abdominal segments reddish-brown, with fine interrupted markings. On the eighth segment is a double setiferous hump, between which and the anterior spine **C** is a white elongated spot as in **F**, centrally constricted, and marked with pale red lines. On the eleventh segment is a smaller hump, between which and that on the eighth is a V-shaped white spot (**E**), opening posteriorly. Terminal segment without feet, forked, as at **D**, and usually elevated.

Fig. 6.

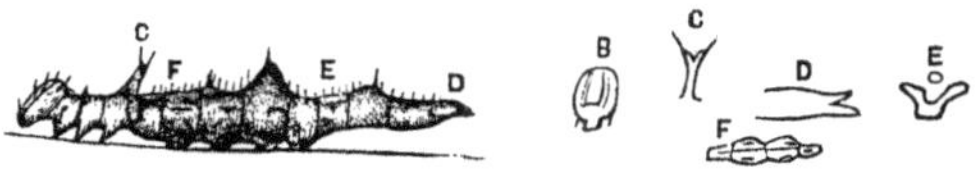

As the larva eats into the margin of a leaf, it extends its body along the excised portion following the curve, holding the edge between its feet, and in this position, from its color and peculiar outline, it can with difficulty be distinguished from the leaf.

The larva has also been found by me on choke-cherry (*Prunus Virginiana*), apple, and on plum (July 28th, one-half inch in length).

This species has proved very difficult to rear, as it usually dies within the cocoon, before assuming the pupal state in the spring.

Platycerura furcilla *Packard.*

Larva eating the leaves of pine (*Pinus strobus*). Length at maturity one inch and five-eighths. *Head* round, of about the diameter of the body, red with conspicuous markings upon the front of lighter red, somewhat in the form of a script *x*, and less distinct reticulations of the same. *Body* presenting a peculiarly mottled appearance from its irregular and broken stripes; its general color dull red; on each segment an irregular band of brighter red; a whitish vascular line within a broken gray stripe; a better defined lateral stripe just above the stigmata, within which, on each segment from the third to the eighth inclusive, are four black depressed spots arranged in a right angle, the upper three in line, the largest of which rests on the crown of the segment, with two behind it and one before; the substigmatal fold is white on the anterior portion of each segment and red on the remainder; rows of tubercles from which clusters of red hairs of unequal length proceed, which, on the anterior segments, incline to yellow; on the first, second, fourth and eleventh segments each, superiorly, are two pencils of red hairs nearly one-fourth of an inch in length, darker at the tips and slightly feathered. (These pencils made their appearance after the last molting.) Stigmata encircled with brown. Legs red.

In the accompanying illustration (Fig. 7) is represented the habit and attitude of the larva in feeding. With its terminal pair of legs clasping the leaves at the sheath, it extends its body along a leaf until it commences to bend, when, by detaching successively the first and following pairs of prolegs, it forces the leaf through its legs until its tip is held between them, in the attitude shown in the figure. When the leaf has been eaten from its tip downward, as far as the contracted segments of the larva will permit, it moves to another leaf, and feeds upon it after the same manner.

FIG. 7.

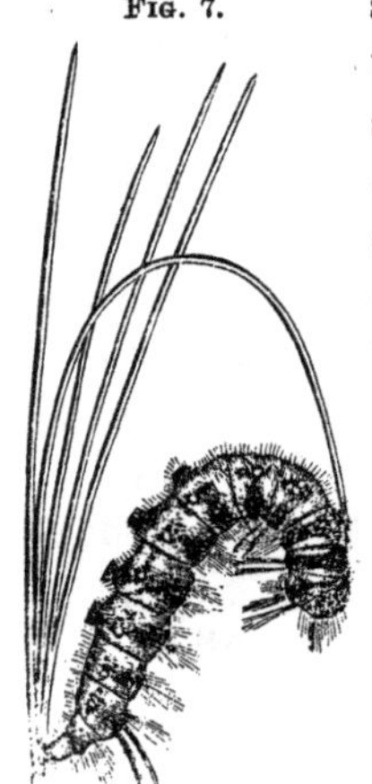

I have observed the same method of feeding in the nearly full-grown larvæ of *Ellema Harrisii* Clemens.

Two of the larvæ above described underwent their last molting September 17th (1859), and a larva farther advanced spun its cocoon beneath leaves lying in the bottom of its feeding cage. A second one spun up on September 25th. An imago emerged June 12th.

The above larvæ were collected at Schoharie. Examples have also been found on pine at Bethlehem, near Albany. It has not been hitherto described.

The imago is much more rare than the larva, and is seldom met with by the collector.

Dryocampa rubicunda (*Fabr.*).

Length of mature larva .70 inch. *Head* reddish-brown, with the ocelli on a lunate spot. *Body* cylindrical, apple-green, closely dotted with minute, whitish, acute granulations; a darker green, narrow, dorsal stripe, and broader subdorsal, lateral and stigmatal stripes, the stigmatal less distinct than the others. Segment one, with four black tubercles on the collar, the central ones transversely oval, the outer ones subtriangular; a spine in front of the stigma and another at the base of the leg. Segments two to eleven, with a substigmatal row of acute, prominent, black spines pointing backward; a lateral row of shorter ones on the inferior margin of the lateral stripe; a subdorsal row of still shorter ones on the superior margin of the subdorsal stripe, marked with whitish at base superiorly; a ventral row on and in range with the external basal portion of the legs and prolegs, those of the prolegs (segments six to nine) quite small, the other seven (segments one to five, ten and eleven) nearly as long as the substigmatal ones, except those on the terminal pair of legs, of which there are two on the base of each, which are quite minute. All of the above spines black, the three superior ones in range transversely on the anterior portion of the segment, and the substigmatal ones on the middle of the segment. In addition to the above, from the fourth to the twelfth segment inclusive, there is a row of whitish, black-tipped short spines on the inferior margin of the subdorsal stripe. Segment two has the two subdorsal spines replaced by two black, blunt, spinous horns one-eighth of an inch long, placed below the subdorsal stripe. The four superior spines of segment eleven, more prominent than the corresponding ones of the other segments. Segment twelve, with a slightly bifurcate spine on the dorsal line, the size of the lateral ones of the preceding segment; another in range with the stigmata, the size of the substigmatal ones, and having a small branch posteriorly; a small intermediate one ranging with the lateral line; another of the same size ranging with the substigmatal line; and a ventral one similar to and ranging with the ventral spines of segments ten and eleven, making nine spines on this segment, nearly ranging transversely. Anal shield triangular, margined externally with eight black spinules, of which the six anterior ones are short, the two terminal ones green at base, larger and pointing backward. Stigmata black. *Legs* tipped with black, the anterior pair with a transversely subelliptical black spot on the base posteriorly, the second pair with a dot similarly placed.

The larva feeds on sugar maple (*Acer saccharinum*). It has been taken for several consecutive years at Schoharie, frequently, on a fence beneath a row of maples, during the latter part of July.

Some examples of the larvæ entered the ground for pupation on the 9th of August.

Tolype velleda (*Stoll*).

Larva at maturity, two and three-fourth inches long, four-tenths of an inch broad, exclusive of lateral fringes. *Head* small and flat, and nearly concealed beneath the two projecting tufts of the first segments. *Body* of a bluish-gray color above, with numerous faint paler longitudinal linings; on the third segment, superiorly, a black band, which is more conspicuous when the caterpillar is in motion; ventrally pale red. On each segment above are two warts, with short black hairs, of which the two on the third segment, anterior to the band, are more elevated. Some short, black and gray hairs, scarcely visible without a lens, are sprinkled over the body more abundantly at the extremities and on the sides. The lateral tufts, proceeding from warts nearly one-tenth of an inch long, are composed of light gray and a few black hairs of unequal lengths, the longest measuring one-fourth of an inch, and some ending in a fan-shaped tip.

When in repose, both extremities of the larva are closely appressed to the surface on which it rests.

The larva was taken July 23, 1861, feeding on elm. On its body were several parasitic eggs, white, acutely oval, and flat beneath, which were carefully removed with the point of a knife-blade. It did not eat after its capture, but remained nearly motionless, with but two or three changes of place, until the 26th, when it spun its cocoon, an inch and a half long and very flat, against the side of the box in which it was secured.

The imago, a female, was disclosed thirty-five days thereafter, on August 30th.

Another larva, found (in 1871) feeding on the leaves of a young apple-tree, made its cocoon on the 10th of July. The imago emerged August 14th, giving it a pupation of the same length as the preceding one.

IV. DESCRIPTIONS OF THE LARVÆ OF SOME NOCTUIDÆ.

Acronycta Americana *Harris MS.*

September 1st, found at Center, Albany county, N. Y., resting on the upper side of an oak leaf, a caterpillar of this species, differing in some particulars from the description given in the *Entomological Correspondence of T. W. Harris*, p. 313.

Head of larva dark brown, almost black. *Body* black, of a silky luster. On the dorsum, a series of thirteen subelliptical raised spots, their transverse length nearly equaling the diameter of the body, of a pale yellow color, orange at the ends, with a black depressed line dividing them on segments four to nine and eleven; these spots, with the exception of those on segments two, three, ten and twelve, have black spatulate bristles springing from a mamilliform base in the lateral orange portion of the spot; those of the first segment project over the head and are six in number (three on each side), of which the two anterior ones are .13 in. long, and the four posterior .28 in. long; on the abdominal segments they are two in number, .13 in. long; those on the eleventh segment measure .16 of an inch, and are directed posteriorly nearly in line with the body; those on the terminal segment are but .06 in. long, and consist merely of a slender stipe, seeming as if the spatulate tip had been broken off. These dorsal spots have also two short bristles anterior to the spatulate ones, and two additional ones on the spots not furnished with the spatulate bristles. Above the legs and prolegs a row of convex shining black crescents, giving out a few hairs; beneath, behind and above each stigma is an elevated glossy black spot, each with one or more short black hairs.

Length of the larva when at rest, from 1.15 in. to 1.25 in.

The larva above described did not produce its moth, nor am I able to obtain any information of it from any of my correspondents with whom I have communicated in relation to it. Mr. F. G. Sanborn, from whom was obtained the example figured by Packard,* informs me that the larva has been familiar to him for some years, but that

* *Guide to the Study of Insects*, p. 305, f. 236.

he has not been able to rear it to the imago. Mr. L. Trouvelot writes me that he had taken the larva several times, but had never been able to raise the imago. On one occasion he had found it in August, fully developed, on a willow bush growing on a stream, in the White Mountains. Mr. Packard states (loc. cit.) that it "is figured in the Harris Correspondence as *Acronycta acris?* var. Americana," but I do not find any reference to it in the Correspondence as *acris*. As the imago was bred by Harris (he states that it appeared June 28th), it may possibly be identified in the Harris collection.* It may, therefore, be presumed to have been correctly referred generically, for although the larva differs remarkably from all other known American forms of Acronycta, yet we may recall the great diversity existing among the European Acroonyctas in their forms and especially in their garniture, perhaps exceeding that in any other genus. Some of these are described as having a few short, isolated, fine hairs; some have quite long, soft, silky hairs covering the entire surface; in others the body is adorned with long diverging pencils, and others present short, stiff, brush-like tufts. Guenée says of *ligustri* and *brumosa*, "on ne compte plus qu'un seul poil, et ce poil est chez l'*alni*, renflé à l'extrémité en manière de rame ou de massue." The hair mentioned in the last species is probably very similar to those which characterize the *Americana* Harr. MS., which I have designated as bristles, although (from memory) they are flattened and lack rigidity.

This species should not be confounded with *Apatela Americana* of Harris, which is *Acronycta hastulifera* (Sm.-Abb.) Guen., an entirely different insect.

*I have since received a communication from Mr. Sanborn, in which he writes me as follows:

"I visited Boston yesterday, and hunted up the species of Acronycta about which you inquire. It is in the Harris cabinet, together with its puparium, numbered 287 (new No). I cannot describe it from memory sufficiently well to enable you, in all probability, to identify it; but if you take an *A. occidentalis* Grote, and suffuse very darkly the inner third of the fore-wings, and deepen the tints of the costal spots, you will have a fair idea of it. It is totally unlike the *Americana*. It reminds one also of the figure of *Microcœlia vinnula* Grote, in the Proc. Ent. Soc. Phil., vol. II, pl. 9 [now *Acronycta vinnula* Grote; Bul. Buf. Soc. Nat. Sci., I, p. 78]."

The above comparisons of Mr. Sanborn should give a good idea of the imago, but I am unable to refer it to any species with which I am acquainted. Now that the preservation of the Harris specimen of the bred imago is known, we shall be able to ascertain what it is, although not in season, I regret, for the present publication.

The larva has also been taken at Schoharie, N. Y., September 9th, feeding on hickory, and by Mr. Meske, at Sharon Springs, N. Y., feeding on beech (*Fagus ferruginea*).

Acronycta morula *Gr.-Rob.*

The larva of this species was taken at Schoharie, N. Y., September 26th, at rest, on some threads spun over a scar on the trunk of a young apple tree, in which position, from its colors and markings, it could scarcely be distinguished from the bark. Length (mature) one inch and a half. *Head* black on the sides and top, and whitish in front, appressed to the stem when at rest. *Body* light brown, with a pale brown median line between two dark brown stripes which, on the middle of each segment, curve outwardly around a wart; on the fourth, seventh and eleventh segments these warts are larger and are bordered without with black; the lateral rows of tubercles are pale brown, with white hairs radiating from them; the hairs of the two lower rows are long, as are those which project over the head; the dorsal hairs, especially those on the warts, are short, appearing as if closely trimmed; above, and running backward from each stigma, is a dark brown dash; whitish dots, each bearing a hair, are sprinkled over the body. Legs black; prolegs greenish.

The habit of the caterpillar seems to be to rest on the bark during the day, after the manner of the Catocalas, feeding only at night.

It spun a thin cocoon, on the 20th of September, in an angle of a box beneath some pieces of bark. The imago emerged June 7th, (1861).

Ceramica picta (*Harris*).

Head small, rounded, pale red. *Body* conspicuously marked with three broad black stripes; the dorsal one is velvety black, with marginal indentations, two of which, near the posterior portion of each segment, are larger than the others; within the stripe, on the crown of the segment, are small, white, transversely oval spots, arranged in a square of four, or with one or two obsolete; between this stripe and the lateral one is a narrow stripe of gamboge-yellow. The lateral stripe is broad, with numerous transverse white markings, appearing blue by contrast with the black, breaking it into lines resembling **IVNW**, etc.; a regularity is traceable in these characters, for example, the stigmata of the central segments are situated in a semi-oval black spot in the base of a **V** character, followed by another **V** and preceded by an inverted one (**Λ**). On each side of the above stripe is a narrow gamboge-yellow one, of which the superior has a setiferous black spot within it near the hinder part of the segments, and the inferior one a corresponding spot but smaller, and a few others in its lower margin. Beneath this, a white stripe mottled with black spots and lines, among which is a black spot beneath a broad **V** over each proleg, and another nearly as large over the base of the anterior leg of the **V**. *Legs* and ventral region tawny-red. Length at maturity two inches.

Entered the ground October 12th (1859) for pupation. In 1857 the larvæ were found abundantly, feeding on turnip. About thirty were collected for rearing, but although they were carefully supplied with fresh leaves, they all died in their larval stage. In 1868 (September 19th) they occurred very abundantly at Schoharie, in a field of cut buckwheat, from which hundreds could have been easily collected. They were also found resting on willows and on various shrubs bordering the field.

The larva in confinement has been observed to eat with great rapidity and to rest frequently from feeding. It increases rapidly in size. Its peculiar markings and bright contrasting colors make it one of our most beautiful caterpillars.

Cucullia convexipennis *Gr.-Rob.*

Larva feeding on the leaves of the golden rod (*Solidago Canadensis*), nearly full-grown, measuring one inch and a half in length; ground color of the body shining black; on the first segment, a small black hump, in which are four short white marks and two white dots and a conspicuous oblong red mark; on the eleventh segment a larger black hump, and between the two a brick-red dorsal stripe; from the hump, extending over the anal segment, black, inclosing one red and eight small white spots. On the sides a broad yellow stripe, shading into white on its borders, and broken transversely by black lines into markings like the Roman letters **NMIVW**. Within the lower portion of this stripe are the stigmata, each resting on a black character running upward into a point and bearing at its apex a short black hair; a few other short hairs may be seen with a glass at several points on the body. Below the yellow band, on the substigmatal fold, is a narrow red stripe, of a darker shade than the dorsal one; above the yellow band are four delicate blue lateral lines, of which the lower one is not continuous. The ventral region is yellow, divided in stripes by black lines and markings; whitish medially. Legs black.

The caterpillar is a very active feeder, eating at first the leaves, and subsequently, when nearly mature, the blossoms. It enters the ground for pupation, where it constructs an earth cocoon, the grains of which are spun together with silk, similar to that of *C. intermedia.** The imago escapes from its cocoon through a round opening made at one end. It is very alert in the breeding cage, rendering it difficult to pin it without the aid of chloroform. It has been captured by Mr. Meske, at Sharon Springs, N. Y., on the 21st of July.

Several of the larvæ were collected at Schoharie, in 1857, nearly mature, in the early part of October; taken also September 8th, 1859; and also September 1st, nearly full grown, feeding on the blossoms of *Solidago Canadensis*, and in 1873, on the same food-plant, in Albany.

Cucullia asteroides *Guenée.*

Head subrotund, flattened in front, green (shade of leaf of food-plant), with paler green reticulations; clypeus bordered with greenish-white, and a lateral curved spot of the same color in which are the five ocelli; labrum and palpi pale green; a few short white hairs.

Body subcylindrical, tapering moderately at the extremities, smooth, shining, with minute white hairs visible with a lens in the usual locations; conspicuously striped in green and yellow, as follows: ground color green; a broad dorsal stripe of bright yellow extending from the head to the anus, and a somewhat narrower substigmatal one of duller yellow, approaching orange, margined beneath with white; on the sides are five green stripes defined by six black lines, of which the stigmatal line is interrupted at and near the incisures, and so inflated upon the stigmata as sometimes to coalesce with the corresponding portions of the suprastigmatal black line; of the five green stripes, the second and fourth are of a yellow-green shade, the first (subdorsal) of a deep green, and the third and fifth of a paler hue. Ventral region with a median line of greenish-white, having two yellow-green lines on each side.

Legs and *prolegs* green, the terminal pair long and extending backward. Stigmata white, acutely elliptical, having their inferior half lying within the yellow substigmatal stripe.

* *Twenty-third Report on the N. Y. State Cabinet*, 1873, p. 214.

Length of larva at maturity two inches; diameter .22 of an inch. Taken at Albany, September 1st, feeding on Solidago.

Another larva, taken on the same food-plant, September 24th, was, in all probability, the same species, although presenting a marked difference in appearance from the one above described. The two superior lateral stripes were in this nearly black, especially on the abdominal segments, apparently resulting from the thickening of the bordering lines and the extension of the interior ones over most of the green ground.

The larva, when captured, was found to have attached to its surface a black oval parasitic egg-shell. It fed sparingly for several days, when it died, and was transferred to alcohol and placed in the State Museum collections.

Catocala —— *sp ?*

Larva taken at Albany, N. Y., resting on the trunk of horse chestnut, June 6th. Length 2.25 inches, diameter on eighth segment .35 in., elongated, attenuated at the extremities, quite flat beneath when resting on a plane surface, bearing dorsally near the posterior margin of the eighth segment a moderately elevated broad wart directed backward, and having the posterior margin of the eleventh segment slightly raised and projecting backward in a hood-like form; the following demi-segment has also its margin similarly projecting, but in a less degree. *Head* .15 of an inch in diameter, subquadrangular, flattened, slightly bilobed, gray with lighter mottlings, surrounded laterally with a black band, which passes over the vertex and anterior to the eyes; the anterior portion of each lobe paler, projecting, bearing each two black points giving out a short black hair; a similar point on the cheek behind the band, and four microscopic ones on the paler bordering of the clypeus; clypeus depressed, nearly half the length of the head, slightly rounded at the apex, with a brown medial line, and (under a magnifier) six papillæ bearing each a short white hair; in front of the eyes two larger black papillæ with white hairs, and also some smaller ones behind the eyes; on each side of the apex of the clypeus is a conspicuous transversely elongated black spot. The collar bears superiorly a double row of four pale papillæ with black hairs.

Body with a few short white hairs laterally and a line of fleshy filaments ranging with the legs. At rest, the second and third segments are closely wrinkled, while segments four to ten are wrinkled only on their posterior half. Ground color pale gray, with brown markings. Vascular line composed of brown dots. Subdorsal line brown, with its margins darker brown, the darker shade of which, on the posterior wrinkles of the segments, presents a maculate appearance. Laterally are irregular linings of brown dots, interrupted on the fourth segment, giving to that ring a whitish appearance by contrast. Trapezoidal spots (of Guenée) inconspicuous but discernible with a magnifier. Stigmata moderately oval, annulated with brown. Legs long, white, striped with pale brown, bearing a few white hairs. Prolegs long, gray; those of the eighth and ninth segments with a fuscous longitudinal line, and with a black line on the crown of each planta. Anal shield with a transverse row of four setiferous papillæ anteriorly, and four on its posterior curve, margined below with black. Ventral region conspicuously marked with a subelliptical black spot of about one-half the diameter of the body, on the posterior portion of each segment.

The larva was apparently full grown when captured. It refused the food which was offered it, not even tasting of the leaves, and died without undergoing transformation.

As so small a number of the larvæ of our numerous species of Catocalas have been described, the above description is presented in the hope that it may be identified by some collector.

V. NOTES ON SOME NEW YORK BOMBYCIDÆ.

In this and the following paper several notes are given which were made a number of years ago, a few of which are accompanied with their dates. It is with hesitancy that some of these notes, descriptive of larvæ, are presented. They are not offered as descriptions, being too incomplete to serve as such, but simply as contributions toward a knowledge of the natural history of our Lepidoptera, of which we possess, as yet, so little information, that the most simple fact observed in relation to them can hardly fail of being of sufficient value to entitle it to record and publication. Even if anticipated, its independent observation gives it confirmatory value, with perhaps the additional value of its occurrence under different conditions of locality, season, food-plant, etc.

Callimorpha Lecontii *Boisd.*

Larva feeding on spearmint (*Mentha viridis*). Length at maturity one inch; tuberculated, bearing fascicles of stiff hairs; dark brown, with yellow spots. It made a cocoon just beneath the surface of the ground July 1st, from which the moth emerged July 24th.

A number of the moths were captured July 28th, beside a small stream in a ravine where spearmint was growing abundantly.

An interesting *militaris* variety of this moth, a female, taken August 8th, lacks entirely the brown dorsal stripe on the abdomen. The thoracic mesial stripe is inconspicuous, and the brown costal and internal margins and the two cross lines of the primaries is limited to lines not exceeding one-twentieth of an inch in width; the spot resting on the median nervules is large.

In two other examples, a male and female, in lieu of the abdominal dorsal stripe is a series of brown spots resting on the anterior margin of the segment and extending three-fourths of its length, narrowing posteriorly; the spots narrower and less conspicuous in the ♀. On the secondaries, near their outer margin, are four brown spots in the ♀ (three in the ♂) of which the largest is transversely elongated, rests on the first median nervule (vein 2) and extends nearly to the median fold, in length equaling the space between the fold and vein

2; on each side of the submedian, nearer the margin than the preceding spot, is a very small and obscure one (in the ♂ the one before the nerve is absent); resting on the discal nervure (vein 5), and wholly within cell 5, is a small spot, slightly larger than the two last mentioned. The brown primaries have eight white spots, surrounded with brown, except the apical one and that at the internal angle, in which the white is continued on the costa and over the fringe; the two spots on the outer margin above the larger one at the internal angle, are quite small.

Arctia Arge (*Drury*).

Larva found in the road, on a warm and sunny day on the 25th of February.

Color dark brown, head and prolegs black, legs tawny. Body with three flesh-colored stripes, one dorsal and two lateral; substigmatal fold colored as the stripes; the hairs, proceeding from tubercles, are long, brown dorsally and tawny laterally; on the segments anteriorly is a small tubercle on each side of and near to the dorsal stripe, and a larger one on the posterior of the segment near the lateral stripe

The caterpillar fed sparingly, for a few days, on a cactus leaf, and commenced the spinning of a slight cocoon on the 1st of March, within which it transformed to a pupa on March 4th.

The moth emerged on the 23d, after a pupation of nineteen days.

Spilosoma virginica (*Fabr.*).

Head black. Body tawny-red, darker on the four anterior segments; a lateral row of broken, irregular black spots; a pale red line below the stigmata; from the tubercles long hairs proceed (the longest of which measure three-fourths of an inch) which are black on the first and second segment and on the sides of the two following, and red over the central and posterior portion of the body. Stigmata white. Exterior basal portion of legs black, the remainder red.

In another example, the hairs were yellow, the dorsal ones approaching to red; the body yellow, darker superiorly above the lateral maculated stripe; incisures superiorly, dusky. Head red. (Schoharie, 1859.)

An interesting sexual characteristic observable in the male of this species and in *S. acrea*, but not in *S. latipennis*, is the process given off by the subcostal nervure for the support of the frenulum, *clothed with black scales*, and connected with a broader black spot resting on the costa.

Spilosoma latipennis *Stretch.*

This species is described and figured by Mr. Stretch,* "from one imperfect broken ♀ (wanting the body) received from Mr. James Angus, of West Farms, N. Y., without any definite locality attached to the specimen."

Its description (loc. cit.) is as follows: "♀, white. Head, thorax and patagia white. Eyes black. Palpi brownish, white beneath. Legs white, with the *coxæ* and *femora* of the anterior pair *bright pink* inwardly; tibiæ and tarsi of the same pair black inwardly, white outwardly. All the wings are pure silky-white, immaculate. The costa of the primaries is decidedly convex from the base to the apex. Expanse of wings, ♀, 1.75 in.; length of body, 0.70 inch."

In the collection of Mr. O. Meske is a ♂ and ♀ of this rare species, captured at Center, N. Y., June 19th, 1872. They show the following features in addition to those above mentioned: The antennæ are white above, black beneath, and with black pectinations of about the same length as in *S. virginica.* Compared with that species, the wings are thinner scaled; the thoracic hairs are longer and finer, readily floated in every direction by the breath, hiding the patagia; abdomen of the ♂ not carinated.

The femora in these are not "bright pink," but of a peculiar bright red shade, between an orange and a vermilion; the coxæ of the ♀ of a paler red, and of the ♂ of a yellowish-red with brown hairs superiorly beneath the head. Palpi of the ♂, white inwardly and black outwardly; the ♀ has only a few fuscous hairs outwardly. Expanse of wings of the ♂ 1.75 in.; of the ♀ 1.80 in. Length of the body of the ♂ .65 inch; of the ♀ .60 inch.

Mr. Grote informs me that a specimen of this species is in the collection of the Buffalo Society of Natural Sciences, which was captured in the vicinity of Buffalo.

Euchætes Oregonensis *Stretch.*

In the collection of Mr. O. Meske, of Albany, is an example of this species, a male, taken on the wing at Center, near Albany, on the 13th of June; another example, also a male, was taken by him in the same locality May 25, 1869, and is now in the cabinet of Mr. C. V.

* *Illus. Zyg.-Bomb. N. Amer.*, I, p. 133, pl. 6, f. 5.

Riley. I have critically compared the former with the description of the type,* and find it to agree in every particular, even to dimensions, so that there can be no doubt of their identity.

Mr. Stretch remarks: "For the type of this species I am indebted to the kindness of Lord Walsingham, who captured the single specimen above referred to in Oregon, during his recent trip to the Pacific coast. In form it approaches nearest to *E. egle*, from which it differs not merely in the color of the wings, but also by the slenderer abdomen and the bright yellow head. Were it not for these latter differences, it might be considered an albino of *E. egle*, though the typical form of that species is yet unknown from the Pacific coast."

I fully concur with Mr. Stretch in his recognition of this form, as distinct from *E. egle*. In addition to other important differences, in both of these eastern examples, the long, slender, cylindrical abdomen is in marked contrast with the short (from .4 in. to .5 in.), thick and conical form observed in the male of *E. egle*.

Of the albino form of *E. egle*, referred to by Dr. Packard † and Mr. Riley, ‡ and accepted by Mr. Stretch upon the testimony of "eastern entomologists," I have no knowledge. I believe that all such examples of "a white variety," will, on critical examination, resolve themselves into *E. collaris* (Fitch), or *E. Oregonensis*.

In consideration of this new habitat of *Oregonensis*, the name selected for it proves to be an unfortunate one; it also presents an argument against the derivation of specific names of insects from the locality of their first observation, especially while so small a portion of our continent has been thoroughly explored,§ our knowledge of geographical distribution so very limited, and the necessary comparison of our fauna with that of Europe not yet made.‖

* *Illus. Zyg.-Bomb. N. Amer.*, 1872–3, I, p. 187, pl. 8, fig. 7 ♂.

† *Proc. Ent. Soc. Phil.*, 1864, III, p. 108.

‡ *Third Rep. Ins. Mo.*, 1871, p. 133.

§ In a recent paper on the Phalænidæ of California, in which thirty-three new species are described, ten of the number bear the name of *Californiaria* or *Californiata*. It is hardly possible that all of these species will prove to be peculiar to that State or even to the Pacific slope, for more extended observations are continually showing us the identity of many of the species of the Pacific coast with those of the Atlantic States.

‖ The species described not long since as *Depressaria Ontariella* from Canadian examples, proves not only to be a common species in the State of New York, but identical with the *Tinea heracleana* of Europe, described by De Geer more than a century ago.

I shall be conferring a favor upon such entomologists as may not have seen the admirable work of Mr. Stretch, * above referred to, in commending it to their notice. Its design is "to furnish good colored illustrations of all the species of Zygænidæ and Bombycidæ found in North America, north of the Mexican boundary, with accompanying letter-press, in which it is intended to embrace everything of interest in relation to each species, which may have appeared in print, with such additional information as may be secured by the author from original sources."

The work is in course of publication, appearing in parts, of which about one-third of the contemplated number are now before the public, as will be seen from the transcript of its title-page herewith given. Volume I contains 242 pages octavo of letter-press, and ten plates, numbering 167 figures. Depicting, as they do, the two families in which are comprised the most beautiful forms, the greatest variety of pattern, the most artistic effects, and the richest coloring of our entire insect fauna, the plates are particularly attractive. The coloring, so far as we have the means of comparison, is very good, for the temptation to exaggeration, for the sake of effect, is not found in the material under representation.

The number of new species contained in this volume (twenty-six), the large number which are for the first time figured, and the very great convenience of a compilation, in a single work, of all that is at the present known of these interesting families, will render it indispensable to all who are engaged in the study of our American moths.

Enchætes collaris (*Fitch*).

A single example of this species has been taken by me at Center, N. Y., and is now in my collection, but without the date of its capture. It is a ♀, having an expanse of wings of 1.62 in.; length of body .48 in. It has also been received by Mr. Meske from Mr. E. L. Graef, of Brooklyn, L. I., labelled as *Spilosoma fulvicosta*, and reported as abundant in the neighborhood of Brooklyn.

This is undoubtedly the species which has been, by some, regarded as an albino form of *E. egle*. Mr. Stretch states † that "specimens, differing in nothing but somewhat inferior size [compared with the example described and figured by him], were forwarded from Penn-

* *Illustrations of the Zygænidæ and Bombycidæ of North America*, by RICHARD H. STRETCH, Vol. I, Part 1 to 9. [San Francisco, Cal.] July, 1872, to Dec., 1873.

† *Illus. Zyg.-Bomb. N. Amer.*, I, p. 188, pl. 8, f. 5.

sylvania by H. Strecker, Esq., in response to a request for the white variety of *E. egle*."

According to Dr. Packard,* "From the same brood of larvæ Mr. Shurtleff has raised both the typical form [of *E. egle*] and a white variety, which agrees well with Dr. Fitch's description of *Hyphantria collaris*."

If it were shown, as it is not, that the variety raised by Mr. Shurtleff was identical with the *H. collaris* of Fitch, still it would fail to prove specific identity of the two forms; before this could be established it would remain to be shown that the "brood of larvæ," from which they were obtained, was the product of a single deposition of eggs. In the event, which may be presumed frequently to occur, of two broods of congeneric larvæ feeding simultaneously on the same plant, the two might very easily become intermingled, and the liability to mingle would be increased in species closely resembling one another. Two such instances of association of larvæ of different species, which would seem to be explicable only through mistaken recognition of one another, have come under my observation, as follows:

In September, of 1869, I collected from a poplar (*Populus tremuloides*) at Bath, N. Y., two folded leaves filled with Ichthyura larvæ, to the number, probably, of sixty. From these I obtained, the following spring, nearly that number of *Ichthyura inclusa* Hübn., together with a single example of *Ichthyura vau* (Fitch), a species which I had not previously met with, but which Dr. Fitch represents as being more frequently taken in his vicinity than either *albosigma* or *Americana* [*inclusa*]. It is quite different from *inclusa*, and the two have not, I believe, been suspected of being the same.

In the other parallel instance, a group of perhaps fifty full grown larvæ of *Clisiocampa sylvatica* Harr., was observed at rest on the trunk of a maple tree in the door-yard of my residence at Schoharie, and scattered among them were several of the larvæ of *C. Americana* Harr. At this time, numbers of this latter species were traveling about on fences, walks and buildings, preparatory to their pupation.

From a company like either of the above, of forms with which we were not familiar, distinct species might be presented to us, with a claim for specific identity resting on the plausible ground of having been reared "from the same brood of larvæ."

I learn from Mr. C. V. Riley, that he has recently been breeding *E. collaris* from the larva, and that he finds it to be very distinct

* *Proc. Ent. Soc. Phil.*, 1864, III, p. 130.

from *E. egle.* His observations on the species will be given in his forthcoming (sixth) Annual Report.

Halisidota caryæ (*Harris*).

The cocoons of this moth were found at Schoharie, N. Y., during the fall of 1856, in large numbers, attached to the under-surfaces of stones, which had been thrown together in a pile extending for several rods along the borders of a wood. A thousand could easily have been secured in a few minutes of time. From one stone twenty-three cocoons were taken, from an area of about five by eight inches, of which fifteen were clustered in a space of fifteen square inches.

The cocoons were kept in a warmed room. An ichneumon emerged December 2d, and the first imago December 15th. A few only of the cocoons were ichneumonized.

Although the larva of this species is not rare, the moth is seldom taken by the collector. The exposed habitat of the cocoon, beneath stones, usually ensures its destruction during the winter months from some of the many predaceous enemies which resort to it for food. Numbers of the cocoons are met with in the spring, in localities where the larvæ abound, with an opening through the sides, and the debris of the pupa within. One that has survived the perils of its hybernation beneath a stone or piece of wood is of very rare occurrence.

Orgyia leucostigma (*Sm.-Abb.*).

A female imago of this species had emerged, August 4th, within its breeding cage standing in a large apartment about ten feet from an open door. At dusk (half-past seven o'clock) males commenced to fly in the room, and precipitate themselves against the gauze front of the cage, moving in every direction over its surface with legs, wings and antennæ in rapid motion, in a persistent effort to force an entrance in the cage. Several attempted to enter through the small crevice left by the imperfectly fitting door at the rear of the cage. Three or four moths were often on the gauze at the same time, whence they could be plucked with the thumb and finger. During the hour that this exhibition continued, forty moths were taken and pinned, from at least a hundred that entered the room.

The larvæ had been more abundant than usual during the season (of 1861, at Schoharie).

Empretia stimulea *Clem.*

On the 30th of August, two of the larvæ were received from Peekskill, N. Y., where they were taken feeding on Indian corn. September 1st, one of the larvæ spun up in its cocoon beneath a leaf.

The cocoon is oval in form, of a reddish-brown color, of a parchment-like texture, and measures .32 in. by .5 in. The imago was not obtained.

My efforts to rear the Cochlidiinæ have been attended with nearly as many failures as with the Ptilodontinæ. Mr. Meske has been quite successful in rearing these and many species usually regarded as quite difficult to mature, by the aid of a uniform supply of moisture during the period of pupation. The pupæ are placed on the surface of the ground, half filling a box, and covered with an inch or two of light moss. On the moss are laid strips of bibulous paper (ordinary blotting-paper is suitable), which, being dipped daily in water, supply the requisite quantity of moisture to keep the moss and ground in a moderately damp state, during the winter months, in a cool apartment. Later in the spring, with an increased evaporation, a second dipping of the paper during the day is required.

Phobetron pithecium (*Sm.-Abb.*).

A larva taken at Bath, near Albany, feeding on hazel (*Corylus Americana*). On September 16th, it spun its cocoon fastened to a twig on which it had been feeding, inserted in some damp sand, on the surface of which the cocoon rested. The cocoon is of an elliptical form, slightly-flattened on the sides, and measures three-tenths of an inch by four-tenths. Its exterior was wholly covered with grains of sand, and in its upper portion were interwoven some of the peculiar curved lateral appendages which impart so singular an aspect to the larva. A good representation of the larva may be found in the *American Entomologist*, vol. II, p. 25.

Another larva was brought to me on September 9th, feeding on pear leaves. It made its cocoon between two leaves September 12th.

The larva is recorded as feeding also on apple, plum, cherry, wild cherry, Siberian crab, white and red oak.

Lithacodes fasciola (*Her.-Sch.*) *Pack.*

Eight of the larvæ were taken during the early part of September, feeding on the leaves of a young plum-tree, three feet in height.

They were of a uniform green color, and without hairs or spines. Their small, oval, brown cocoons were spun between leaves.

A male and a female imago emerged on the 11th of May. They were found hanging from the top of the box inclosing them, with the body curved upward toward the head, so that the tip was directed perpendicularly to the surface on which they rested. This peculiar posture, frequently observed also in *Asopia farinalis*, did not appear to be owing to the inverted position on the box cover, for, on turning it carefully over in different directions without alarming the moth, no change of posture occurred.

The imago has been taken at Schenectady, N. Y., on July 3d, attracted by lights.

Nadata gibbosa (*Sm.-Abb.*).

Larva found on the ground; feeds on maple. Length at maturity 1.75 in.; diameter .35 in.

Head large, flattened in front, a shade of green darker than the body, ocelli black; mandibles yellow, with black on their inner edge. *Body* grass-green ventrally and laterally, and greenish-white dorsally; with paler granulations; smooth, segments rounded, and incisures deep; laterally a line of transversely elongate whitish spots. Anal shield rounded and yellow bordered. The larva is represented in Fig. 8.

FIG. 8.

Its pupation commenced September 20th, beneath a leaf fastened by some threads to the ground. The imago emerged June 1st. It is quite rare in this vicinity.

The following note probably refers to another example of the same species:

Caterpillar taken on a fence under a row of maples, October 20th; length one inch. Head apple-green, with yellow mandibles. Body yellow-green, with a yellow stripe on the side, and transverse interrupted markings of the same color. Anal plate apple-green, semi-elliptical, with yellow dots and border. Legs apple-green.

The head of the larva was abnormally large, perhaps from having recently molted, or possibly from parasitic attack. The season was too far advanced to permit of its being supplied with proper food, and it did not mature.

Notodonta ———?

Found, September 19, 1868, at Schoharie, feeding on willow, a remarkable looking larva, a figure of which is annexed. (Fig. 9.) The first horn-like projection is on the third segment, the second and longest on the fourth segment, the large bifid hump on the eighth segment, and a terminal one on the eleventh segment. The markings of the larva are exceedingly delicate in whitish and various shades of brown, finely reticulated. The long flexible horn of the fourth segment is capable of considerable motion, and is sometimes directed backward. The terminal segments are at times carried in an elevated position. When at rest, the central segments are contracted as represented in the figure, and its head and anterior segments are turned sideways. When extended in feeding it presents the appearance shown in outline in Fig. 10.

FIG. 9.

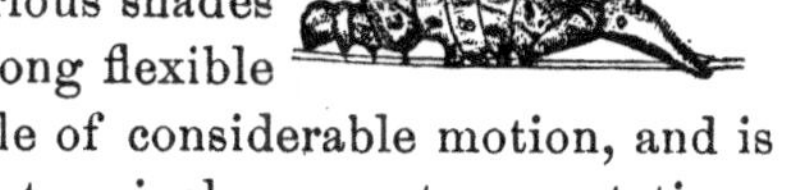

FIG. 10.

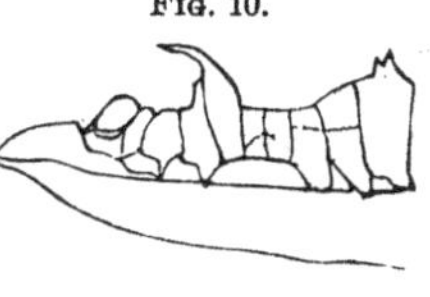

I have met with the larva on this occasion only. Without taking a description of it, I forwarded it to Prof. Glover, of the Agricultural Department at Washington, for figuring and description, and was informed by him that he had seen a single example of it, and had already given a representation of it in one of his plates. The imago was unknown to him.

Edema albifrons (*Sm.-Abb.*).

Late in November, twelve naked pupæ of this species were found lying in a cluster on some leaves beneath a stone. Kept in a warm room, the moths emerged between January 9th and February 11th, all during the night time.

The larva is described in the Harris Correspondence, p. 304.

Cerura borealis (*Boisd.*).

Mature larva feeding on willow, September 11th, represented in Fig. 11. Head small, red. Body apple-green, red dorsally from the fourth segment backward, reaching nearly halfway down the side on the sixth segment, and after a contraction expanding to the stigmata on segment seven, and thence contracting posteriorly; the red margined below with yellow. Anal extremity forked, and extensible

FIG. 11.

at pleasure in two long slender tail-like projections, annulated with red and green, and at their greatest extension disclosing a white ring.

Some young larvæ, collected August 22d on willow, apparently prior to their first molting, were found quite delicate to rear. Although furnished with fresh leaves twice a day, only one was matured by inclosing in its box the end of a twig of a growing plant. The larva is of slow growth, requiring about six weeks for it to mature.

The larva makes a flattened cocoon, three-fourths of an inch long, and quite firm in its texture, from the quantity of sawdust-like bits which it gnaws from the wood upon which its cocoon is placed and weaves together with its thread, excavating the wood, in some instances, to the depth of a tenth of an inch. Composed almost wholly of the gnawed material, and but slightly elevated above the surface, it is very liable to escape observation.

When engaged in the construction of its cocoon, the larva is very active, moving its head with remarkable rapidity. A cocoon was made August 22d, and two others September 1st.

Within a warmed room the imago has emerged April 1st and April 7th.

Telea Polyphemus (*Linn.*).

A female which mated June 18th–19th, deposited three hundred and fifteen eggs on the 19th. Of these, two hundred and seventy-eight produced their larvæ June 29th; fifteen died in the act of emerging from the shell; seventeen containing larvæ died within the shell, not fully developed, and five were probably unfertilized.

The eggs are round as seen from above, quite flattened on the upper and lower surfaces, white, with an intermediate brown band .04 in. broad, on which at two opposite sides is a small elongated white spot, and sometimes, between the two, an obscure whitish line.

From another bred example of the moth, two hundred and thirty-three eggs were deposited, and twelve additional were found within the abdomen, making two hundred and forty-five in all. The moth was somewhat dwarfed from its artificial feeding.

Actias Luna (*Linn.*).

As an illustration of the occasional extraordinary abundance, for a single season, of some of our usually rare insects, the following statement, made to me by Mr. Otto Meske, is worthy of record:

At about the commencement of his entomological studies, in 1864 or 1865, and not long after his arrival in this country, a fine specimen of the above moth was brought to him by a friend who had captured it at Sharon Springs, N. Y., where any desired number of the same, it was stated, could be collected. Charmed with the beauty of the "fair empress of the night," now for the first time seen by him, and desirous of procuring other examples for himself and for his European friends, as soon thereafter as his engagements permitted, he hastened to Sharon Springs. The day following his arrival there, he visited the hickory groves in the vicinity where the moth was represented as occurring. The season had too far advanced, by several weeks, (August) for the moth; but on almost every tree, pendant, fruit-like, from the lower branches, on leaves drawn downward by their heavy burden, were found one or more of the caterpillars—in their matured garb of transparent green enameled in dots of silver and pearl, so beautiful to the lepidopterist, and not unattractive to the unscientific eye. Sixty-four of the larvæ—the utmost capacity of his collecting case—were carried to his hotel, as a portion of the trophies of the morning ramble.

Cocoons were obtained from the entire number; for a readines to spin themselves up at any time, after their fourth molting, under the slightest provocation of a temporary withdrawal or an inferior supply of food, is a characteristic of the species (as has also been observed by European entomologists of *Aglia tau*); from this habit undoubtedly results the frequency with which crippled specimens of the moth are met with when artificially reared.

The following spring when the moths emerged, while they were hanging in profusion from curtains and from the walls about him, Mr. Meske proposed to Dr. Speyer, of Germany, with whom he was in correspondence, to send to him such a number of the cocoons as would serve to test the practicability of the acclimatization of the species in Germany; not doubting but that the ensuing season would be equally prolific with the preceding. Dr. Speyer was delighted in the prospect of so beautiful an addition to the insect fauna of Europe, and expressed himself as impatient to undertake the experiment. From that period to the present, although Mr. Meske has passed each intervening summer at Sharon, and brought from that superior collecting ground most valuable entomological contributions, not over a half dozen of the cocoons have been sent to Germany, toward the fulfillment of his promise; and in a letter lately received from

Dr. Speyer, after an enumeration of scores of our American Lepidoptera, of which he needs no more examples either for his cabinet or for study, he reminds Mr. Meske that his cabinet is still deficient in a perfect female Luna.

Hemileuca Maia (*Drury*).

A crippled imago emerged July 8th, 1872, being the first from a small number of larvæ carried to pupation the previous year. The usual white band on the upper surface of the primaries was interrupted in the middle; beneath it was continuous and broader than above. Upon opening its abdomen, one hundred and fifty-two eggs were obtained therefrom, of a uniform reddish-brown color.

On the 17th of July a second one (a male) emerged. September 7th, a third was observed just as it escaped from its puparium. It not being convenient to entrust it to the care of any one, and desirous of securing it in a perfect condition, I removed it, with the utmost care, to a small box, which I carried in my hand to the railroad train for which I was on the point of leaving. While in the cars, the box was held in position to subject it to as little motion as possible. When examined in the evening, after a three hours' ride, the moth was found with its wings dried but entirely unexpanded, and with its abdomen retaining the elongate form and sutural extension with which it emerged from the puparium. The motion of the cars had caused an entire arrest of its last stage of development.

I am informed by Dr. Hagen that this species was quite common, in 1872, at Detroit, Mich., and in Maine, where the caterpillar was observed feeding on *Spiræa salicifolia*, as noticed by Prof. S. Smith, at Norway, Me., in 1865. In Massachusetts it was not rare in its occurrence.

Gastropacha Americana *Harris*.

Larva feeding on birch (*Betula lenta*), August 18th, nearly mature, measuring two inches in length.

Body slate-gray, mottled with black, beneath flattened and greenish; on the sides, beneath the stigmata, a series of tufts of reddish hairs, three-sixteenths of an inch long; on the incisure of the second and third segments, a scarlet band superiorly, divided by a black line and black at the ends, only observed when the larva is extended or in motion; on the first segment, two small tubercles on each side, and one on each side of the following segments; from the tubercles are given out tufts of gray hairs mingled with white ones which are clavate at the tip; the lateral fringe with numerous gray clavate hairs.

Legs black; prolegs ash-color, with a black spot between each pair.

The caterpillar made its slight cocoon between two leaves, enveloping it in a wool-like substance. Another example, occurring on maple, spun up between some leaves on September 12th.

It is quite liable to parasitic attack. A pupa which did not develop, was opened, and found to contain the puparia of nine Tachinæ.

Clisiocampa Americana (*Fabr.*).

Some larvæ of this species which made their cocoons on the 6th of June, completed their transformation and appeared as moths on July 6th.

Young larvæ have been observed, just disclosed from the egg-belt on the 18th of April.

(ZYGÆNIDÆ.)

Ctenucha virginica (*Charp.*).

Larva taken on grass, upon which it may be presumed to feed, as it has also been found thereon by Dr. Packard, in Maine. Head large, shining black. Body reddish-brown dorsally, darker shaded on the incisures, black laterally, with two light cream-colored stripes on each side; its short, brush-like fascicles of hairs proceeding from tubercles and nearly covering the body, are black on the back, and dusky and black intermingled on the extremities and sides; intermediate ones (?) ochre-yellow.

An imago was obtained June 4th from a cocoon found a few days previously, attached to the leaves of a cedar seedling, about two inches above the ground.

The cocoon is oval, .75 in. in length, composed of the hairs of the caterpillar, which are gray and black, and about one-fourth of an inch long, and, under a lens, show distinct feathering. Through the hairs could be seen the inclosed dark brown pupa.

The moth has frequently been observed in a grove of pines and cedars at Schoharie, in 1859. It is not readily alarmed when at rest, and its flight is slow and steady, permitting its easy capture. It was unusually abundant in 1861, at the same locality, being frequently seen about dwellings and in gardens. On the 23d of June numbers were observed in a ravine, beside a brook bordered with deciduous trees.

Scepsis fulvicollis (*Hübn.*).

Two males and two females of this usually rare species were collected at Bethlehem, Albany county, on September 14th, 1870, resting on or flying about the blossoms of *Solidago* at mid-day.

This accords with an observation of Doubleday:* "I took it in September, in Illinois, on flowers, especially on the different species of *Solidago*, flying by day." He also adds, "I took it in Florida by night; for they used to fly to my lamp. I do not remember to have taken one by day there."

I have only taken the species (and its allies) by day, and I know of no other instance of its capture at lights.

A perfectly fresh specimen was taken at Schoharie (the only instance, during several years, that it came under my notice there), resting on a window pane within a room which it had entered through an open door.

I have previously noticed the attractiveness of the *Solidago* to *Lycomorpha pholus* (Drury),† six individuals having been observed by me regaling themselves on the blossoms of a single plant, while a hundred or more could have been collected at the time from the same locality (a hill-side at Schoharie, August 16, 1859). Melsheimer states that the larva of this species is found on the lichens growing on the trunks of hickory trees.

* *Entomological Correspondence of T. W. Harris*, 1869, p. 122.

† *Twenty-third Ann. Rep. on the N. Y. St. Cab. N. H.*, 1873, p. 193.

VI. NOTES ON SOME NEW YORK NOCTUIDÆ, ETC.

Diphtera deridens *Guenée.*

Larva resembling an Arctia in form, somewhat narrowed anteriorly and broadest toward the posterior segments, as represented in Fig. 12. Head white, with black markings as seen at *a.* Body white, segments rounded, smooth, but from the points where in an Arctia the tubercles are located, soft white hairs, one-fourth of an inch long, radiate, as fine as the finest silk spun by caterpillars, which curve at their tips and interlace, entirely enveloping the body. Length at maturity, 1.25 in.; diameter at broadest portion, .25 in.

Fig. 12.

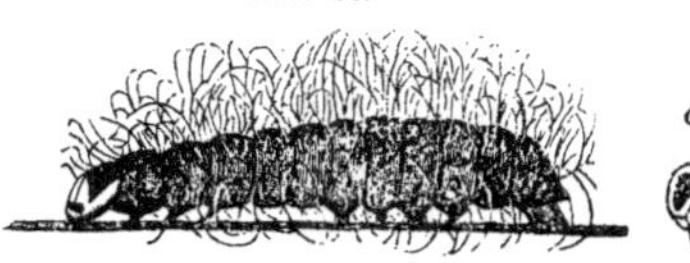

On September 4th, it made an oval cocoon, of uniform texture throughout, of fine silk.

On the 10th, it had undergone its pupal change; the dark-brown pupa could be distinctly seen through the delicate cocoon.

The imago was disclosed May 25th (1862).

Acronycta Americana *Harris MS.*

Since the printing of the notes on this species, on page 135 of this Report, I have been permitted to see colored figures of the larva and imago of *Acronycta alni* Linn. The larva, in its ground color, dorsal series of yellow spots, shape and comparative length of bristles etc., represents our larva so well, that, at the first glance, it might be taken for an accurate representation of it. It has the four long and two short bristles on the first segment, with none on the following two segments, as in ours. The single bristle shown in the figure on segments ten and eleven and the three on segment twelve, are probably inaccuracies of delineation through carelessness of the artist, as evidently are the placing of a bristle on each of the incisures of the sixth segment, and the location of several of the bristles elsewhere than in the lateral portions of the spots. The spots are shown as being marked with a number of irregular black lines, like Chaldaic letters, unlike the single, transverse, impressed line in our larva. In the accompanying text the larva is said to be rare, and to feed on oak, willow and wild plum.

The imago is represented of the size and shape of wings of *A. psi* figured beside it (with which our *A. occidentalis* Grote was for a long time confounded), the discal spots are more deeply marked, and the inner third of the primaries is brown, conforming in these particulars to the features indicated by Mr. Sanborn, as noticed in the Harris specimen. Other marked features are, the brown of the internal margin continued as a median band across the wings, separating the discal spots, and a distinctly defined brown border on the white secondaries.

A. alni is figured in Wood's Catalogue of the Lepidoptera of Great Britain, pl. 14, fig. 314; and following it, as fig. 315, is *A. psi.*

From the above, I think it highly probable that our species will prove to be identical with *A. alni* of Europe; if not the same, it is certainly very closely allied to it.

The figure of *Acronycta funeralis* Gr.-Rob. (*Proc. Ent. Soc. Phil.*, vi, pl. 3, f. 8), resembles very closely that of *A. alni* above referred to, differing mainly in the diffused border of its inferiors, their fringe cut with black instead of simply dotted, and in the absence of the submarginal black streak of the superiors behind the cell. Mr. Grote informs me that he has seen *A. alni*, and that *A. funeralis* and also *A. connecta* Gr. (*Bull. Buff. Soc. Nat. Sci.*, I, p. 79) resemble it.*

Acronycta hastulifera (*Sm.-Abb.*).

Larva feeding on horse-chestnut; two inches long, covered with gray or light red hairs. Body cream-color, with two dark interrupted stripes on each side, and two on the back. Head, feet, anal seg-

* To the kindness of Mr. Sanborn I am again indebted for a second visit made by him to the Boston Soc. of Nat. Hist., in compliance with my request for a critical comparison of the Harris specimen of "*A. Americana*" with the description and figure of *A. funeralis* Gr.-Rob., and for information just received from him, that the two are, without doubt, the same species.

Now that both the larval and perfect stages of *A. funeralis* are described and figured, a satisfactory comparison may be made with *A. alni*, to determine the question of their identity which has arisen from the marked resemblance in the representations of their unique larvæ.

The *habitat* of *A. funeralis* is evidently quite an extended one. The larva has been taken by Mr. C. V. Riley, at Portland, Me., on elm (*Ulmus Americana*), and by Prof. Bessey, as Mr. Riley informs me, at Ames, Iowa. Mr. Grote's type is from Ohio.

ment, eleventh segment superiorly, and under side of body, black. Two dorsal pencils of black hairs one-third of an inch long, on the fourth segment, two on the sixth segment, and one on the eleventh; the pencils have a black spot at their base.

Another larva feeding on the linden (*Tilia Americana*) had the body pale green, with yellowish hairs dorsally and white laterally.

When confined in a box, the larva spins a firm cocoon in one of the angles, in which it interweaves bits of the material bitten from the space inclosed by the cocoon.

Harris describes the cocoon (*Apatela Americana*, in Insects of New England, second edition, p. 338) as having "the half-oval web of silk, intermixed with the hairs of its body." Of six cocoons constructed by larvæ collected by me, only one contained intermixed hairs.

The female moth has the upper surface of the wings darker, and the under surface less shining than in the other sex.

Acronycta oblinata (*Sm.-Abb.*).

Larva feeding on the blossoms of smart-weed (*Polygonum punctatum* Elliot), September 15th. Length one inch and one-fourth. Velvety black, with a tawny red substigmatal stripe. Segments with tubercles, from which clusters of short hairs radiate, which are red on the upper part of the first four and last two segments, and white on the intermediate ones; from the tubercles on the terminal segment, long hairs proceed. Stigmata white.

Spun a cocoon between some leaves which it drew together.

Agrotis tricosa *nov. sp.*

I have for some time had set apart in my collection, three distinct forms of "*Agrotis subgothica* Haworth." Now that Mr. Grote, in correction of some former determinations, has recently pointed out, beyond question, the true *Agrotis subgothica*, and shown it to be the species redescribed by Guenée as *A. jaculifera*, and has also described as *A. herilis* a second form which Guenée had regarded as a variety of the former (var. B, not A), it only remains in order to clear up the confusion so long existing among these forms, to indicate the third species, which is easily to be distinguished from the other two.

A. tricosa is between *subgothica* and *herilis*, approaching nearer to the former in its antennal pectinations, the general coloring of its primaries, the pale subcostal nervure with brown linings, and in the form, color and marking of its orbicular. In size and in the dark coloring of its secondaries it is nearer to *herilis*.

It is readily distinguished from *subgothica* by its smoky-brown secondaries, broadly dark outwardly and paling gradually inwardly, in marked contrast with the distinctly brown-bordered white wings of that species, especially in the ♂. It has not the red reniform of *subgothica*.

Its most prominent differential features, compared with *A. herilis*, are the following: While the antennal structure is the same in the three species, in this the pectinations are stronger than in *herilis*. Both the subcostal and median nervures are pale gray, bordered with brown lines. The orbicular is concolorous with these nervures, more broadly open above, almost or completely uniting superiorly with the reniform, and with an interior brown line forming nearly a triangle, in continuation of the two lines bordering the subcostal; in *herilis* the orbicular is less open, approaching a U, and in some instances contracted above into a suborbicular form, and separated by some space from the reniform. It also differs from *herilis* in the acute extension of the internal tooth of the anterior median line, connecting with or nearly approaching the posterior median; in the interspaceal forked black rays behind the cell and between the median nervules; in the greater distance of the posterior median line from the reniform; in its less distinctly marked posterior median line (in *herilis* usually continuous and composed of interspaceal crescents); in its paler subterminal region; in its better defined subterminal line preceded by sagittate spots; in the distinct marginal black crescents interspaceally; in its paler costal region and general dark brown shade of the wing, instead of blackish; in its paler tegulæ, and collar less prominently marked with a transverse black line.

Material under examination in the above comparisons; *A. subgothica*, 17 ♂'s, 12 ♀'s; *A. tricosa*, 9 ♂'s, 5 ♀'s; *A. herilis*, 14 ♂'s, 5 ♀'s.

Measurements of the expanse of wings of the three species give the following results:

		Average.	Min.	Max.
A. subgothica......	10 spec.	1.55 in.	1.37 in.	1.62 in.
A. tricosa.........	9 "	1.70	1.61	1.81
A. herilis.........	14 "	1.60	1.37	1.69

A. tricosa appears to be as abundant as it allies in this portion of the State, appearing contemporaneously with them. We have no knowledge of its occurrence, or of *herilis*, in England, where probably *subgothica* was alone introduced from this country. Dr. Boisduval reports *A. jaculifera* (*subgothica?*) among California collections.

The following is the synonymy of the above species:

Agrotis subgothica *Haworth.* Lepidop. Britan., 1810, Part —.
A. subgothica Stephens. Illus. Brit. Ent., 1829. Haust. II, p. 25, pl. 22, f. 3.
A. subgothica Wood. Illus. Cat. Lep. Ins. Gr. Brit., 1833-8, p. 36, pl. 9, f. 149.
A. jaculifera Guenée. Spec. Gen. Lep., 1852, V. p. 262, pl. 5, f. 4.
A. subgothica Fitch. 1st-2d Rep. Ins. N. Y., 1856, p. 314, pl. 3, f. 1.
A. jaculifera Riley. 1st Rep. Ins. Mo., 1869, p. 82, pl. 1, f. 11.
A. subgothica Grote: in Bul. Buf. Soc. Nat. Sci., 1873, I, p. 99.

Agrotis tricosa *nov. sp.*
A. jaculifera var. A. Guenée Spec. Gen. Lep., 1852, V. p. 262.
A. subgothica Riley. 1st Rep. Ins. Mo., 1869, pp. 81–2, f. 29 *b* (not *a*).
A. subgothica Packard. Guide Stud. Ins., 1869, p. 306, f. 238 (right hand fig.).

Agrotis herilis *Grote.*
A. jaculifera var. B. Guenée. Spec. Gen. Lep., 1852, V, p. 262.
A. herilis Grote: in Bul. Buf. Soc. Nat. Sci., 1873, I. p. 99.

Hadena lignicolor (*Guen.*) *Grote.*

A larva of this species was found lying in a cell beneath a stone, on the 18th of May. It changed to a pupa during the night. The imago emerged June 29th.

A number of examples of the moth have been taken by me, but I have never known it to occur abundantly.

Hadena adjuncta (*Boisd.*) *Grote.*

Caterpillar feeding on blossoms of golden rod (*Solidago Canadensis*). Length one inch and one-fourth. Body pale apple-green; a narrow vascular stripe bordered by darker green lines; on each segment superiorly, a semicircular dark-green line, concave anteriorly; fourth, fifth and eleventh segments, marked with olive-green, the last segment elevated in a hump; a pale green stigmatal line, giving off diagonally a line to the back of each proleg.

Buried in the ground for pupation, and made a cell just beneath the surface. Imago emerged April 2d. When disturbed, drops upon its back and lies motionless for several minutes. Schoharie, 1857.

Cucullia florea *Guen.*

Through the kindness of Mr. A. R. Grote, of Buffalo, I have had the privilege of examining a Cucullia, which he regards as the above species. It is, unfortunately, in quite poor condition, having lost many of its scales, its fringes and its wings somewhat injured, etc. The single example from which Guenée's description was drawn, was also "assez mauvaise," and his diagnosis consequently is quite brief and incomplete. To add to the perplexity, Guenée gives us a figure (*Noctuélites*, II, pl. 7, f. 9) so entirely at variance with his description, in coloring, markings and size, that it can only serve to mislead.

The description (l. c., p. 134) is as follows: Primaries of the form of the preceding [*postera*, *asteroides* and *asteris*], of a uniform, deep bluish ash-gray, without a light or ferruginous shade, with the costa and the internal border blackish. The two median spots very vague, but distinguishable, surrounded and filled with blackish spots (groupes). Tooth of the internal border single, concolored, followed by an internal shade, surmounted itself by a straight blackish line. Extrabasilar line slightly visible, with rounded angles. Secondaries a little nacreous, with border broadly blackish, and with nervures deeper. 40 mm.

While the example before me does not wholly conform to the above description, the differences are such as may result from the imperfect condition of the specimen. It may, therefore, without much risk of error, be accepted as the *florea* of Guenée, now for the first time, it is believed, recognized in this country. The following are some of its features:

The primaries are bluish-gray, giving a very decided blue reflection when viewed obliquely. The costal margin seems as if it may have been suffused with blackish. The internal margin is blackish above a slender black line running from the basilar curve of the wing, to the outer margin. A black line, interrupted at the nervules, rests on the outer margin. There were apparently white and black submarginal streaks in the interspaces. The condition of the wings does not permit the tracing of the discal spots. The white mark (tooth) at the internal angle is crescentiform, preceded and followed by blackish, with the two black lines outwardly as above mentioned by Guenée, and as shown in his figure.

The secondaries are somewhat hyaline, tinged with brown, and with a lustrous brown border, quite narrowed toward the internal angle, and at the apex occupying nearly one-fourth of the wing. The veins are clothed with dark scales.

The tegulæ are gray, with a few intermingled black scales.

Wings beneath, a lustrous smoky-brown. The tuft supporting the frenulum is rust-red. There are no cellular lunules, and the marginal band of the secondaries is obsolete.

Expanse of wings 1.85 in.; length of body, .75 in. In the figure of Guenée, the alaric expanse is represented .28 in. in excess of that given in the description (40mm.).

This species seems to be the most rare of our Cucullias, the above example being the only one of which I have knowledge. *C. postera* appears to be nearly as rare. A single pair is in the collection of Mr. Meske, and, from their photographs, Mr. Strecker has identified an example in his possession, taken at Falls of Schuylkill, Philadelphia, Pa.

? Chariclea exprimens (*Walker*).

Caterpillar feeding on rose leaves; length one inch and one-eighth, head red, body green with yellow lateral stripes, along which a number of black spots are sprinkled. On the first segment, dorsally, are four black spots, on the second are two, and on the terminal (?) segment are four; these are of a larger size than the remaining ones, of which there are four on each of the other segments, forming a trapezoid in which the two anterior are considerably nearer the mesial line than the two posterior.

Taken August 4th, 1859, at Schoharie, and on the 7th August changed to an imperfect pupa, which did not develop.

It is believed to be identical with other larvæ from which *C. exprimens* has been reared.

Chamyris cerintha (*Treits.*).

From a cocoon made in an angle of a box, the moth emerged May 12th. When disturbed, it runs rapidly about the box, without taking wing. Upon suddenly opening the box, it has, in several instances, been observed to drop upon its back and lie in that position, with folded limbs, for several minutes, counterfeiting death.

It has been captured in the vicinity of Albany, and by Mr. Meske at Sharon Springs.

Plusia balluca (*Hübn.*).

A moth of the above species emerged from a cocoon July 6th (1861), which was taken about the 20th of June from a hop vine. The cocoon was attached to the under side of a leaf, and through its thin and loose threads, like a spider's web, the white pupa could plainly be seen, with some black marks on its back. After the escape of the imago, the puparium was of a light horn color, with small black spots surrounding the stigmata and with black lines on the incisures of the four anterior dorsal segments, and with two black spots at the base of each wing-cover. The tongue-case projected above an eighth of an inch, over three of the abdominal segments. Anal spine short and straight.

This beautiful moth has been frequently taken at Schoharie, in some seasons not having been at all rare.

Plusia æroides *Grote.*

A moth of this species emerged from cocoon July 8th (1861). The larva was taken while sweeping some low plants of violets, etc. No description was made. It was of a delicate apple-green color, without hairs, with rather deep incisures. It is believed to have had some yellow markings upon it.

Scoliopteryx libatrix (*Linn.*).

The moth emerged August 3d, from a cocoon which had been found lightly spun within a willow leaf. There are probably two annual broods of this species, as I have taken it in the early part of May.

I learn from Dr. Speyer that the brief description which I have given of the larva in the *23d Report on the State Cabinet*, p. 695, corresponds with the European form, and that they agree, also, in their habit of pupation.

Catocala parta *Guen.*

Larva found on the willow July 7th. It spun some leaves together, and twenty-two days thereafter the moth appeared.

Mesographe stramentalis *Hübn.*

Larva measuring from seven to eight-tenths of an inch in length. Head small, glossy black. Body spindle-shaped, slate-colored dorsally, dull green ventrally; on the dorsum are two rows of small white dots; just above the stigmata a broad yellow stripe, and whitish markings below the stigmata; collar glossy black; several rows of black tubercles having on two sides of most of them a white dot, and in each tubercle a short black hair.

The larvæ occur abundantly on leaves of the horse-radish, at Schoharie, during the latter part of October and first of November, living on the under surface, and consuming nearly the entire portion of the leaves except the principal ribs. When taken in the hand and held loosely, they usually succeed by their rapid contortions in dropping to the ground, and by their quick movements in finding a hiding place.

I have not obtained the imago from the above larva, but presume that it is correctly referred in the *Harris Correspondence*, page 322, to *Pionea eunusalis* Walker, which, according to Zeller,* is but one of the forms of the very variable *Pionea stramentalis* (Hübn.) Guen.

Zeller also claims that the genus Mesographe of Hübner was so well defined, that there was no propriety in the erection of the new genus Pionea from it by Guenée, and that consequently *stramentalis* should continue to be known as a Mesographe.

The moth appears to be as variable in this country as it is said to be in Europe.

Nematocampa filamentaria *Guenée.*

Larva found suspended by its thread from a maple tree (*Acer saccharinum*) on the 1st of July. It was placed in a box with some leaves to feed upon, and on the 4th it inclosed itself for its transformation, within three small pieces of a leaf which it had cut from the edge, and spun together with a few silken threads. The imago emerged on the 14th, after a pupation of ten days.

The larva is described in the *Harris Entomological Correspondence*, page 322, and a figure of the larva in the peculiar attitude which it assumes in repose, is given on plate 3, fig. 5.

Ennomos magnaria *Guenée.*

Larvæ feeding on lilac (*Syringa vulgaris*). Slight cocoons were spun between leaves August 29, and the moths appeared September 14th.

In the *Harris Entomological Correspondence*, page 320, the larva is recorded as feeding, in the months of August and September, on *Tilia*.

* *Beiträge zur Kenntness der nordamerik Nachtfalter*, 1872, p. 75.

Amphidasys cognataria *Guenée.*

Larva a looper, with ten feet; two inches in length. Head forked, light red. Body with two brown tubercles on the first segment; laterally, on the eighth segment, two transversely-elongate brown warts; on the eleventh segment, two small red warts on a brown patch; two white dots near the anterior portion of each segment dorsally, and two similar ones below the stigmata of the eleventh segment.

The larva feeds on maple. Entered the ground for pupation August 11th (1859). The imago emerged the latter part of May.

Abraxas ribearia *Fitch.*

Larvæ taken on currant bushes, buried in the ground for pupation July 4th. The first imagines appeared ten days thereafter.

Note on the Season of 1858.

Pyrameis Atalanta (Linn.) has been rare, last year quite abundant. *Pyrameis cardui* not observed, but abundant last year. *Pieris oleracea* (Harris) has abounded for two years, but, previous to that time, I had been able to collect but a single example. *Papilio Turnus* has been unusually numerous; early in the season it was as frequent as *Colias Philodice*, while two years previous not one specimen was observed throughout the entire season. Not one *Grapta interrogationis* has been seen, and a very few of *Grapta comma* or *G. Progne.* Not a single specimen of Catocala has been collected, while the previous year several species were obtained.

Note on the Season of 1859.

A very unfavorable season for collections, in marked contrast with the abundance of insect life the preceding year. Some of the most common Lepidoptera have not appeared at all, and others have only occasionally been seen. The following is a statement of the comparative abundance or absence of some of our Diurnals:

Abundant.	Few.	None.
Colias Philodice ...	Papilio Turnus	Papilio Troilus.
	Papilio Asterias.......	Pieris protodice.
Pieris oleracea.....	Danais Plexippus	Argynnis Myrina.
	Argynnis Aphrodite...	Argynnis Cybele.
	Argynnis Bellona.....	Grapta int'rogationis.

Abundant.	Few.	None.
Satyrus Alope . . .	Grapta comma	Grapta J-album.
	Grapta Progne	Limenitis misippus.
Satyrus Nephele.	Melitæa tharos.	Limenitis Arthemis.
	Vanessa Milbertii	Limenitis Ursula.
	Vanessa Antiopa	Pyrameis huntera.
	Pyrameis Atalanta	Pyrameis cardui.
	Lycæna comyntas	Eudamus Tityrus.
	Chrysophanus Americana.	

But one Catocala was observed during the season. Very few Noctuidæ were attracted by light on windows at night, except of the species of *Agrotis subgothica*, which occurred in great abundance.

VII. DESCRIPTIONS OF NEW SPECIES OF CUCULLIA.

Cucullia Speyeri *nov. sp.*

Palpi gray beneath, brown above. Front, with three transverse rows of projecting scales, gray at tips and black at base. Collar edged below with a sharply defined black line, pale gray in front, with a paler band just before its middle, bordered with lines of brown of which the posterior is the broader; when elevated, the apex and hinder part show brown hairs. Thoracic hairs brown. Tegulæ light gray, paler than the primaries, with a few scattered black scales near their superior margin as in *C. asteroides* Guenée. Abdomen acutely pointed in the male (Fig. 13), with long terminal hairs, and in the female (Fig. 14) ending with a long flattened tuft; whitish shading into gray terminally, more inclined to gray in the female, and in this sex interspersed with brown scales beneath; four brown dorsal tufts of about the size of those of *C. intermedia* Speyer; on the sides of the terminal segment of the female is a small spot of dull ochrey-yellow hairs, and a few projected from the incisure beneath. Tibiæ concolorous with the tegulæ, with a slender black line superiorly.

Fig. 13.

Primaries straight on costal margin, rounding to the apex as in the European *lucifuga* W.-V., and more curved than in *intermedia;* posterior margin slightly dentate, regularly sloping to the internal angle; interior margin nearly straight; breadth about equal to *intermedia*, exceeding *asteroides;* narrower and more acuminate in the male than in the female, as in all species of this genus; of a pale gray color, somewhat darker on the costal and internal margin and with a silvery reflection intermediately. Demi-line indicated by a short, very oblique black streak resting on the costa, bordered behind with white and a black dot nearer the base, above the subcostal nervure. Median transverse

Fig. 14.

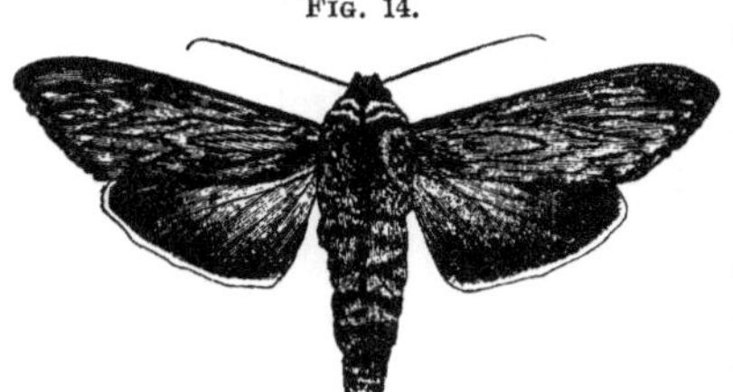

lines black, relieved in front by pale gray, better defined and more acutely dentate than in *intermedia;* the anterior line five-toothed, obscurely geminate, the interior line more distinctly defined above the inferior portions of the subcellular teeth, the two subdorsal teeth obscure, the cellular tooth slightly overlapping the orbicular spot, the medial one bisected by the slender black basilar line, and the internal one quite acute; the posterior transverse line traceable only below the cell, except as obscurely continued in a whitish zigzag shade across the nervules, and marked in its inception by a short, oblique, black line, followed by a white dot resting on the costa just beyond the bifurcation of veins 6 and 7. Reniform spot indicated only in its lower portion by a black line proceeding (in the female) from near the origin of the first median nervule (vein 2), curving over veins 3 and 4 near their bifurcation, and terminating in the black streak in cell No. 4. Orbicular spot visible in its lower portion as a short, slightly curved black line above the median nervure, at the point of projection of vein 2, its outline doubtfully traceable with a lens, contracted from its normal quadrate form, as seen in *C. postera* Guen.,* to that of a figure 8, being almost bisected

* In *C. postera* (a ♂ and ♀ example), the "orbicular" spot is quadrate, resting above and below on the subcostal and median nervures, angularly concave interiorly and less angularly convex posteriorly; the interior and posterior margins are black, interrupted just above their middle, appearing as four short lines; interior of the spot gray, with two elongate quadrangular markings, of which one is above and the other below the cellular fold. In this species the "orbicular" is better defined than in any other of the species under my observation.

In *asteroides* (three ♂'s and three ♀'s), in the strongest marked male, the spot is not defined in outline, but is represented by three black dots beneath the subcostal and three above the median nervure; four additional black dots, in range with these, pertaining to the reniform (making two rows of five spots each), give to the discal region the punctuated appearance observed in the European species next noticed. In other examples, the middle one of the three lower spots is wanting, and in others the middle of the upper ones also; this more frequently in the female.

In *C. absynthii* Linn., of Europe (one ♂), the orbicular assumes a very interesting form, from which, together with corresponding characters in the reniform, it has received the name of *punctigera* (*Berl. Mag.*, III, 100), and *la pointillée* by Engramelle. It is composed of six black subquadrate spots, of which three lie in a row beneath the subcostal and three above the median nervure, the middle spot in each row surrounded with white scales which sometimes extend over a part or all of one or both of the contiguous spots.

In *asteris* W.-V. (three ♂'s, two ♀'s), the female has the spot four-punctated and of the normal form, with traces of the interior transverse lines. In the male, the marginal lines form the four oblique sides of a hexagon, with a trace of the superior transverse interior line only.

by the extended acute cellular tooth of the anterior median line, which in this species surpasses the same tooth in *intermedia* and its allies, and nearly equals that of *chamomillæ*. Nerves and nervules clothed with black scales, and faintly relieved by white ones; the latter bordered with white scales as they approach the margin, which are continued on the fringe, extending nearly one-half across it. In the interspace (cell 4), below the disco-central nervure (vein 5), a black line running from the outer margin of the reniform, half-way to the margin, whence its continuation may be traced, with a lens, beneath the interspaceal white streak; in cells 5 and 6 are indistinct black streaks beneath the white streaks, not reaching the margin; in cell No. 3, a short black streak extending from outer third of interspaceal streak to the margin; in cell No. 2 a corresponding black streak but shorter and broader; in cell No. 1 b, a black line, having a white streak before and behind it, extends from the marginal termination of vein 2, obliquely to the median fold, increasing in breadth before reaching it, where it runs into a black line resting on the fold, which is broadest at the point of contact and loses itself just before reaching the posterior median band; this oblique line is more prominent that in *intermedia*, *lucifuga* and *umbratica*, but less conspicuous than in *postera*, *asteroides* and *asteris*.* Terminal margin with an obsolete black line, interrupted by the nervules.

In *lucifuga* W.-V. (one ♂, one ♀), an arc of the superior and another of the inferior portions of the margin are visible, leaving the outline of the spot not defined.

In *intermedia* (four ♂'s and four ♀'s), close scrutiny with a lens reveals, in the more perfect specimens, a faint, fine, curved, black line above the median nervure, representing the inferior margin of the spot.

In *umbratica* Linn. (two ♂'s and four ♀'s), two dots, of which the outer one is sometimes elongated, mark the lower corners of the spot, visible in all the specimens. In the most distinctly marked female, two oblique lines below the subcostal, running the one outwardly and the other inwardly, indicate the superior portion of the spot.

In *chamomillæ* W.-V. (one ♂ and one ♀), the spot is reduced to two black dots, one at each end of a pale elongate spot beneath the subcostal.

In *convexipennis* Gr.-Rob. (one ♂ and two ♀'s), the spot is obsolete, having its position indicated only by two black dots beneath the subcostal ; a third dot, and occasionally a fourth outside of these, pertain to the reniform.

In *lichnitides* Ramb., *lichnitis* of Guenée (two ♀'s), are two black dots above the median, and over these a semi-elliptical pale spot bordered with brown, deeper at the ends, lying between the subcostal and the cellular fold.

In *scrophulariæ* W.-V. (three ♂'s), the same features appear as in *lichnitides*, but not quite so well defined.

* The engraver has omitted these interspaceal black streaks from the figures, and failed to represent correctly the associated white ones.

Fringe squamose, long, equal in length to the space between veins 2 and 3 on the margin; basal third brown, and a narrow brown line through its middle; outer scales clavate, and, under a lens, white, centered with pale brown.

Secondaries abruptly rounding into the apex, which is slightly acute, less so than in *asteroides;* outer margin convex, excavate between the subdorsal and median nervules and prominently angulate on the submedian; in the male, white, hyaline, with a lustrous brown border, shading paler from the margin inwardly, broadest apically where it occupies one-sixth of the length of the wing, narrowing regularly to the internal angle, and terminating between the submedian and internal nervures. The subdorsal and median nerves and nervules with brown scales, which almost cover the nervules, especially in the subterminal region; the submedian with a marginal spot of a few brown scales. Fringe white, with a few pale brown scales on the superior half of the wing. In the female, wing whitish (smoky-white) basally and slightly hyaline; border a lustrous brown, of not quite so dark a shade as in *asteroides*, broader than in that species, about equal to *intermedia*, occupying nearly one-third of the wing apically, its inner margin tolerably well defined, and is continued indistinctly along the internal nervure nearly to the base: costal region also shaded with brown above the s. c. nervure and extending into the cell in the basilar region. The submedian and its branches more heavily clothed with brown scales than in the ♂. Fringe white, traversed with brown in its superior half.

Beneath, primaries lustrous pale brown, the costa gray basally, the ♀ with ochreous hairs supporting the frenulum. Secondaries without the discal spot, which is also lacking in *asteroides*, evident in *asteris* and conspicuous in *umbratica*, *lucifuga* and *intermedia;* white in the ♂ with brown scales in the costal and apical region and a few on the terminal margin; the terminal and subterminal regions give a creamy reflection in a certain light. In the ♀, more numerous brown scales costally, and with a lustrous brown border nearly as broad and as prominent as on the upper surface, traversed by the paler nervules.

Expanse of wings of ♂, 1.98 in.; of ♀, 2.07 in. Length of body, including anal tuft of ♂, 1 in.; of ♀, .95 in.

I take pleasure in naming this species after Dr. Speyer, in recognition of the study that he has bestowed on the perplexing species of the genus, as shown in two excellent articles on some closely allied and for a long time confused European species, published in the *Stettiner Ent. Zeit.*, in 1858 and 1859; also in his valuable paper published in the same journal in 1870, and a translation given in the 23*d Ann. Report on the N. Y. State Cabinet*, but for which paper our *intermedia* might yet have been regarded identical with the quite dissimilar *umbratica* of Europe.

The detection of this interesting species is due to the indefatigable zeal in collecting, of Mr. Otto Meske, of Albany, and to his ready perception of new Lepidopterous forms or features. A pair, in perfect condition, are in his cabinet, from which the above description is drawn, and, as they are the only examples which have come under my observation, it is probably quite rare. The female was taken by him, at Albany, on the 6th of June; the male was found at Sharon Springs, N. Y., on the 15th of August, at rest upon a fence, and was recognized, before being pinned, as differing from *intermedia*.

The description has been carefully drawn and extended, perhaps, to an unusual degree of minuteness, for the reason that the genus presents us with species which can only be separated from one another, in the imago state by very careful discrimination, and requiring for their identification, a faithful delineation of inconspicuous features. M. Guenée, in his *Species Général des Lépidoptères*, remarks, that in Cucullia "it frequently occurs that caterpillars the most different, it might almost be said, the most opposite, produce moths so very near, that it is only by great care that they can be distinguished." In remarking on *C. lucifuga* Rœsel, he says: "The *lucifuga* of Treitschke appears to me a *lactucæ;* that of Duponchel is a *chamomillæ;* that of Borkhausen seems an *umbratica*, as also those of Stephens and Esper." The greatest confusion has existed in regard to *C. blattariæ*, for, according to Gueneée, the *caninæ* of Rambur and the *thapsiphaga* of Duponchel are identical with the true *blattariæ* of Esper; the *blattariæ* of Her.-Sch. is *thapsiphaga* Treitschke; the *blattariæ* of Boisduval is *scrophularivora* Rambur; the *blattariæ* of Duponchel is *prenanthis* Boisduval. *Thapsiphaga* figures as synonyms of three different species. *Umbratica* Linn. appears as *lactucæ* of Fabricius ?, Hübner, Treitschke, Haworth and Stephens, the *lucifuga* of Esper and Borkhausen, and the *taneceti* of Stephens. *Lychnitis* Engramelle and *scrophulariæ* W.-V. so strongly resemble one another that several authors have believed them to be identical. Esper figures both as the same species, but according to Speyer, there can be no doubt that they are distinct, he having reared numbers of the imagines of both species from larvæ presenting constant differential features. On the authority of the same author, the females of *lactucæ* and *umbratica* are with difficulty separable.

The genus is also very numerous in species. Guenée records forty-four European species. Up to the present, only five* American species had been described, but in all probability a number of others will hereafter be detected.

Of the species with which I am able to institute a comparison, *Speyeri* stands between *intermedia*, *lucifuga* (nearer to the latter in the gray and more distinct markings of the primaries) and *asteroides*. The female bears a stronger resemblance to *lucifuga* than does the male. In the white color and hyalescence of its secondaries, its abdomen and brown scales of the tegulæ, it approaches *asteroides*, but lacks the angulated white line near the internal angle characterizing that species, *asteris*, *postera*, *florea*, *convexipennis*, etc. The resemblance of the secondaries to those of *asteroides* is particularly marked, only that in the latter species the border is less conspicuous.

Having been informed by Mr. Herman Strecker, of Reading, Penn., that he had in his cabinet an undetermined Cucullia, near to *intermedia*, I submitted to him a photograph of the above species, requesting its critical comparison with the species in his possession. As these pages are passing through the press, I learn from him that the two are identical, and that he has two examples of it which were taken at Falls of Schuylkill, Philadelphia. It had also been taken, he states, near Reading, and it did not appear to be more rare in that vicinity than *intermedia*.†

*These are *asteroides* Guen., *postera* Guen., *florea* Guen., *intermedia* Speyer, and *convexipennis* Gr.-Rob. *Chamomillæ* W.-V. is credited by Walker to the State of New York and Hudson's Bay, but its occurrence in this country has not, that I am aware, been confirmed. A species described as *C. Yosemitæ* by Mr. Grote, has subsequently been stated by its author not to pertain to the genus.

Dr. Boisduval, in his *Lépidoptères de la Californie*, 1869, p. 89, credits the European *asteris* to California, erroneously regarding it as identical with *asteroides*, stating of it, "élevée de la chenille par M. Lorquin sur le *Solidago Canadensis*. M. Guenée en a fait une espèce à part sous le nom *d'Asteroides*." The two species differ so much in their features, that the above error could not have occurred if examples had been placed side by side for comparison.

† Through the kindness of Mr. Strecker, I am in the receipt of one of the above examples (a female), whereby I am able to verify his determination of the species. It conforms to the typical examples as above described, with the single exception that the anal tuft, instead of being flattened, is contracted to a point nearly as acute as in the male, although not so long.

A few copies of these papers have been accompanied by two photo graphic plates containing figures of all the species noticed in this paper, viz.: Plate I, *postera*, *asteris*, *asteroides*, *Speyeri* and *intermedia*, in each sex; Plate II, *lucifuga* ♂, *absynthii* ♂, *chamomillæ* ♂ ♀, *umbratica* ♂ ♀, *convexipennis* ♂ ♀, *scrophulariæ* ♂ and *lichnitides* ♀.

In consideration of the far greater accuracy ensured in the representation of the above species by the aid of photography than it is possible to attain by any other method, it is to be regretted that the expense of the production of such plates prevents their presentation in the entire issue of this Report.

Cucullia serraticornis *nov. sp.*

Antennæ biserrated; the serratures, which are less conspicuous at the base and tip, as seen from above, consist apparently of a row on each side beneath of conical projections, bearing rows of curved whitish cilia on their lateral margins beneath, which increase in length from the base to the apex. Palpi nearly horizontal, gray with brown scales, third joint short and rounded. Front gray with some black scales. Collar yellowish-brown below the prominent black transverse line; above gray with the usual paler line bordered with darker scales, and still darker ones on the upper margin of the collar. Tegulæ concolorous with the wings. Thorax fuscous; an abdominal series of similar colored tufts on the first four segments. Abdomen gray, paler basally.

Primaries straight costally or slightly concave from the folding over of the marginal nervure, gently curving to the apex, which is obtuse; outer margin entire, sloping moderately to the inner margin, which is long and nearly straight. Color pale ash-gray, darker on the inner margin. Anterior median line blackish, teeth acute, preceded by a white shade, beyond which some blackish lines almost geminate it. Posterior median line obsolete, except in cell 1 *b*, where it is bidentate; the teeth preceded by an elongate-oval, brown bordered white spot on the submedian fold, extending to the anterior median line; followed by a white angulated line (the "tooth" of the internal angle), from the concavity of which a black streak (the usual interspaceal streak of the internal angle) runs obliquely to the first median nervule (vein 2). Costal margin over the place of the orbicular, with a diffuse brown shade and two oblique brown streaks at the inception of the posterior median. Basilar line black, fine; a slender black line on the internal margin. Reniform indicated by a row of black dots anterior to the discal cross-vein; orbicular only visible as a central pale shade and four outer brown dots. Nerves and nervules clothed with black scales; on the interspaces intermediately are brown scales, with a white streak centrally, beneath which, on the subterminal margin, are the usual black streaks in all the interspaces, the most conspicuous of which is that in cell 4 (farther removed from the margin than the others). Fringe white, cut with brown on each side of the nervules, opposite the interspaceal lines of brown scales; these brown ciliary scales of each interspace joined by a brown marginal line.

Secondaries acute, excavated opposite the cell, slightly dentate; white, hyaline. Nerves and nervules heavily marked with black scales, especially toward the margin; no distinct marginal border, but in place thereof the extreme margin is brown, with some brown scales extending a short distance therefrom, and farther in cells 1 *b* and 4; some brown scales on the costal nervure apically.

Beneath, primaries pale brown with an æneous reflection; a conspicuous brown spot on the discal cross-vein. Secondaries, with brown scales on the nerves and nervules and marginally as above, though less abundantly; sprinkled with brown scales costally, and in the cell above the fold; cellular fold and discal cross-vein above it broadly covered with brown scales, diffuse on the latter, giving a conspicuous cellular spot; these features seen in transparency from above.

Described from two ♂'s, differing materially in size; the larger and better specimen, from which the features are mainly drawn, measures two inches expanse of wings, length of body .86 in.; the other 1.70 in. expanse, length of body .72 in.

In addition to disparity in size, the two examples differ somewhat in shape of wings, those of the smaller being narrower and more acute, to the degree that the male usually varies from the female in the several species of this genus; yet the two examples are undoubted males, as is shown by their frenulum examination. In all other particulars, so far as they are traceable, the two are identical. They are unfortunately in poor condition, and the description above given may require correction.

Habitat, California. From Mr. James Behrens (No. 5), through Mr. Grote.

The species can at once be separated from all other described Cucullias, by its serrated antennæ (in the ♂), it being the only species known in which this form exists. Conforming in other respects to the typical forms of the genus, the simple difference of antennal structure does not seem to warrant its separation, but simply a modification of the generic diagnosis as given by Guenée; "to antennæ [usually] cylindrical and entirely smooth in both sexes." It may be recognized by the double interspaceal brown ciliary cuttings, the prominent cellular spot and the brown cellular line of the secondaries beneath, extending from the base to the discal cross-vein. The latter feature will probably be found to be less conspicuous in the ♀; it is feebly represented in one ♂ example of *C. intermedia* in my possession, and still more indistinctly in a ♂ of *C. lucifuga.*

In the presence of the "tooth" of the internal angle of the primaries, the hyalescence of the secondaries and general coloration, the species seems allied to *asteroides* and *florea.* I regret that I am unable to give a comparison of shape of wings, owing to the variation, as above stated, in the examples before me.

VIII. OBSERVATION OF SOME NEW YORK RHOPALOCERA FOR THE YEAR 1871.

A calendar of the occurrence of the Rhopalocera, upon the plan presented in the Report for 1870, was commenced for the following year, but was necessarily suspended early in the month of July, after the following records had been made:

Papilio Turnus *Linn*..............	May 30; June 1, 8, 16; July 7.
Papilio Troïlus *Linn*	June 1, 8, 16.
Papilio Asterias *Drury*............	May 19, 30; June 1, 8, 13.
Pieris rapæ (*Linn.*)...............	March 14; May 2, 12, 16, 23, 30; June 1; July 7.
Colias Philodice *Godt*.............	May 2, 12, 16, 19, 23, 30; June 1, 8, 13, 16; July 7.
Danais Plexippus (*Linn.*)..........	June 1; July 7.
Argynnis Aphrodite *Fabr*..........	July 7.
Argynnis Cybele *Fabr*............	June 13.
Argynnis Myrina (*Cram.*)..........	June 8, 13, 16; July 7.
Phyciodes tharos (*Drury*) *Hübn*....	June 1, 8, 13, 16; July 7.
Phyciodes Batesii *Reak*............	June 8, 16.
Charidryas Nycteis (*Doubl.*) *Scudd*..	July 7.
Melitæa Phaeton *Fabr*............	June 16; July 7.
Grapta J-album (*Boisd.-Lec.*)	June 1.
Vanessa Antiopa (*Linn.*)..........	May 2, 12, 16, 19, 30; July 7.
Limenitis misippus (*Fabr.*).........	June 1, 8, 13, 16; July 7.
Neonympha Eurytus (*Fabr.*)........	June 1, 8, 13, 16.
Thecla Irus (*Godt.*)................	May 2, 12, 16, 19, 23, 30; June 8.
Thecla Augustus *Kirby*............	May 2, 12, 16, 19.
Thecla Niphon (*Hübn.*)............	May 2, 12, 16, 19.
Thecla Melinus (*Hübn.*)............	May 12.
Thecla Edwardsii *Saund*...........	July 7.
Lycæna violacea *Edw*.............	May 12, 19.
Lycæna neglecta *Edw*.............	May 16, 19, 23, 30; June 1, 8, 16.

Lycæna comyntas (*Godt.*)........ ..	May 2, 12, 16, 19, 23, 30; June 1, 8; July 7.
Lycæna Scudderii *Edw*............	May 30; June 8, 16; July 7.
Chrysophanus Americana (*Harr.*)...	May 16, 19, 23, 30; June 1, 8, 13; July 7.
Thorybes Pylades *Scudd*...........	June 1, 8, 13, 16; July 7.
Epargyreus Tityrus (*Fabr.*)........	May 30; June 8, 16.
Nisoniades Juvenalis (*Fabr.*).......	May 12, 16, 19, 23, 30; June 8.
Nisoniades Persius *Scudd*..........	May 12, 16, 19, 23, 30.
Nisoniades Lucilius *Lintn*.........	June 1.
Nisoniades Martialis *Scudd*........	May 16, 19, 23, 30; June 8, 16.
Nisoniades Ausonius *Lintn*.........	May 12.
Nisoniades Brizo *Boisd.-Lec.*......	May 2, 12, 19.
Nisoniades Icelus *Lintn*...........	May 19, 23, 30; June 8, 16.
Ancyloxypha Numitor (*Fabr.*) *Feld.*	June 13.
Amblyscirtes vialis (*Edw.*) *Scudd*...	June 8, 16.
Ocytes metea *Scudd*..............	May 16, 19, 23, 30; June 8.
Atrytone Zabulon (*Boisd.-Lec.*) *Scudd.*	May 30; June 1, 8, 16, 13.
Polites Peckius (*Kirby*) *Scudd*......	June 1, 8, 13.
Limochores Mystic (*Edw.*) *Scudd*....	June 8, 13, 16.
Limochores bimacula (*Grote-Rob.*)...	June 13.
Limochores Taumas (*Fabr.*)........	June 13, 16; July 7.
Limochores Manataaqua *Scudd*.....	June 16; July 7.
Lerema Hianna *Scudd*............	May 23, 30; June 8, 16.

IX. DATES OF COLLECTION OF SOME NEW YORK HETEROCERA FOR THE YEAR 1872.

Sphingidæ.

Sesia gracilis *Gr.-Rob*		June 4.
Thyreus Abbotii *Swains*		June 6.
Darapsa Chœrilus (*Cram.*)	June 19,	June 27.
Deilephila lineata (*Fabr.*)		June 18.
Deilephila chamænerii *Harr*		June 20.
Philampelus Pandorus (*Hübn.*)		June 27.
Smerinthus geminatus *Say*		June 8.
Smerinthus excæcatus (*Sm.-Abb.*), larva		Oct. 5.
Daremma undulosa *Walk*		June 1.
Sphinx chersis (*Hübn.*)		June 27.
Sphinx drupiferarum *Sm.-Abb*		June 19.
Sphinx kalmiæ *Sm.-Abb.*, larva on ash		Sept. 24.
Sphinx Gordius *Cram*	May 29,	June 27.
Agrius eremitus *Hübn*		June 27.
Agrius eremitus, larva on mint		Sept. 24.
Ellema Harrisii *Clem.*, larva on pine	Sept. 3,	Sept. 19.

Ægeridæ.

Ægeria tipuliformis (*Linn.*)	June 16,	June 23.

Zygænidæ.

Scepsis fulvicollis *Hübn*		Sept. 24.
Ctenucha virginica (*Charp.*)		June 12.

Bombycidæ.

Euphanessa mendica (*Walk.*)	July 8–	July 21.
Hypoprepia miniata (*Kirby*)		July 28.
Utetheisa bella *Hübn*		Sept. 24.
Arctia arge (*Drury*)		July 28.
Spilosoma virginica (*Fabr.*)	May 30–	June 29.
Orgyia leucostigma (*Sm.-Abb.*)		July 9.
Parorgyia parallela *Gr.-Rob.*, larva on pine		Sept. 24.

Ichthyura albosigma (*Fitch*), larva on poplar.... Sept. 5, Sept. 19.
Actias Luna (*Linn.*)........................ June 7, June 12.
Platysamia Cecropia (*Linn.*)........................... July 2.
Hyperchiria Io (*Fabr.*)............................... June 12.
Eacles imperialis (*Drury*), larva on pine............... Sept. 8.
Anisota senatoria (*Hübn.*)............................ June 14.
Anisota stigma (*Sm.-Abb*)., larva on oak............... Sept. 24.
Tolype laricis (*Fitch*)............................... Sept. 6.

Noctuidæ.

Acronycta occidentalis *Grote*................ May 25, June 29.
Acronycta oblinita (*Sm.-Abb.*), larva on smart-weed...... Sept. 1.
Leucania pallens (*Linn.*)..................... June 7, June 17.
Leucania unipuncta *Haworth*................ June 6, Aug. 31.
Micrococlia diphteroides *Guen*........................ June 25.
Hydrœcia nictitans (*Linn.*).................. July 14, July 28.
Hydrœcia lorea *Guen*...................... June 24, July 11.
Hydrœcia sera *Gr.-Rob*..................... July 5, July 14.
Hydrœcia immanis *Guen*............................... Aug. 25.
Nephelodes violans *Guen*............................. Sept. 6.
Hadena lignicolor (*Guen.*) *Grote*............. July 4, July 11.
Hadena arctica *Boisd*...................... June 27– July 28.
Hadena dubitans (*Walk.*) *Grote*............... July 7, Aug. 19.
Hadena devastator (*Brace*) *Grote*............. July 7– Aug. 14.
Apamea iaspis *Guen*................................. July 7.
Apamea finitima *Grote*................................ June 17.
Celæna herbimacula *Guen*................... June 17– July 21.
Agrotis suffusa *W.-V.*..................... June 17, July 26.
Agrotis venerabilis *Walk*................... Sept. 15, Sept. 24.
Agrotis subgothica *Haworth*................. July 21, July 30.
Agrotis herilis *Grote*................................ Aug. 14.
Noctua clandestina *Harr*............. July 4, July 14, Sept. 23.
Noctua bicarnea *Guen*................................ July 14.
Noctua augur *Fabr*.......................... July 4, July 13.
Xanthia circillaris *Naturf*,........................... Sept. 19.
Tæniocampa instabilis (*Rœs.*)................. April 4, May 8.
Aplecta herbida *W.-V.*................................ July 7.
Hyppa xylinoides (*Guen.*)................... Aug. 14, Sept. 19.
Mamestra chenopodii (*Albin*).................. May 29, Aug. 14.
Xylina cinerea *Riley*............. April 20, Aug. 4, Sept. 26.

Xylina Bethunei *Gr.-Rob* .... Sept. 5.
Cucullia intermedia *Speyer* .... July 14.
Cucullia postera *Guen* .... July 10.
Cucullia asteroides *Guen* .... Aug. 12.
Cucullia asteroides, larva on Solidago .... Sept. 7.
Cucullia convexipennis *Gr.-Rob.*, larva on Solidago ...... Sept. 7.
Chariclea exprimens (*Walk.*) .... June 30.
Heliothis armigera *Hübn.* .... June 29.
Melaporphyria immortua *Grote* .... June 1.
Chamyris cerintha (*Treits.*) .... June 19.
Erastria nigritula *Guen* .... June 4, June 14.
Erastria carneola *Guen.* * .... June 4–Sept. 23.
Erastria muscosula *Guen* .... June 24, July 8.
Leptosia concinnimacula *Guen.* † .... May 16–May 29.
Placodes cinereola *Guen* .... July 28.
Plusia precationis *Guen* ...... June 1, July 13, Aug. 8, Sept. 22.
Plusia simplex *Guen* .... June 7, Aug. 14.
Plusia ærea (*Hübn.*) .... June 28, Aug. 17, Sept. 30.
Plusia æroides *Grote* .... July 7, July 14.
Plusiodonta compressipalpis *Guen* .... July 19.
Deva purpurigera *Walk* .... June 29.
Anomis xylina *Say* .... Sept. 25.
Amphipyra tragopogonis (*Linn.*) .... July 13.
Amphipyra pyramidoides *Guen* .... Sept. 30.
Catocala ultronia *Hübn* .... Aug. 4.
Catocala relicta *Walk* .... Aug. 19.
Catocala concumbens *Doubl* .... Sept. 5, Sept. 11.
Catocala amatrix *Hübn* .... Sept. 5, Sept. 11.
Catocala cara *Guen* .... Sept. 6, Sept. 19.
Catocala piatrix *Grote* .... Sept. 6, Sept. 19.
Catocala desperata *Guen* .... Sept. 11.
Catocala habilis *Grote* .... Sept. 14, Sept. 19.
Catocala retecta *Grote* .... Sept. 19, Sept. 24.
Catocala ilia *Cram* .... July 15.
Catocala polygama *Guen* .... July 13, July 18.
Catocala cerogama *Guen* .... July 30.

* This species appears to have successive broods during the season, as it was observed on June 4, 24; July 17, 21, 28; August 5, 19; September 8, 19, 23.

† More abundant than previously observed, having been captured May 16, 21, 24, 26 and 29.

Drasteria erechtea (*Cram.*).................. Sept. 1, Sept. 23.
Euclidia cuspidea *Hübn*............. May 21, June 12, Aug. 1.
Poaphila quadrifilaris *Hübn*......... May 29, June 4, June 19.

PHALÆNIDÆ.

Eutrapela transversata (*Drury*)............. July 24, Aug. 3.
Sicya truncataria *Guen*........................... July 12.
Angerona crocataria (*Fabr.*)...................... June 12.
Hyperetis alienaria (*Her.-Sch.*).................. June 12.
Nematocampa filamentaria *Guen*.................... June 25.
Endropia hypochraria (*Her.-Sch.*)................. June 6.
Endropia homuraria *Grote*......................... June 12.
Endropia serrata *Drury*........................... June 25.
Ellopia fiscellaria *Guen*................... Sept. 18, Oct. 5.
Caberodes phasianaria *Guen*....................... July 3.
Caberodes majoraria *Guen*......................... July 3.
Ennomos magnaria *Guen*............................ Sept. 28.
Cleora pulchraria *Minot*................... Sept. 6, Sept. 24.
Tephrosia disconventa *Walk*....................... June 6.
Acidalia enucleata *Guen*................... June 12, July 20.
Acidalia sinuata *Pack*............................ July 16.
Stegania pustularia *Guen*......................... Aug. 8.
Macaria bisignata *Walk*........................... June 12.
Macaria 4-signata *Walk*........................... July 4.
Corycia albata *Lef*............................... June 7.
Corycia semiclarata *Walk.*=*Bapta viatica* Harvey...... June 7.
Fidonia bicoloraria *Minot*................. May 25–June 12.
Fidonia Faxonii *Minot*..................... May 21–June 7.
Hæmatopis grataria (*Fabr.*) *.............. June 9–Sept. 24.
Aspilates coloraria (*Fabr.*)............... May 16, June 4.
Phasiane mellistrigata *Grote*..................... May 29.
Lozogramma defluaria *Walk*................. May 14–June 4.
Abraxas ribearia *Fitch*........................... July 15.
Melanthia albicillata (*Linn.*).................... May 29.
Melanthia ruficillata *Guen*................ June 24, July 21.
Melanippe gothicata *Guen*......................... June 24.
Coremia propugnata *W.-V*................... May 20, Oct. 28.
Coremia ferrugata *Alb*..................... July 30, Aug. 14.
Camptogramma fluviata *Hübn* †.............. July 28, Aug. 1.

* Observed June 9, 12; July 4, 14; August 19, 25; September 1, 8, 24.

† *Camptogramma gemmata* Hübn., is the ♀ of *C. fluviata*, as has been recently ascertained through rearing both forms from a single oviposition.—SPEYER.

Cidaria diversilineata *Hübn*			July 3.
Cidaria gracilineata *Guen*		Aug. 1–	Sept. 8.
Eupithecia interrupto-fasciata *Pack*			Sept. 25.

(COLLECTED PRIOR TO 1872.)

Boarmia gnopharia *Guen*	Schoharie	July 30.
Boarmia intraria *Guen*	Center	May 21.
Boarmia humaria *Guen*	Schoharie	June 15.
Boarmia indicitaria *Walk*	Schoharie	July 1.
Tephrosia disconventa *Walk*	Center	June 7.
Tephrosia spatiosaria *Walk*	Schoharie	June 5.
Paraphia subatomaria *Wood*	Bethlehem	June 17.
Aplodes latiaria *Pack.**	Schoharie	June 5.
Aplodes approximaria *Pack*	Center	June 9.
Cabera intentaria *Walk*	Schoharie	June 21.
Cabera erythemaria *Guen*	Bethlehem	June 25.
Halia subcessaria *Walk*	Schoharie	July 26.
Corycia semiclarata *Walk*	Center	May 21.
Eumacaria brunneata *Pack*	Center	May 25.
Caripeta divisata *Walk*	Schoharie	July 26.
Larentia perlineata *Pack*	Schoharie	May 3.
Eupithecia interrupto-fasciata *Pack*	Schoharie	April 29.
Eupithecia vernata *Pack*	Albany	May 25.
Cidaria rigidata *Walk*	Schoharie	May 14.
Fidonia truncataria *Walk*	Center	May 23.

Deltoidæ.

Chytolita morbidalis (*Guen.*) *Grote*		June 12.
Zanclognatha lævigata *Grote*	July 9–	July 23.
Philometra longilabris *Grote*		Aug. 24.
Phalænostola larentioides *Grote*		Aug. 1.
Rivula propinqualis *Guen*		June 19.
Helia phæalis *Guen*		Aug. 18.
Epizeuxis lituralis *Grote*		July 3.
Epizeuxis strictilinealis *Grote*		July 25.
Epizeuxis Americalis (*Guen.*)		Sept. 1.

*This species was erroneously determined in the 23d Ann. Rep. St. Cab. N. H., p. 196, line 23, as *A. mimosaria* Guen.

Pangrapta decoralis *Hübn.*=*Hypena elegantalis* Fitch ... June 4.
Phalænophana rurigena *Grote* .... May 22.
Palthis angulalis *Hübn* .... June 24.
Hypena evanidalis *Rob* .... June 28, July 7, Sept. 9.
Bomolocha abalienalis (*Walk.*) .... June 18.
Plathypena scabra (*Fabr.*) *Grote* .... Sept. 6– Sept. 23.
Meghypena velifera *Grote* .... July 13.
Macrhypena deceptalis (*Walk.*) *Grote* .... July 26.
Lomanaltes lætulus *Grote* .... May 25.
Tortricodes bifidalis *Grote* .... June 10.

PYRALIDÆ.

Botis terrealis *Treits* .... June 1.
Botis plectilis *Gr.-Rob* .... June 5.
Botis thesealis *Led* .... July 10, Aug. 1, Sept. 24.
Botis marculenta *Gr.-Rob* .... July 6.
Eurycreon chortalis *Grote* .... May 21, June 4.
Asopia olinalis (*Guen.*) .... July 9.
Asopia farinalis (*Linn.*) .... July 7, July 28.
Asopia fimbrialis *W.-V.* .... July 7.
Cataclysta opulentalis *Led* .... June 21.
Nomophila noctuella *Hübn* .... Aug. 24, Sept. 24.
Scoparia centuriella *W.-V.* .... June 23.

TORTRICIDÆ.

Nolophana malana (*Fitch*) .... June 25.
Nolophana (Asisyra) Zelleri *Grote* .... June 5.

TINEIDÆ.

Cryptolechia Schlagæri *Zeller* * .... June 12.
Depressaria heracliana *De Geer* .... Aug. 12, Aug. 19.
Crambus chalybirostris *Zeller* .... Aug. 26.
Crambus girardellus *Clem* .... July 21.
Crambus laqueatellus *Clem* .... June 24.
Pterophorus tenuidactylus *Fitch* .... June 30.
Pterophorus marginidactylus *Fitch* .... June 30.

* Seven examples of this beautiful moth were collected at this time. As it sits at rest on the upper surface of a leaf, its peculiar form and singular combination of colors render it almost undistinguishable from a deposit of bird excrement. A simulation so nearly perfect cannot fail of giving it, while in repose, almost entire immunity from its enemies.

X. DESCRIPTION OF A CONVENIENT INSECT CASE.

[From the Fifth Annual Report on the Insects of Missouri, 1873.]

For beauty and security, and the perfect display of the larger Lepidoptera, I have seen nothing superior to a box used by Mr. Lintner, of Albany, N. Y. It is a frame made in the form of a folio volume, with glass set in for sides, and bound in an ordinary book cover. The insects are pinned on pieces of cork, fastened to the inside of one of the glass plates; and the boxes may be set on end, in library shape, like ordinary books. For the benefit of those who wish to make small collections of showy insects, I give Mr. Lintner's method, of which he has been kind enough to furnish me the following description: *

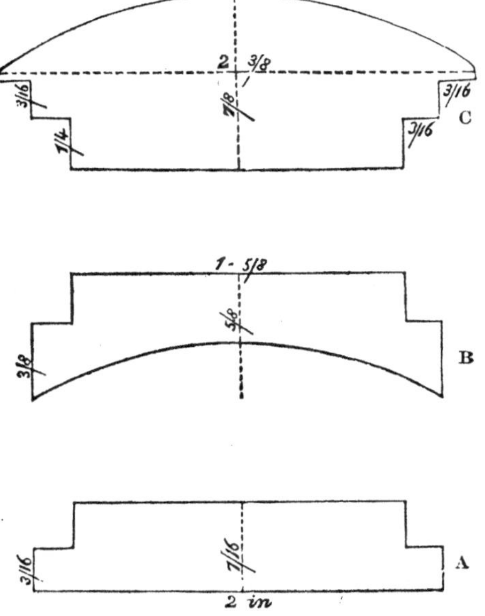

Figures A, B and C represent, in section, the frame-work of the volume; A showing the ends, B the front, and C the back. The material can be prepared in long strips of some soft wood, by a cabinet-maker (if the collector has the necessary skill and leisure for framing it), at a cost of sixty cents a frame, if a number sufficient for a dozen boxes be ordered. Or, if it be preferred to order them made, the cost should not exceed eighty cents each. Before being placed in the hands of the binder, the mitering should be carefully examined, and any defect in fitting remedied, so that the glass, when placed in position, may have accurate bearings on all the sides. The interior

* The description was originally intended for one of the State Museum Reports, but, at the request of Mr. Riley, State Entomologist of Missouri, it was sent to him for publication in his Fifth Report, and by his permission it is here republished.

of the frame is covered with tin-foil, made as smooth as possible before application, to be applied with thoroughly-boiled flour paste (in which a small proportion of arsenic may be mixed), and rubbed smoothly down to the removal of the blisters which are apt to appear. The tin-foil can be purchased, by weight, at druggists, and the sheets marked off and cut by a rule in strips of proper width, allowing for a trifle of overlapping on the sides. Its cost per case is merely nominal.

First quality single-thick glass for sides must be selected, wholly free from rust, air-bubbles, veins or any blemish. Such glass can be purchased at fifteen cents a pane. The lower glass, after thorough cleaning, especially of its inner surface, with an alkaline wash, and a final polishing with slightly wetted, blank printing paper, is to be firmly secured in its place by a proper number of tin points; the upper glass is but temporarily fastened. The binder must be directed to cover the exposed sides of the frame with "combed" paper, bringing it over the border of the permanent lower glass and beneath the removable upper glass.

The covers of the volume are of heavy binder's board (No. 18), neatly lined within with glazed white paper. On the inside of one of the lids may be attached, by its corners, a sheet with the numbers and names of the species contained in the case, or these may be placed on the pin bearing the insect. If bound in best quality of imitation morocco, with cloth sides, lettered and gilded on the back, the cost (for a dozen cases) need not exceed $1 each. If in half Turkey-morocco, it will be $1.50.

The lettering and ornamentation of the back will vary with the taste of the individual. The family designations may be permanently lettered, or they may be pasted on the back, on a slip of paper or gum-label, as are the generic names, thus permitting the change of the contents of a case at any time, if desired.

The bits of cork to which the insects are to be pinned are cut in quarter-inch squares from sheet-cork of one-fourth of an inch in thickness. If the trouble be taken to trim off the corners, giving them an octagonal form, their appearance will be materially improved, and much less care would be required in adjusting them symmetrically on the glass.

The cement usually recommended for attaching the cork to the glass is composed of equal parts of white wax and resin. My experience with this has not been favorable, for, after the lapse of a few years, I have invariably been subjected to the serious annoyance of being compelled to remove the entire contents of the case, clean the glass and replace the corks with new cement. From some cause, inexplicable to me, a gradual separation takes place of the cork with its cement from the glass, first appearing at the angles of the cork, and its progress indicated by an increasing number of iridescent rings which form within until the center is reached, when, if not previously detached, the insect falls with the cork, usually to its injury and that of others beneath it.

A number of years ago I happened to employ, in attaching a single piece of cork in one of my cases, a cement originally made for other purposes, consisting of six parts of resin, one of wax and one of venetian red. Several years thereafter, my attention was drawn to this piece, by finding it as firmly united as when at first applied, and at the present time (after the lapse of twelve years) it is without the slightest indication of separation. Acting upon this hint, I have of late used this cement in the restoration of a number of my cases, and with the most satisfactory results. It is important that the cement, when used, should be kept heated (by a spirit lamp or gas flame) to as high a degree as it will bear without burning. An amount sufficient to cover the bottom of the flat metal vessel containing it to the depth of an eighth of an inch will suffice, and prevent the cork from taking up more than its requisite quantity. It should occasionally be stirred to prevent the precipitation of its heavier portions. The cork may be conveniently dipped by the aid of a needle inserted in a handle, when, as quickly as possible, it should be transferred to the glass, for the degree of adhesion seems to depend upon the degree of fluidity of the cement. From some experiments made by me, after the corks had been attached as above, in heating the entire glass to such a degree as thoroughly to melt the cement until it spreads outward from beneath the weight of the cork, and then permitted to cool—the glass meanwhile held horizontally that the corks might not be displaced—the results appear to indicate that the above cement, applied in this manner on glass properly cleaned, will prove a *permanent* one; it is scarcely necessary to state that this method is not available where the glass has been bound as above.

Preparatory to corking the glass for the specimens assigned to it, the spaces required for them are to be ascertained by arranging them in order on a cork surface or soft-wood board. On a sheet of paper of the size of the glass, perpendicular lines, of the number of the rows and at their proper distances, are to be drawn, and cross lines equal in number to the insects contained in the rows. The distances of these lines will be uniform, unless smaller specimens are to occupy some portion of the case, when they may be graduated to the required proportion. With the sheet ruled in this manner and placed beneath the glass, the points where the corks are to be applied are indicated by the intersection of the lines. The sheet, marked with the family of the insects for which it was used, and, with the numbers designating its divisions, may be laid aside for future use in the preparation of other cases for which it may be suitable. In a series of unbound cases in my collection, in which the glasses measure 11 × 14½ inches, I have used for my Lepidoptera and laid aside the following scales, the citation of which will also serve to show the capacity of the cases: 3 × 8, Catocalas; 2 × 7 and 3 × 9, Sphingidæ; 4 × 11 to 4 × 14, Bombycidæ; 5 × 13 to 6 × 16, Noctuidæ; 8 × 16 and 8 × 20, Lycænidæ and Tortricidæ.

The unbound cases above referred to are inexpensive frames, made by myself, of quarter-inch white wood or pine, the corners mitred, glued and nailed with three-quarter inch brads, lined within with white paper (better with tin-foil), and covered without with stout manilla paper. The glasses are cut of the exact size of the frame, and, when placed in position thereon, are appressed closely to it by laying upon them, near each corner, a heavy weight (the weights used by me are four-inch granite cubes, weighing nearly seven pounds each, which are sufficient to overcome the curvature of the glass); strips of an enameled green paper cut to the width of one inch, are pasted over the edges of the glass, extending a little beyond the thickness of the frame, and brought downward over the sides of the frame. On its front, two gum-labels, indicating the insects inclosed, are placed at heights respectively of seven and twelve inches, when, if all has been neatly done, the cases present a tasteful appearance upon a shelf. When there is reason to believe that the case will need to be opened for the change or addition of specimens, it will be found convenient to employ, for the fastening of the upper left-hind side of the upper glass, paper lined with a thin muslin, to serve as a hinge when the other sides have been cut.

Should it become desirable to bind these cases, outside frames may be constructed after the plans above given, with the omission of the inside quarter-inch (the equivalent of these frames), in which these may be placed and held in position by two or three screws inserted in their sides.

INDEX

TO

ENTOMOLOGICAL CONTRIBUTIONS.

ENTOMOLOGICAL CONTRIBUTIONS—NO. IV.

BY J. A. LINTNER.

I. ON MERMIS ACUMINATA RUDOLPHI.

A PARASITE OF THE LARVA OF CARPOCAPSA POMONELLA.

In April, 1875, an example of this entozoan was received at the State Museum of Natural History, from the Hon. Joseph D. Friend, of Middletown, Orange Co., N. Y., with the request for information in relation to it. Its examination, in connection with the circumstances under which it was found, proved it to be of so much general interest, that the following reply was made to him, through the columns of the *Albany Evening Times*, of April 12th:

DEAR SIR: The "new apple-worm" which you submitted to us a few days since as having been found in a few instances in Orange county, N. Y., during the present season, coiled about the heart of the apple, bore so strong a resemblance to the Gordius, or "hair-worm" as it is ordinarily called, that on its presentation by you I unhesitatingly (but erroneously) referred it to that genus. The brief notice of the new worm, which has appeared in some of our journals, has excited no little apprehension, lest so diminutive a creature, the diameter of a horse-hair, five inches in length, and of the color of the apple pulp—might be introduced in its living state unnoticed into the stomach, and continue its existence as an internal parasite within some of the organs of the body.

The Gordius is not an uncommon animal. Under its popular name of hair-worm or hair-snake (either appellation being given to it, as the particular species or sex noticed approaches its minimum length of about four inches, or its maximum length of twenty-six inches), it has probably come under the observation of most persons living in the country. They are occasionally met with in turning up damp soil, where little groups of several individuals are sometimes found knotted together, occupying a cell in the ground. More frequently they occur in standing water by the roadside and in wagon ruts, in drinking troughs, in old wells, and in small pools on the banks of creeks or rivers. In color, shape and size, they bear so strong a resemblance to a hair from the mane or tail of a horse, as partially to excuse the very general superstition which prevails in relation to them, that they have actually originated from such hairs, and that if a horse-hair be placed in a barrel of rain-water, it will in due time be converted into a living hair-snake. Of course, the more intelligent portion of the community need not be told of the utter impossibility of such a transformation, by which a body devoid of animal life can become a living being. It is a law of nature, without exception, that all animal existence, the lowest as well as the highest, commences with an egg.

The Gordius belongs to that division of the animal kingdom known as the Entozoa, embracing animals which pass a portion of their existence at least, within the bodies of other animals. Our common grasshoppers are frequently infested with Gordii, and I once was so fortunate as to discover an individual in the act of emerging from the head of a grasshopper. They have also been found in crickets, in some of the butterflies, in various species of beetles, in aquatic larvæ of insects as of caddis-worms and May-flies, in the honey-bee, etc.

Much of the history of the Gordius remains unknown. Dr. Leidy has observed the operation of its laying its eggs, in a long, thread-like string, broken asunder in several places, but aggregating the extraordinary length of ninety-one inches, — more than ten times the length of the worm extruding it. The entire number of eggs contained in this oviposition was, by a careful calculation, computed at nearly seven millions (6,624,-800). The young Gordius, microscopic in size, and very unlike its parent in form, has been observed, and its entrance followed into the body of water larvæ, through the thin integuments at the joints of the legs. Their subsequent development during their condition of internal parasites is unwritten.

Upon submitting the specimen with which you favored us, to microscopic examination, it was seen to differ in its internal structure and in its more pointed extremities, from the species of the genus GORDIUS above referred to. As these forms have been so little studied and so little is known in relation to them, it was deemed proper before naming it for you in accordance with your request, to submit it to the eminent authority, Dr. Leidy of Philadelphia, who has given special study to the Entozoa. Having examined it, he returns the following very interesting information: "The worm is a species of MERMIS, a parasite of the larva of *Carpocapsa pomonella*, or apple-worm moth, which accounts for its presence in the apple itself. A similar specimen was referred to me a short time since, for an account of which see the next or forthcoming number of the Proceedings of the Academy of Natural Sciences [Philadelphia]."

MERMIS is a genus closely allied to GORDIUS. Leidy states that he has frequently seen specimens of it, which he calls "the white hair-worm," within insects—in one instance crawling out of a Carolina grasshopper which was struggling in a ditch. Siebold describes *Mermis albicans* of Europe (two to five inches long, of a whitish color) as parasitic in the drones of the honey-bee.

This new phase of parasitism of the Mermis upon a caterpillar living within the apple, at its core, and often in its younger stage within the seeds, is so remarkable and interesting an announcement, that we shall anxiously await the promised paper, for the explanation of much that seems mysterious to us. In what manner, and at what time, does the Mermis effect its entrance in the body of the Carpocapsa apple-worm? The eggs of the apple-moth are deposited on the blossom end of the apple, where the skin is the thinnest, at various periods during the summer months. Hatching within a week, the young caterpillar passes directly into the apple, eating its channel as it proceeds toward the core. Here it remains until it has completed its growth, when emerging from the apple it crawls down the branches, or drops itself to the ground by its thread, to seek some safe place of shelter in which to construct its cocoon. At this time and even during its subsequent hybernation in its larval form within its cocoon, it is exposed to parasitic attack; but this cannot be the period of the entrance of the Mermis, for its presence within the apple indicates its previous existence in the worm. It could not have taken possession of the worm while the apple was attached to the tree, for its structure would not admit of its ascending the trunk and branches of the tree. Should a wonderful instinct lead it to seek its prey through the closely packed excremental matter filling the worm-holes of the "wind-falls" lying on the ground, then the worm in its exit from the fruit, usually very soon after the apple falls, would carry its guest away with it, instead of leaving it behind to excite our wonder and perhaps alarm several months thereafter. Damp cellars would seem to be an appropriate habitat for the Mermis, but its abode within the stored apples would naturally be more brief in this case than in the preceding, the latest worms at this time having attained or being near their maturity. Without the labor of penetrating the apple, its prey could much more conveniently be found in the unchanged larvæ hidden often in immense numbers between the boards of the apple-bins or beneath the barrel hoops within a cocoon too slight to offer any material resistance to the entrance of so thread-like an organism.

The interesting inquiry also arises, is this the first state of parasitism in the cycle of the Mermis' history, or, as would seem more probable, has it already undergone its first transformation from its larval to its perfect form within the body of some other animal, totally unlike our apple-worm? It is to be hoped that Dr. Leidy's observations have enabled him satisfactorily to solve these several enigmas.

As an aid to the development of the history of the interesting animal, it will be of service if as full statements as possible be obtained of the conditions under which it has occurred in each instance in Orange county the present season—the first knowledge we have of its presence in the apple, or any other fruit. Was it found invariably in worm-eaten apples; and if so, had the fruit been much or little eaten? Were any remains of the Carpocapsa-worm noticed as associated with it? Were any living apple-worms seen in the apples eaten during the winter? Where and in what manner were the apples stored in which the Mermis was present? Can it be ascertained if the infested apples were hand-picked or "wind-falls?"

Any replies to the above, or such additional information as may promise to be of service, will be thankfully received at the State Museum.

The great importance of a knowledge of the history of the internal parasites, and the interest connected with them, from the fearful results following the introduction of some of the class within the human body, as for example, the *Trichina spiralis*, is my apology for replying to your inquiries at some length. The detection of the Mermis in the apple, in a few instances, recently, need not, we think, occasion alarm. It is possible that in eating an uncooked apple without the proper mastication, a living Mermis might be introduced into the stomach, for of the Gordius (a closely allied genus as above stated) Leidy says, "It is perhaps the hardest or most resistant to the feel of any of the order, and it is tough and elastic. It is very tenacious of life, and when cut into several pieces will continue to live and move for some time afterward." But should it escape the ordeal of the teeth and pass uninjured into the stomach, there is reason to believe that the action of the gastric juice and other conditions to which it would be there subjected, would deprive it of life before it could pass into the intestines or penetrate the integuments of the body.

Subsequent to the above communication, I addressed another to Mr. Friend, asking for such additional information as he might possess, or be able to obtain in relation to the interesting parasite. In reply he sent me a letter which he had received from Mr. James T. King, of Middletown, stating under date of April 16, 1875, as follows:

I regret to be able to give you but a very meagre report on the apple-worm. Two or three years have passed since the party who gave me the scanty information below, found the first specimen left with me.

The apple in which it was found was a fine looking fall pippin, appeared to be sound, was blown from the tree during a violent wind-storm at night, and picked up the next day. The worm was coiled up in the fleshy part of the apple, about midway between the skin and the core. It was white, or of the same color as the pulp, and when uncoiled measured seven inches in length, and about one-fiftieth of an inch in diameter. It remained quite active for several hours, and was then placed in alcohol.

The specimen delivered to you was found this winter, but the person who discovered it and brought it to me could give me no definite information in regard to it. * * * *

Accompanying the above, Mr. Friend also writes: —

I send you herewith a letter which I have received from Mr. James T. King, Druggist and Chemist of Middletown. Mr. King forgot to state in his letter, that specimens of the worm have been found in stewed apples, by a family residing in Middletown, somewhat broken up, but in no other way sensibly affected by the heat to which they had been subjected.

Now, that public attention has been so widely drawn to this subject by the publication, in so many of the newspapers of the State, of your recent letter, I think it safe to predict that during the coming summer and autumn much more satisfactory information will be gathered respecting this curious and hitherto unknown parasite.

The above prediction was not verified, and it may therefore be inferred that the parasite is not increasing rapidly, and that its presence in fruit cannot be expected to be of frequent occurrence.

But a single instance of its detection has since been brought to my notice. Prof. J. H. Comstock, of the Department of Entomology, in Cornell University, has informed me that he has in his possession an example of *M. acuminata*, taken in January 29th, of the present year [1876], from a worm-eaten "Seek-no-further." The apple was grown in the vicinity of Ithaca, and had been stored in a bin in a cellar. The fruit had been shaken from the tree, but not allowed, it is believed, to lie upon the ground for any length of time.

Prof. Comstock proposes, during the coming season, to communicate with a large number of pomologists with a view of learning of the distribution, abundance, and such additional facts as may add to our knowledge of this interesting creature.

The first published notice of the detection of the Mermis in fruit (unknown to me at the time of my communication to Mr. Friend), appeared in the Gardener's Monthly, for May, 1872, a periodical published in New-York. A reference to this notice is made in Prof. Riley's Fifth Report on the Insects of Missouri, p. 49, in connection with descriptions of two additional parasites (Hymenoptera) of the apple-worm (*Carpocapsa pomonella* larva), discovered by Prof. Riley.

Mr. P. H. Foster, of Babylon, N. Y., communicates to the Gardener's Monthly, as follows:

I discovered a parasite on the above worm [*Carpocapsa pomonella*] in the year 1869. I sent a specimen to Mr. B. D. Walsh, of Illinois, which he calls a species of hair-snake (Gordius). I also found one last summer imbedded in the apple-worm in the center of a large pear. This Gordius is white. * * * * Prof. Leidy, of Philadelphia * * * describes several, and mentions one which he calls the white hair-worm (Mermis), which is the only one that corresponds with the specimen I have reference to.

Prof. Riley informs me that subsequent to his reference, above cited, he had obtained two specimens of Mermis from Carpocapsa larvæ found in fruit, and two other examples from larvæ taken from beneath bandages placed around the trunks of apple trees, to serve as a place of retreat for the larvæ during their transformations, from which they could be taken and destroyed. He had also taken a similar specimen from the posterior part of the brain of an owl.

The specimen taken by Prof. Riley from the brain of the owl may be presumed to be the same or closely allied to those described by Prof. Wyman,* which he has found so common, in the brain of the snake-bird or water-turkey, in Florida (in seventeen out of nineteen specimens shot), that their presence might be presumed to be the normal condition of the bird. Prof. Wyman finds them to correspond so closely to the *Eustrongylus papillosus* of Diesing, that he thought they might prove to be identical. In every instance they were coiled up on the back of the cerebellum, in numbers varying from two to eight. Figures of them are given, showing the male and female, their position on the cerebellum, enlarged views of their extremities, and the development, within the oviduct, of the egg to the free young embryo. Nothing is known of their transfer from the oviduct, through some other animal probably, to the brain of another bird.

In a subsequent communication to the *American Naturalist* (Vol. VI, p. 560), Prof. Wyman presents very interesting additional observations upon these parasites, and, upon the bird in which they have their habitat, and designates them as *Filaria anhingæ*.

The communication of Prof. Leidy to the Philadelphia Academy, to which reference has been made, in which he describes *Mermis acuminata*, is reported in the Proceedings of the Academy of Natural Science, of Philadelphia, for February, 1875, as follows:

Professor Leidy remarked that Mr. Thomas Meehan had submitted to his examination some worms which had been

---

**Proceedings of the Boston Society of Natural History*, October 7, 1868. See, also, *American Naturalist*, vol. III, p. 41, 1870.

found in an apple. They consisted of one entire individual and the anterior half of a second, and apparently pertain to the *Mermis acuminata*, a long thread-worm which has been discovered infesting the larvæ of many insects. Among others, it is parasitic in the larvæ of the fruit-moth of the apple, which readily accounts for its presence in the fruit. Twenty-five years ago (Proc. 1850, p. 117) he had described a worm, belonging to the collection of the Academy, and labeled as having been obtained from a child's mouth, which was evidently the same species. It having been in a child's mouth is probably to be explained by supposing that the child had eaten an infected apple.

The characters of the present specimens of the worm, both females, are as follows: Body filiform, pale fuscous, narrower anteriorly. Head conical, truncate, with the mouth simple and unarmed. Caudal extremity thicker than the head, obtusely rounded, and furnished with a minute spur-like process. Length, five inches eight lines; cephalic end at mouth $\frac{1}{12}$ mm.; a short distance below $\frac{1}{5}$ mm.; middle of body $\frac{3}{8}$ mm.; near caudal end $\frac{1}{4}$ mm.; mucro $\frac{1}{12}$ mm. long, $\frac{1}{80}$ mm. thick.

We transcribe for comparison the description of the example obtained from a child's mouth, to which reference is made above, together with the accompanying interesting remarks, which show the apprehension entertained by Dr. Leidy, at that time, at least, of serious results which might follow the introduction into the human system through the mouth, of the Mermis and allied species of Entozoa.

Filaria hominis oris.—Body white, opaque, linear, thread-like; mouth round, simple, posterior extremity obtuse, furnished with a short, curved, epidermal hooklet $\frac{1}{500}$ inch in length, by $\frac{1}{2000}$ inch in diameter at base. Length, five inches seven lines, greatest breadth $\frac{1}{66}$ inch; breadth at mouth $\frac{1}{250}$ inch; at posterior extremity $\frac{1}{80}$ inch.

Remarks.—The description is taken from a single specimen preserved in alcohol, in the collection of the Academy, labeled "obtained from the mouth of a child."

Is it a young individual, or perhaps a male of the *Filaria Mendinensis*, or Guinea-worm? The latter, as is well known, infests the human body, often growing to an enormous length, several yards or more, in the inter-tropics of Asia and Africa. It is frequently brought in the body of negro slaves from Africa to America, where no entozoon of the kind has ever been noticed to be parasitic in man, as an indigenous production. From some late observations on the course of life of entozoa, helminthologists have been led to suspect that most and probably all entozoa pass different stages of their existence in different animals. If such be the fact, may not the

Filaria Mendinensis owe its introduction into the human body from the custom which prevails in those countries where the worm is found, of using insect food. Insects are well known to be infested with Filariæ, probably more than any other class of animals. In Egypt, Arabia, etc., the locust is eaten; in Guinea, etc., the larger coleoptera, in the raw state; and in this condition Filariæ may often be swallowed, and reach a higher development of their existence in the human body.

In the same paper, Dr. Leidy describes two additional species of similar Entozoa, — the one (*Filaria canis cordis*) as indicated by the name given it, taken from the heart of a dog. The two examples were white, opaque, linear, nearly uniform throughout, posteriorly subulate, pointed; mouth simple, round. Length ten to ten and a half inches; greatest breadth $\frac{2}{5}$ of a line, anteriorly $\frac{1}{5}$ of a line. The other species (*Filaria boæ constrictoris*), was found in the areolar tissue, in an irregular or tortuous position, between the muscles of the ribs and the integument of a boa constrictor. This was a more robust form, ten inches in length by $\frac{4}{5}$ of a line broad, of a white color and longitudinally striated.

Dr. Leidy has also recently found* the common house-fly (*Musca domestica*, it may be presumed) to be infested with a thread-worm, of about a line in length, which takes up its abode in the proboscis of the fly. From one to three worms occurred in about one fly in five. The parasite was first discovered in the house-fly in India, by Carter, who described it as *Filaria muscæ*, and suggested that it might be the source of the Guinea-worm in man.

In view of these unwelcome suggestions, that it may be a necessary section of the life history of several of these entozoa that they should be introduced into the human body through the food of which we partake, there to undergo their final development, it is much to be regretted that the entire history of all the species to which man is exposed is not yet known. Much attention has been paid to them, but their study has proved a difficult one. A monograph of the Hair-worms, by M. Villet, has recently been published, of which we know nothing beyond the information given in the American Naturalist for December, 1874, to the effect that it was then being published in the "Archives de Zoölogie Expérimentale." The author had found the larvæ encysted in the larvæ of CHIRONOMUS (be-

---

* *American Naturalist*, Vol. IX, p. 247. 1875.

longing to the Diptera and aquatic in their habits), and afterward, in September, in the mucous lining of the intestines of fishes, thereby, in conjunction with the previous labors of Grube, Leidy, and Meissner, clearing up their metamorphoses. The larvæ are tadpole-shaped. The habits of GORDIUS seemed quite distinct from MERMIS found living in insects.

Dr. Speyer communicates to me the information that the occurrence of the Gordiacæa in the body of insects has frequently been observed in Europe, notices of which may be found in several interesting communications from Von Siebold, in the *Stettiner Entomologische Zeitung* for the years 1842, -43, -48, -50 and -54. The species which infests most frequently the Lepidoptera is *Mermis albicans*. They are found in both the larvæ and the perfect insects, oftener in the former. They occur in the larvæ which feed on tall trees, as well as those which live on plants and low shrubs. Wet seasons seem to be more productive of the parasitism, and Dr. Speyer recalls, a number of years ago, during an unusually wet season, his having met with several of such instances. From an example of *Hadena adusta* he had a Mermis emerge, of the length of eight and a half inches, and another from *Hesperia lineola* after it had been pinned. Prof. Von Siebold suggests that a heavy dew may moisten the trunks of trees sufficiently to enable the Mermis to ascend them.

The Mermis parasite (species not stated by Dr. Speyer) also infests the *Carpocapsa pomonella* larvæ in Europe.

These pages have been for some months in type. In the meantime, Dr. Packard's *Report on the Rocky Mountain Locust*, in Hayden's Geolog. and Geograph. Survey of Colorado, for 1875, has been received. In it (pp. 663–667) he gives an account of the several species of GORDIUS and MERMIS occurring in the United States, transcribing from the paper above cited (see full title below *), the descriptions of the Gordii and also the history of *Gordius aquaticus* as given by Villot, carrying it beyond its encysted state in the intestines of fishes, to its free and aquatic state the following spring. This stage it attains by boring through the cyst into the intestinal cavity of the fish, thence passing with the fæces into the water, where material changes take place before it assumes the active stage. For a more full account of these several transformations, see p. 665 loc. cit.

---

*Monographie des Dragonneux (Genre GORDIUS Dujardin), par A. Villot. (Archives de Zoölogie expérimentale et générale, tome 3, No. 1, 2. 1874. Paris.)

II. THE NEW CARPET-BUG — ANTHRENUS SCROPHULARIÆ.

During the summer of 1874, notices appeared in various newspapers of the ravages of a carpet-bug, quite different in its appearance and in the character of its depredations from the well-known carpet-moth, *Tinea tapetzella*, which for so long a time had been the only known insect depredator on our carpets. Its *habitat* was stated to be beneath the borders of carpets where nailed to the floor, eating in those portions numerous holes of an inch or more in diameter. Occasionally it located itself in the crevices left by the joinings of the floor, following which, entire breadths of carpet would be cut across as by scissors. In several instances carpets had been destroyed — new ones as readily as older — and it was questioned whether their use could be continued, in view of a prospective increase of the alarming ravages.

The insect was new to every one, and no one could form a rational conjecture as to what order of the Insecta it belonged. It was described as a small ovate object, about one-eighth of an inch in length, thickly clothed with numerous short bristle-like hairs, and terminating in a pencil of these, forming a tail. It was exceedingly active in its motions, and when disturbed in its concealment would glide away beneath the base-boards or some other convenient crevice so quickly as in most instances to elude capture for its closer inspection. They were found only during the summer months.

In 1876 it was reported in many dwellings in Schenectady, and in the month of July examples of it, for the first time, came under my observation, taken, upon search having been instituted, under the carpets of my residence at Schenectady, where its presence had not been suspected. It was evident, on the first inspection, that it was the larva of a beetle, and in all probability a member of the very destructive family of DERMESTIDÆ, which comprises several of our most injurious depredators on animal substances.

A number of the larvæ were secured and fed upon pieces of carpet in order to rear them. In September they had evidently matured, and had assumed their quiescent pupal state within the skin of the larva, first rent by a split along the back for the escape of the perfect insect. At this stage they presented characters which led me to refer them, in all probability, to the genus ANTHRENUS.

In October, the first perfect insect emerged. Being entirely new to me, they were sent to Dr. LeConte, the distinguished coleopterist of Philadelphia, for determination. He returned answer that they were the *Anthrenus scrophulariæ* — a species well known in Europe for its destructiveness, but now for the first time detected in this country.

Notice of the discovery was communicated by me to the Albany Institute at its meeting of October 17th, 1876, and a report of the same published in the *Albany Argus* of October 21st. Owing to the interest attached to the introduction in our country of another addition to the already formidable list of injurious insects of European origin, the paper, or extracts therefrom, appeared in several of the journals of this and adjoining States. Through the publicity given it, I became informed of the presence of the insect in many localities in New York and other States. Examples of a beetle, believed to conform to the brief description which I had given of *A. scrophulariæ*, and known to possess the like habit of feeding upon carpets, were sent to me by Mr. A. S. Fuller of the *Rural New-Yorker*, for comparison. The species had been in his cabinet for some time, under the name of *Anthrenus lepidus* Le Conte, having received the first examples from Oregon in 1871 or 1872. Later, in 1874, specimens referred by him to the same species were found abundantly in a dwelling in Market street, New York, and thereafter in various parts of the city and neighboring localities. The examples reared by Mr. Fuller from larvæ taken in New York city were clearly identical with *A. scrophulariæ*. Upon informing Dr. Le Conte that examples of this species were in cabinets under the name of *A. lepidus* and requesting an explanation, he wrote me that the latter name had been given by him to a form which he had found on flowers at San Francisco and San Jose in 1850; * that it differed from the *A. scrophulariæ* of Europe in

* *A. lepidus*, breviter ovatus, supra niger, thoracis lateribus albo-squamosis, gutta nigra inclusa, elytrus fasciis tribus angustis suturaque albo-squamosis, macula antica suturali aureo-squamosa ornatis, basi parce albo-squamosis. Long. .11 in San Diego, Cal. — *Proc. Acad. Nat. Sci. Phila.*, 1854, p. 112.

its sutural line being white instead of red; but that in all probability it should only be regarded as a variety of the European species.

Dr. Le Conte suggests that it may have been imported into California from Southern Europe during the Spanish occupation of that country. The eastern invasion of the insect, he believes to have been within a few years through the importation of carpets at New-York.

The accompanying figures, very faithfully drawn by Prof. Riley, represent *A. scrophulariæ* in three of its stages, viz., *a* the larva, *c* the pupa, and *d* the imago or beetle. At *b*, the skin of the larva, after the beetle has emerged from the fissure on the back, is shown. The figures are enlarged — the lines beside them representing the natural size.

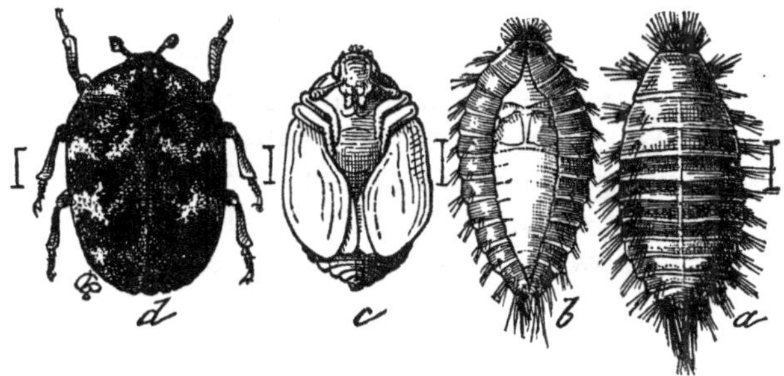

The larva — the form in which it is usually found when pursuing its ravages beneath the carpets — measures, at maturity, about three-sixteenths of an inch in length. A number of hairs radiate from its last segment in nearly a semicircle, but are more thickly clustered in line with the body, forming a tail-like projection almost as long as the body: this terminal pencil of hairs is not shown in its full extent in the figure, doubtless taken from an immature individual. The entire length of the insect, including the pencil of hairs, is, in the largest specimens, nearly three-eighths of an inch. Measured across the body and the lateral hairs, its breadth just equals the length of the body. An ordinary magnifier will show the front part of the body, where no distinct head is to be seen, thickly set with short brown hairs, and a few longer ones. Similar short hairs clothe the body — somewhat longer on the sides, where they tend to form small tufts. Towards the hinder end may be seen on each side three longer tufts (thrice as long) projecting laterally; but these are not always visible, as the insect by the aid of a peculiar muscular arrangement, has the power of folding them out of sight along its sides. The body has the appearance of being banded in two shades of brown — the darker band being the central portion of each ring, and the lighter, the connecting portion of the rings, known as the incisure. By turning it upon its back, the six little legs, of which it makes such good use, can be seen, in vigorous efforts to regain its former position — its struggles while in this condition sometimes producing a series of jumps of about an eighth of an inch in length.

Having attained its full growth, it prepares for its pupal change without the construction of a cocoon or any other provision than merely seeking some convenient retreat. Here it remains in a quiet state, unaltered in external appearance, except somewhat contracted in length, until it has nearly completed its pupation, when the skin is rent along its back, and, through the fissure, the pupa is seen. A few weeks having passed, the pupal skin in its turn is split dorsally, and the brightly colored wing-covers of the beetle are disclosed. Still a few additional days of repose are required for its full development, when the now fully matured beetle crawls from its protective coverings of pupal case and larval skin, and appears in its perfect form — its final stage.

The earliest beetles emerge in the month of October, and continue to make their appearance during the fall, winter and spring months. Soon after their appearance probably, they pair, and the females deposit their eggs for another brood of the carpet-eating larvæ.

The beetle is quite small — smaller than would ordinarily be expected from the size of the larva — being only about one-eighth of an inch long by one-twelfth broad. An average of five examples before me gives, length .125 inch, breadth .085 inch. Its form is almost a perfect ellipse as seen from above ; its back and under surface are quite rounded. When turned upon its back, it often for a few moments counterfeits death, with its legs so closely folded to the surface as scarcely to be seen, and in this state the ordinary observer might be at a loss to know the lower from the upper side.

It is a beautifully marked little insect in its contrasting colors of white, black and scarlet, arranged as follows : The edge of each wing-cover, where they meet on the back, is bordered with red (forming a central red line), with three red projections from it outwardly — one on the middle of the back, and one other toward each end. Take a straight line and divide in four equal parts by three cross lines, and we have nearly the position of these projections. At the extreme tip of the wing-covers is a widening of the bordering line, making almost a fourth projection from it. The first projection, near the head, is connected with a white spot, running upwardly on the middle of the front border of the wing-cover. On the outer border of the wing-covers are three white spots nearly opposite the red projections. The intermediate spaces are black. The segments of the body beneath are covered with pale red scales, and the thoracic region (which bears the legs) with whitish scales.

The above description, although not presented as a scientific one, will suffice for the identification of the beetle when met with.

The detection of this insect adds to our fauna another species of the dreaded genus of ANTHRENUS, which there is reason to fear will equal in its destructive agency the well-known museum pest, the *A. varius* (formerly known as *A. musæorum*), the obtrusive guest of all our collections of natural history, whose ravages it seems impossible fully to guard against, and so exceedingly difficult to control.

It does not confine itself wholly to carpets, but it also infests and injures various articles of wearing apparel, hanging in closets or lain away in drawers. An instance has also been stated, but awaits confirmation, of its preying upon cotton fabrics—a habit not attaching to either the clothes or carpet moths.

It is known to have become almost ubiquitous in a house which had been for some time occupied by it, notwithstanding the persistent efforts made for its extirpation. Its exuviæ were encountered in trunks, boxes, tied up packages, drawers, beneath floor oil-cloths, etc. Late in the season (October) clusters of twelve or more of the full-grown living larvæ were disclosed, to the disgust and dismay of the housekeeper, in turning over a paillasse, the borders of which they had selected as a safe retreat on which to undergo their final transformation.

I have this present year found that a convenient place in which to discover the beetle, is upon the windows of the infested rooms during the day. In the latter part of April examples were taken upon the windows of my residence at Schenectady. After the middle of May, a systematic search instituted for them, gave several examples each day. In the six days from May 17th to 22d, forty-four specimens were taken from the three windows of two upper rooms. Should investigation show that the beetle is drawn to the windows before the deposition of its eggs, their ready capture and destruction at this time will offer an easy method of preventing their increase.

Should this insect continue to increase until its complete naturalization shall make it as common as *A. varius* (a dozen or more of which may sometimes be seen feeding on a single flower), it is difficult to conceive how, under such a visitation, the comfort of carpets can still be indulged in within our homes. Even now, when it has barely commenced its ravages, it is reported as having inflicted very serious pecuniary losses in several instances, where carpets have been entirely ruined; and such terror has its presence imparted, that not a few prudent housekeepers have already abandoned the customary nailing of their carpets to the floor, that frequent examinations may be made during the summer months for the discovery and destruction of the unwelcome guest.

The remarkable invasion of a dwelling in Cold Spring, N. Y., in the summer of 1874, after a twelve months' absence of the family in Europe, was by a larva of Anthrenus (as determined by Dr. Packard), which is now believed to have been this species. According to the statement made, "they took complete possession, from the cellar to the attic, in every nook and crevice of the floors, under matting and carpets, behind pictures, and eating every thing in their way." From this account we may infer an almost incalculable capability of increase if left to itself, and draw the lesson of the absolute necessity of combatting its invasion by every means in our power.

It will unquestionably prove an exceedingly difficult pest to dislodge. The ordinary applications of camphor, pepper, tobacco, turpentine, carbolic acid, etc., are powerless against it. It has even been asserted that it "grows fat" on these substances. An effectual means of destruction, and preventive against new invasions, is still to be discovered. The free use of benzine has been recommended in some of our journals, to be used in the saturation of cotton, with which to fill the joinings of the floors and crevices beneath the base-boards. This is to be done during the winter months, at which time the insect will be occupying these retreats, either in its perfect beetle form, or as eggs deposited for another brood; to either of these the direct application of benzine would be fatal. To some of my correspondents I have suggested the pouring of kerosene oil in the crevices of the floors, and filling of all places of retreat with cotton saturated with the oil. I would regard this as less dangerous in its use than benzine, and equally efficient.

The recommendation recently made in several of our newspapers, of the Persian insect-powder for the destruction of the insect, I believe to be of no value. I have not deemed it worth the trouble of experimenting with it, but I have been told by those who have given it a trial that it has been found to be of no avail whatever.

The insect has not as yet become sufficiently abundant in New York to be found resorting to plants for its food. The variety *Anthrenus lepidus*, which was introduced in California sufficiently long ago to permit its complete naturalization, was discovered there, in numbers, feeding upon some of the Compositæ. The *Anthrenus varius* is often found, in its perfect state, taking its food from the blossoms of different plants in the garden or field. I have met with it abundantly on peonies. It has also been found to frequent the rocket flower, *Hesperis matronalis*, a fragrant and showy perennial.

If the plants known to be attractive to the *A. varius* can be introduced into our houses, and made to flower during the months of April and May, I believe that the carpet-beetles would be drawn to them in preference to windows, perhaps as soon as they emerge from the pupæ.

We are unable to give at the present any precise statement of its distribution. It is known in Oregon, as well as in California. It is believed to be distributed throughout most of the western States, and it is known to occur in various portions of the State of New York.

It is announced as having appeared in considerable force in Syracuse. In Utica it has inflicted serious damage in many dwellings. It has occurred at Buffalo, but not in such numbers as to have originated the name sometimes applied to it—the *Buffalo bug*—a name given to it on the Pacific coast probably, from a fancied resemblance to that animal. Its presence has also been detected in Albany, but no serious ravages have been reported. It has occasioned much alarm in several places in the State of New Jersey. Without doubt it is committing its depredations in many localities where its work is ascribed to the carpet-moth, than which it is a far more pernicious insect.

A lady to whom I was relating the destructive capabilities of the new pest, congratulated herself that her carpets were free from it. The following morning her husband brought to me a beetle which he had taken from his face during the night, which proved to be the creature that I had described to her the previous evening—the abundant presence of which in her home, she had not suspected.

From the serious nature of its depredations as above referred to but in part, the secrecy with which it conducts them, the extreme difficulty with any known appliance of eradicating it—it becomes very important, as a preventive against its alarming increase, that it should, from the outset, be combatted by all the means known to be efficacious against its allied forms, or which may give promise of success as against a new foe.

It may be interesting, in connection with the above notice of this last importation, to recall the fact that nearly all of our most injurious insects have been introduced from Europe. Of a long catalogue given by Professor Riley, in one of his valuable reports, a few may be mentioned here:

The Hessian-fly (*Cecidomyia destructor*), the wheat-midge (*Diplosis tritici*), the cheese-maggot (*Piophila casei*), the house-fly (*Musca domestica*), the currant-worm (*Nematus ventricosus*), oyster-shell bark-louse (*Aspidiotus conchiformis*), several species of plant-lice (*Aphides*), the cockroach (*Blatta orientalis*), the croton-bug (*Ectobia germanica*), the meal-worm (*Tenebrio molitor*), the grain-weevil (*Sitophilus granarius*), the bee-moth (*Galleria cereana*), the codling-moth of the apple (*Carpocapsa pomonella*), the cabbage-moth (*Plutella cruciferarum*), the carpet-moth (*Tinea tapetzella*),* the clothes-moth (*Tinea vestianella*), the fur-moth (*Tinea pelionella*),* the currant borer (*Ægeria tipuliformis*), and within the few past years, the asparagus-beetle (*Crioceris asparagi*), and the well-known destructive cabbage-butterfly (*Pieris

*Mr. V. T. Chambers finds differences in these two species from the European ones (*Canadian Entomologist*, 7, pp. 124, 125).

rapæ). All of these, and the formidable list might be greatly extended, we have received from Europe, while very few of our native insect pests have been sent in return. Should our late exportation of the Colorado potato-beetle (*Doryophora decemlineata*), prove as injurious in Europe as in this country, which there is much reason to doubt, we shall still be very far from having made a commensurate return. While the few American species which have been introduced in Great Britain and on the continent have not spread to any great extent, in almost every instance where injurious insects have been brought thence to this country, their number and their ravages have been greatly increased. Thus, while the recent advent of the *Anthrenus scrophulariæ* has brought consternation in many of our homes, we have been unable to find any record of its preying upon carpets, or other woolens, in the Old World, where it has been so long known. Even special inquiry made by me of one of the leading Entomologists of Europe, has failed to elicit any such information. It is said there to infest dried meats and similar substances. Perhaps its fondness for carpets is a new taste which its transportation hither has developed.

III. ISOSOMA VITIS SAUNDERS.

THE GRAPE-SEED FLY.

During the early part of September of the present year (1876), a cluster of grapes, in very bad condition, which had been received at the office of the *Country Gentleman*, of Albany, was submitted to my examination, with a request for information as to the cause of the injury or disease. It was accompanied with the following letter from the gentleman in whose garden, at Plainfield, N. J., the grapes had grown:

EDS. COUNTRY GENTLEMAN.—I send you by mail to-day a single bunch of Walter grape, to ask if you will tell me in the columns of your paper, what is the matter with my grapes. They have been affected in same manner since the first year they bore (now the third year of bearing), but never so badly as this year. The disease attacks all kinds (I have about a dozen), and in different parts of my garden. It shows itself about the time of ripening, and this year promises to destroy the whole crop before they are fairly ready to pick. If this bunch carries well, you will probably notice a berry or two just touched, and you will see that the puncture is very small. Dr. Hexamer thinks it may be sun-burn, but I am sure this is not the trouble, for this very bunch which I send you was cut from underneath heavy foliage, and bunches heavily shaded are just as badly affected as others more exposed. I have watched for bees and wasps, but find scarcely any. Occasionally a single bee will be found sucking the juice of a berry already broken. Neither do I see many birds on the vines, though we have English sparrows in the neighborhood. The vines are very thrifty, and are taken as good care of as I know how to do.

H. R. M.

The nature of the difficulty was so singularly concealed that it was not readily detected by me. A critical examination, however, brought to view the hidden cause, and enabled me to return the following answer:

The injury which threatens to destroy an entire crop of grapes, as above narrated, proceeds from an insect pest which promises to prove very detrimental to the grape-growing interest of our country, and which, therefore, should be promptly met and circumvented by all the means in our power.

The bunch of grapes, as received, contained a large number of shriveled berries, upon which, as also upon nearly every one of the perfect ones, could be seen with the naked eye, a small round dot, in the center of which an elevated roughened surface was visible with a magnifier. The dot marks the spot where a very minute four-winged fly had punctured the skin and deposited its egg. The egg hatching, the larva passes through the pulp into one of the seeds, upon the kernel of which it feeds, and within the empty case undergoes its transformation to its pupal state, having previously provided for the escape of the perfect fly by gnawing an aperture of sufficient size in the seed.

For the detection of this insect, for our knowledge of its habits and transformations, and for its description, we are indebted to Mr. W. Saunders, the able editor of the *Canadian Entomologist*, who first observed the insect, in Canada, in the fall of 1868. It was at first believed by him to be the larva of a curculio, but subsequently was correctly referred by Prof. Riley to the hymenopterous genus ISOSOMA. In the *Canadian Entomologist* for November, 1869, it is described by Mr. Saunders as *Isosoma vitis*. The fly is quite small, being but about one-sixth of an inch in spread of wings ; its head, thorax and abdomen are black, the wings clear and iridescent, and the legs brown and black. The species is interesting from its belonging to the same genus with the destructive joint-worm fly, the *Isosoma hordei* (Harris), which has proved so very destructive to the wheat, rye and barley crops ; and, perhaps, even more interesting from a remarkable difference in the sexes, pointed out by the late Mr. Walker of the British Museum, "one of them representing the carnivorous EURYTOMA, and the other the herbivorous ISOSOMA, and thus one species figuratively combines the diminishers of vegetation and the controllers of such diminution."

In an account of this insect Mr. Saunders says: "On the 20th of August, 1868, we observed that many of the berries in the bunches of a Clinton vine, under our care, were shriveling up. On opening the grapes, we noticed that most of the smaller berries — that is, those which had shriveled earliest— contained only one seed, and that of an unusually large size ;

but some of the larger withered grapes contained two seeds, each having a dark spot upon its surface. On cutting the seeds carefully open, the kernel was found almost entirely consumed, and the cavity occupied by a small, milk-white, footless grub, with a pair of brown hooked mandibles, a smooth and glossy skin, with a few very fine short white hairs. When at rest, it was nearly oval in form, but when in motion, its body became elongated, varying in length from one-fifteenth to one-twelfth of an inch.''

Mr. Saunders found the larvæ unchanged within the seeds, and quite active, in an examination made in the month of February. Early in July they were still soft, but motionless, and they may have been in the pupa state. On the 9th of August, a number of the perfect insects, dead, were found when the contents of the bottle containing the seeds were turned out upon a piece of white paper. They had probably effected their escape during the last half of July.

It is possible that this insect may not prove so destructive as it threatens to do, by spreading from garden to garden, and throughout our vineyards. If it were left unmolested, under circumstances continuing to favor its increase, it is capable within a few years of compelling the abandonment of the culture of the grape in our country. Hidden within the seeds, it could readily be distributed in the transportation of the grapes to distant markets, through the several States of the Union. This is the first instance that we have heard of its appearance within the United States.

Encouraged by the recollection that at different times in the past, when a destructive insect pest has threatened to pursue its ravages without the probability of its arrest by human agency, some kindly parasite or climatic condition has come to our aid — so we shall hope that in the present instance the little *Isosoma vitis* will not find conditions more favorable to its existence in its new habitat, or others that it may select, than it enjoyed in Canada, where its spread seems to have been, for the time, at least, arrested.

No means, however, should be left untried that promise to arrest and destroy it. Knowing the history of the insect, we are able to state that it can best, perhaps only, be destroyed while in the larval or grub state, or previous to attaining its perfect condition. This may be done by burning all the shriveled grapes, as well as those not shriveled, but showing the dot marking the entrance of the inclosed grub — in short, all infested clusters. Or, the clusters might be buried, with a foot of solid ground above them, through which, the perfect insects if developed under such circumstances, would not be able to penetrate.

Regarding so serious an attack upon the grape as of great economic importance, and with our knowledge of it limited to an experience in a single locality, I deemed it proper to request of the gentleman from whom its announcement had been received, a detailed statement for record, of the circumstances and conditions attending it, as an aid toward working out the life history of a dangerous insect pest, of which so very little was known.

He kindly communicated the following statement:

DEAR SIR — Your favor of 17th November is received. Please accept my hearty thanks. I will gladly take pains to learn what I can next year about the insect and inform you.

I only have a small place in New Jersey, where I live, and in my garden I have some twelve or fifteen vines only. The place I purchased five years ago, built my house and made my garden. The lot was a portion of an old neglected apple orchard. The soil is a light sandy loam, with a sub-soil of clear sand and gravel, running down probably twenty feet. Most of my grape vines are set in a border facing the south-west, trained on a post and wire trellis, with an ordinary picket fence about two feet in the rear — so they have an abundance of light and air.

About 150 feet from this trellis, facing it (hence a north-east exposure), is the rear of my house. On a piazza here I have a Croton vine, very thrifty, and not far away, against an out-house, I have a Catawba, with a south-eastern exposure. These two vines bear largely and are the only ones in my yard, apparently free from the attack of this insect.

Where I previously lived, I never had any trouble worth mentioning with grape enemies, and so was not on the look out for this insect. My grapes bore a very little two years ago (first crop), and I do not remember any appearance of disease. Last year they (or some of them) bore quite a crop, and the grapes were badly injured. Gardeners and fruit growers in my neighborhood, thought the trouble came from heavy and continued rains following a dry spell, the sap starting so vigorously as to burst the berry; I therefore gave the matter not much thought. This year, as you know, many of my vines fruited largely — notably the Concord, Martha, Walter, Croton, Catawba, with a smaller fruitage of Delaware, Hartford Prolific, etc. With the exception heretofore mentioned, *all* were attacked. I noticed it first as the grapes reached their full size, and began to ripen, and I suppose I lost quite three quarters of the fruit on the vines thus attacked.

I do not know how widespread the trouble may be. Some of my neighbors have suffered somewhat, but I think none so much as I. I propose to take pains to get the experience of the fruit-growers around me, and will then, as you request, communicate with you further.

Yours respectfully,
H. R. MUNGER,
PLAINFIELD, N. J.

Notices of "grape-rot," as a serious evil prevalent in various parts of the country, have appeared, from time to time, for the past few years, in our agricultural journals. Its cause has been extensively discussed, many speculations have been advanced, earnest study has been given it, but up to the present time, its occurrence, like that of the pear-blight, has received no satisfactory explanation.

Is it possible that it may be but a phase of a formidable *Isosoma vitis* attack? This question arises, when too late to answer it through examinations the present year. It may be that the conjecture, for such it merely is, may at once be dismissed as without foundation, by those conversant with the disease (of which the writer only knows the name), and can recall in it, conditions inconsistent with those attendant upon the insect attack, as above reported. The shriveling of the berry, its discoloration (its partial decomposition, perhaps, of which no mention has been made), and its dropping to the ground, might easily present most of the features of an ordinary decay.

If the question herewith raised of the identity of the two may not be at once authoritatively answered by the grape-grower familiar with all the phases of the grape-rot, then it will remain as a most interesting subject for determination the coming season. Should the suspicion be verified, and the cause of the wide-spread and growing evil be discovered so singularly hidden within the seeds, then there is scarce a doubt, but that by the sacrifice of one or two crops the progress of the evil can be effectually arrested.

IV. LIST OF LEPIDOPTERA.

COLLECTED BY W. W. HILL, IN THE ADIRONDACK REGION OF NEW YORK.

It is with much pleasure that we present the following record of some recent collections of Lepidoptera from one of the most elevated regions of the State of New York. It is, we believe, the first published local list, of any considerable extent, of the Lepidoptera of the State, and in view of the absolute necessity of such lists to an extended knowledge of the distribution of insects, it will be appreciated by the student. The great interest pertaining to the subject of geographical distribution, and its important bearing upon the derivation and modification of species, is illustrated by the admirable chapter on "The Geographical Distribution of the Phalænidæ of the United States," constituting pages 567–594 of Dr. Packard's *Monograph of the Phalænidæ.*

The enthusiasm of the entomologists of an adjoining State, has led them to explorations of a peculiarly interesting field lying beyond the limits of their own State — the White Mountains of New Hampshire. For successive years, the members of the Cambridge Entomological Club have established a midsummer encampment upon the slope of Mt. Washington, by which, through their protracted sojourn for weeks, and opportunity for collecting crepuscular and nocturnal forms, they have been able to enrich their cabinets and those of their correspondents with many rare boreal species, to accumulate much valuable biological information, and to present local lists of Lepidoptera, Coleoptera and Orthoptera which have been received as special contributions to science.

Meanwhile, the extensive Adirondack region with its numerous lofty mountain peaks, its deep gorges, its hundreds of lakes — perhaps second only to the White Mountains in point of interest to the entomologist of any locality in the United States east of the Rocky Mountains — has been permitted, each year, to bury within itself its entire entomological wealth.

Previous to the collections noticed in this paper, scarce an insect had been drawn from it. At the present, nothing has been reported of its mountain insect fauna. Many new species are doubtless to be discovered there, and the first comparison of its fauna with that of other elevated and more northern regions is yet to be made. It is not impossible (although our eastern friends will not admit the possibility) that the naked summit of Mt. Marcy may yet yield to earnest search another locality for that very interesting butterfly of so restricted range—*Chionobas semidea*, while aspirations less lofty, would in all probability be rewarded by the addition of *Argynnis montinus* to our State fauna.

It is sincerely to be hoped that, from the growing interest manifested in entomology, the numerous accessions to the number of its students, the facility for study afforded by recent publications and in several extensive classified collections — the reproach resting on the Entomologists of New-York, may speedily be removed. And while the thorough exploration of any locality can scarcely fail of bringing to light much new material, the ambitious student may have for his incentive the assurance that in the Adirondack region, and especially among the Adirondack Mountains proper, there is open to him an unexplored field where faithful search will assuredly yield him a most abundant return.

For the valuable information embodied in the following List, in its enumeration of species, dates of apparition, comparative abundance of species and of sex, we are indebted to the zeal of Mr. W. W. Hill, of Albany. Although having but recently devoted himself to entomological study, the ardor with which he has entered upon it, the unwearying industry displayed in its pursuit, and the very satisfactory results thus far attained, give every assurance that the science to which he has so earnestly consecrated his available time, will be materially advanced by his labors.

The collections were made in Township No. 4, of Lewis county, at Fenton's, and its immediate vicinity. The elevation above tide has not been computed, but may be given approximately at 1450 feet.

The larger proportion — perhaps three-fourths — of the Noctuidæ were attracted by light, and taken within the Fenton House. Quite a number were captured "at sugar"; the inexperience of the collector in this usually very successful method made it far less remunerative than under other circumstances it would have been, and than it promises to be hereafter.

It is to be regretted that this List could not present the results of an entire season, instead of being necessarily so partial an exhibit of the Lepidoptera of the region, from the brief time to which the collections were limited. As may be seen from the record, they commenced on July 1st (1876), and were continued until the 21st of August. On the part of Mr. Hill, they terminated August 4th, the subsequent additions having been made by friends sojourning at the place. An absence of one week during this time limits the period of regular collecting to four weeks.

Collections of considerable less extent made during the preceding year (1875) from July 1st to August 4th, are also incorporated with the above. Of the species taken at this time no dates were recorded; when, therefore, they were not met with during the following season, they appear in the list simply as taken in July or August, 1875.

RHOPALOCERA.

Papilio Turnus *Linn*	July,	1875.
Pieris oleracea (*Harr.*)	"	17, 19, 27, 29 31.
P. rapæ (*Linn.*)...........	"	19, et al.
Colias Philodice *Godart*	"	12, et al.
Danais Archippus (*Fabr.*).....	"	3, 10, 20, 27, 31.
Argynnis Cybele *Fabr*........	"	19, 31.
A. Aphrodite *Fabr*... ...	"	12 to 29. 12 ♂, 5 ♀.
A. Atlantis *Edw*.........	"	4 to 21. 8 ♂, 15 ♀.
A. Myrina (*Cramer*).....	"	27, 30. 3 ♂.
A. Bellona *Fabr*.........	"	19.
Phyciodes Nycteis *Doubl.*	"	8, 10.
P. tharos (*Drury*).......	"	2, 3, 5, 6, 7, 18. 9 ♂.
P. Batesii (*Reak.*)	"	1875.
Melitæa Phaeton (*Drury*)	"	1875.
Grapta comma (*Harr.*)........	"	14.
G. Faunus *Edw*.........	"	17, et al. 6 ♂, 5 ♀.
G. Progne (*Cramer*)......	"	1.
G. J-album (*Bd.-Lec.*)...	"	1875. Aug. 1. 2 ♂, 2 ♀.
Vanessa Antiopa (*Linn.*)......	"	1875.
V. Milberti *Godart*......	"	1875. Aug. 5.

Pyrameis Atalanta (*Linn.*)....	Aug.	1875.
P. huntera (*Fabr.*).......	July	27, Aug. 1, 5.
Limenitis Arthemis (*Drury*)...	"	5, (15 ♂, 2 ♀) 10, 12, 15, Aug. 1. 19 ♂, 4 ♀.
L. disippus (*Godt.*)......	"	1875.
Neonympha Eurytris (*Fabr.*)..	"	1875.
N. Canthus (*Linn.*)......	"	3 to 19. 8 ♂, 2 ♀.
Satyrus Alope (*Fabr.*)	"	19, 26, 31, Aug. 1. 5 ♂.
S. Nephele (*Kirby*)......	"	15, 17, 19.
Thecla strigosa *Harr*..........	"	1875.
T. Titus (*Fabr.*).........	"	6, 17, Aug. 1, 4. 7 ♂, 3 ♀.
Lycæna neglecta *Edw*.	"	1875.
Chrysophanus Thoe (*Bd.-L.*) ..	"	1875.
C. Americana (*Harr.*)....	"	2, 3, Aug. 1, 5,
Pamphila Zabulon (*Bd.-Lec.*) .	"	1875.
P. Leonardus (*Harr.*)....	"	1875.
P. Peckius (*Kirby*)......	"	18.
P. Mystic *Edw*..........	"	2, 3, 4, 5, 12. 7 ♂, 2 ♀.
P. Cernes (*Bd.-Lec.*).....	"	2, 3, 4, 5, 12. 6 ♂, 3 ♀.
P. Metacomet (*Harr.*)....	"	2, 5, 6, 8, 10, 18. 4 ♂, 2 ♀.
Eudamus Pylades (*Scudd.*)....	"	1875.

HETEROCERA.

SPHINGIDÆ.

Sesia uniformis *Grote-Rob*.........	July	5, 12.
Daremma undulosa *Walk*.........	"	1875.
Smerinthus geminatus *Say*........	"	27.

ZYGÆNIDÆ.

Ctenucha virginica (*Charp.*).......	July	5 to 18. 10 ♂ 2 ♀.
Lycomorpha pholus (*Drury*)......	Aug.	12.

BOMBYCIDÆ.

Callimorpha militaris *Harr*........	July	1875.
Arctia virgo (*Linn.*)	"	"
A. Saundersii *Grote*	"	"
Spilosoma textor *Harr*............	"	"
Orgyia nova *Fitch*................	"	26.
Parorgyia Clintonii *Gr.-Rob*......	"	12.
Nadata gibbosa *Walk*.............	"	7.

Crocota ferruginosa *Walk.* .... July 18, 19.
Dryopteris rosea (*Walk.*)......... " 1875.
Cœlodasys unicornis *Sm.-Abb.*.... " "
Clisiocampa Americana *Harr.*..... " 11, 12, 18, 27.

NOCTUIDÆ.

Pseudothyatira cymatophoroides.. Aug. 12.
P. expultrix *Grote* .... July 4, 6, 9, 27. 3 ♂, 6 ♀.
Thyatira scripta *Gosse* .... " 30.
Diphthera fallax *Her.-Sch.*......... " 11, Aug. 3. 1 ♂, 1 ♀.
Briophila lepidula (*Grote*) .... " 12.
Acronycta lobeliæ *Guen.* .... " 12.
A. Radcliffei (*Harvey*)....... Aug. 26.
A. lepusculina *Guen.*......... July 7.
A. Americana *Harr.* .... " 10, 12, 18.
A. brumosa *Guen.* .... " 29, Aug. 3.
A. superans *Guen.* .... " 11, 14.
A. hamamelis *Guen.* .... " 10.
A. dentata (*Grote*)........... " 11.
Micrococlia fragilis *Guen.*......... Aug. 2.
M. diphteroides *Guen.*........ July 10.
M. obliterata *Grote.* .... " 4.
Agrotis Chardinyi (*Boisd.*)........ " 30.
A. sigmoides (*Guen.*) .... " 3, 9, 10.
A. perattenta *Grote*.......... " 1875.
A. conflua *Tr.*.............. Aug. 1875.
A. perconflua *Grote.* .... " 8.
A. baja (*W.-V.*) .... July 26, 30, Aug. 13, 22.
A. Normaniana *Grote.* .... " 1875.
A. rufipectus *Morr.* .... " 29.
A. haruspica *Grote.* .... " 11, 17, 27.
A. bicarnea (*Guen.*) .... " 13, 28, Aug. 5, 15.
A. subgothica (*Haw.*)......... " 1875.
A. herilis *Grote* .... " 27.
A. tricosa *Lintn.*............ " 3, 27, Aug. 2.
A. plecta (*Linn.*)............ " 9, 14, 21, Aug. 7.
A. redimicula *Morr.*.......... " 30.
A. pitychrous *Grote*.......... " 1875.
A. muræenula *Gr.-Rob.*........ " "
A. ypsilon (*Rott.*) .... " 31, Aug. 1, 8.
A. placida *Grote*............. " 26.

Agrotis	turris *Grote*.	July 30.
A.	pressa *Grote*.	" 8.
A.	prasina (*W.-V.*)	" 17, 27, 29. 4 ♂, 2 ♀.
A.	astricta (*Morr.*)	" 15.
Dianthœcia	lustralis *Grote*	" 9.
Mamestra	purpurissata *Grote*	" 27 to Aug. 8. 4 ♂, 3 ♀.
M.	nimbosa (*Guen.*)	" 1875.
M.	imbrifera (*Guen.*)	" 4 to 15. 9 ♂, 2 ♀.
M	subjuncta *Gr.-Rob.*	" 8.
M.	legitima *Grote*	" 10, 13, 15, 29.
M.	trifolii (*Esper*)	Aug. 8.
M.	lorea (*Guen.*)	July 2, et al. 6 ♂, 5 ♀.
M.	renigera (*Steph.*)	" 11 to 31. 6 ♂, 2 ♀.
M.	olivacea *Morr.*	" 30 — Aug. 21. 10 ♂. 1 ♀.
Hadena	loculata (*Morr.*)	" 3, 11, 16, 30, Aug. 6.
H.	devastatrix (*Brace*)	" 12 to Aug. 8.
H.	lateritia (*Hüfn.*)	" 3, 9, 10.
H.	sputatrix *Grote*	" 8 to Aug. 9.
H.	apamiformis (*Guen.*)	" 8.
H.	impulsa (*Guen.*)	" 10.
H.	arctica *Boisd.*	" 3, 4, 6, 10, 17, 26, 27.
H.	lignicolor (*Guen.*)	" 8 to Aug. 7. 3 ♂, 8 ♀.
H.	verbascoides (*Guen.*)	" 10, 15.
H.	sectilis (*Guen.*)	" 6 to 31. 2 ♂, 6 ♀.
H.	modica (*Guen.*)	Aug. 4.
H.	Hillii *Grote*.	July 26.
Hyppa	xylinoides *Guen.*	" 6, Aug. 15, 19.
Eriopus	mollissima *Guen.*	" 1875.
Phlogophora	periculosa *Guen.*	" 27, 29, 31. 5 ♂, 3 ♀.
Euplexia	lucipara (*Linn.*)	" 1 to 18. 7 ♂, 4 ♀.
Brotolomia	iris (*Guen.*)	" 1875.
Nephelodes	violans *Guen.*	Aug. 3, 19, 22.
Hydrœcia	sera *Gr.-Rob.*	July 12, 14, 16, 17, 27. 5 ♂.
H.	nictitans (*Linn.*)	" 27 to Aug. 19.
Achatodes	zeæ (*Harr.*)	" 30, Aug. 4.
Leucania	pallens (*Linn.*)	" 1.
L.	phragmitidicola *Guen.*	" 1875.
L.	Harveyi *Grote*	Aug. 9, 19.
L.	lapidaria *Grote*	July 9.
L.	adonea (*Grote*)	" 10.
L.	commoides *Guen.*	" 6, 11, 16, 26. 1 ♂, 5 ♀.

Leucania unipuncta *Haw* .... Aug. 13, 21.
Caradrina fidicularia (*Morr.*) .... " 1.
Amphipyra tragopogonis (*Linn.*) .. " 19.
A. pyramidoides *Guen.* .... July 30, Aug. 4, 6.
Orthodes infirma *Guen.* .... " 11.
Tæniocampa oviduca *Guen.* .... " 11, 17.
Tæniosea gentilis *Grote* .... " 9.
T. perbellis *Grote* .... " 6.
Plastenis pleonectusa (*Grote*) ...... " 1875.
Eucirrœdia pampina (*Guen.*) ...... Aug. 19.
Anytus sculptus *Grote* .... " 21.
Cucullia convexipennis *Gr.-Rob* ... July 3.
C. asteroides *Guen.* .... " 3.
C. postera *Guen.* .... " 15, 29.
C. intermedia *Speyer* .... " 18.
Nolaphana Zelleri *Grote* .... Aug. 17.
Plusia æreoides *Grote* .... July 3 to Aug. 7. 13 ♂, 3 ♀.
P. balluca *Guen.* .... " 29.
P. Putnami *Grote* .... " 1875.
P. bimaculata *Steph.* .... " 6, Aug. 3.
P. precationis *Guen.* .... " 28, Aug. 4, 13.
P u-aureum *Boisd* .... " 30.
P. mortuorum *Guen.* .... " 11, 30.
P. epigæa *Grote* .... " 1875.
P. ampla *Walk.* .... " 30.
P. simplex *Guen.* .... Aug. 4, 5.
Erastria carneola *Guen.* .... July 7, 30, 31, Aug. 1, 7, 9.
E. albidula *Guen.* .... " 3.
E. muscosula *Guen.* .... " 3, 6, 11.
Leptosia concinnimacula *Guen.* .... " 1875.
Lithacodia bellicula *Hübn.* .... " 2, 3.
Drasteria erechtea (*Cram.*) .... " 29, Aug. 1.
Catocala ultronia (*Hübn.*) .... " 31, Aug. 1.
C. Ilia (*Cram.*) .... " 8.
C. polygama *Guen.* .... Aug. 5.
C. præclara *Gr.-Rob.* .... July 31.
Pangrapta decoralis *Hübn.* .... " 2, Aug. 17.
Homopyralis tactus *Grote* .... " 8, 10, Aug. 7.

DELTOIDÆ.

Pseudoglossa lubricalis (*Geyer*) .... July 11 to Aug. 1.
Epizeuxis æmulalis (*Guen.*) .... " 1875.

Epizeuxis americalis (*Guen.*)....... Aug. 2.
Litognatha nubilifascia *Grote*...... July 3, 4.
Zanclognatha lævigata *Grote*...... " 16.
Z. ochreipennis *Grote*........ " 11.
Z. marcidilinea *Grote*........ " 14.
Philometra longilabris *Grote*....... " 4, 9, 11, 14.
Rivula propinqualis *Guen*......... " 2, 27
Palthis angulalis *Hübn*............ " 2, 10, 11, Aug. 17.
Renia discoloralis *Guen*........... " 1875.
R. Belfragei *Grote*............. " 15 to 31. 5 ♂, 2 ♀.
Bleptina caradrinalis *Guen*........ " 1, 12.
Hypena humuli *Fitch* .... " 15, 21, 29.

PHALÆNIDÆ.

Chœrodes transversata (*Drury*).... July 27 to Aug. 3. 5, ♂ 9 ♀.
*Ennomos magnaria *Guen*.......... Aug. 12, 15.
Endropia serrata (*Drury*).......... July 1875.
E. bilinearia *Pack*........... " 8, 9, 10, 13, 14. 10 ♂.
E. effectaria *Walk*. .... " 11, 19, 26.
Ellopia fiscellaria (*Guen.*) .... " 1875.
Sicya truncataria *Guen*............ " "
Angerona crocataria (*Fabr.*)....... " 6, 8.
Nematocampa filamentaria *Guen*... " 17, 18, 19.
Boarmia humaria *Guen*............ " 18, 30.
B. pampinaria *Guen*. .... " 6.
Paraphia subatomaria *Guen*. ...... " 1875.
Synchlora rubivoraria *Riley*....... " "
Ephyra pendulinaria *Guen*........ " 31.
Acidalia enucleata *Guen*........... " 1, 13.
†A. quadrilineata *Pack*. ...... " 3, 4, 11.
A. inductata *Guen*........... " 1, 11, Aug. 8.
Stegania pustularia *Guen*.......... " 18, 28, 30, Aug. 4, 11.
Cabera erythemaria *Guen*.......... " 1.
C. variolaria *Guen*. .... Aug. 7, 22.
Eumacaria brunnearia *Pack*. ...... July 28.

*I cannot agree with Dr. Packard in the reference of this form to the *alniaria* of Europe, the two differing so much in the shape of the wings. Dr. Speyer has remarked of *magnaria*, " near to our *alniaria*." Mr. von Meske has recently sent a large number of the eggs of *magnaria* to Dr. Speyer, which will probably furnish the means for such a comparison, in all of the four stages of the insect as will authoritatively decide the question of identity which is raised.

†One example of the species was entirely without lines.

Macaria enotata *Guen.* .... July 8, 27.
M. granitata *Guen.* .... " 3, 8, 11, 18, 27.
Phasiane trifasciata *Pack.* .... " 29.
P. mellistrigata *Grote* .... " 30, Aug. 19.
*Thamnonoma brunneata (*Thunb.*). " 4. 2 ♂, 1 ♀.
Eufitchia ribearia (*Fitch*).......... " 16, 19, 30.
Caripeta divisaria *Walk.* .... " 1875.
Hæmatopis grataria (*Fabr.*)........ Aug. 8.
Aspilates Lintneraria *Pack.* .... July 27.
Heterophleps triguttata *Her.-Sch.* .. " 1875.
Odezia albovittata (*Guen.*) .... " 5, 8.
Scotosia undulata (*Hübn.*) .... " 5.
Melanippe hastata (*Linn.*)......... " 6.
M. sociata (*Bork.*) .... " 3, Aug. 17.
M. lacustrata *Guen.* .... " 1, 3, 6, Aug. 5.
M. basaliata (*Walk.*)......... " 3, 27.
Melanthia ruficillata *Guen.* .... " 4 to Aug. 8. 6 ♂, 5 ♀.
Coremia designata (*Hüfn.*)........ " 6, 30, Aug. 17.
C. ferrugaria *Clerck.* .... Aug. 9.
Cidaria Packardata *n. sp.* .... July 27.
C. testata (*Linn.*) .... Aug. 15.
C. albolineata *Pack.* .... " 15.
C. cunigerata *Walk.* .... July 6—Aug. 17. 7 ♂, 6 ♀.
C. hersiliata *Guen.* .. .. " 7, 14.
C. truncata (*Hübn.*).......... " 4 to Aug. 22. 5 ♂, 4 ♀.
Spargania magnoliata *Guen.* .... " 11, 15.
Oporabia cambricaria (*Curtis*)..... " 11.
O. 12-lineata (*Pack.*) .... " 7.
†Larentia cæsiata *W.-V.* .... " 10.
Eupithecia miserulata *Grote* .... " 11.

PYRALIDÆ.

Asopia farinalis (*Linn.*)........... July 26, 29, 30, 31.
A. devialis *Grote* .... " 12, 13, 17, 18.
A. squamealis *Grote* .... " 9.
Ennychia octomaculalis (*Linn.*) ... " 11.
Desmia maculalis *Westw.* .... Aug. 1876.
Botis theseasalis *Walk.* .... July 6, 8, 16, 29, 30.
B. plectilis *Gr.-Rob* .... " 1, 3, 4, 5, 14, 17.
B. marculenta *Gr.-Rob.* .... " 27.

*Determination of Dr. Packard.
†Determination of Dr. Packard.

Botis badipennis *Grote*	July 30.
B. generosa *Gr.-Rob*..........	" 5.
B. hircinalis *Grote*...........	" 4.
B. subolivalis *Pack*..........	" 5.
Scoparia centurialis *W.-V.*.........	" 28.

MICROLEPIDOPTERA.

Galleria cereana (*Fabr.*)....	Aug. 1875.
Crambus girardellus *Clem*.........	July 10.
Crambus præfectellus *Zell*.........	" 3, 17, Aug. 6.
Tortrix cerasivorana (*Fitch*).......	" 17.
Ditula blandana *Clem*.............	" 7.
Sericoris cæsialbana *Zeller*........	" 4.

In the above list two hundred and fifty-four species are recorded, and represented, as in the detailed memoranda preserved by Mr. Hill, but not convenient to reproduce in these pages, in 796 examples, viz.: of the Rhopalocera, 152 ♂'s and 69 ♀'s; of the Heterocera, 306 ♂'s and 269 ♀'s.

The preponderance of the captures of male Rhopalocera is very marked, being 120 per cent in excess of the females: of the Heterocera, the males exceed the females by only 12 per cent.

The paucity of the Sphingidæ and Bombycidæ reported, is to be explained by the late date at which the collections commenced. Numerous sphinges had been observed, attracted to light, during the month of June. The period of greatest abundance of most of the Diurnals both in species and in individuals had also passed. On the 5th of July, *Limenitis Arthemis* was still abundant, but in worn condition. At this date, Mr. Hill captured of this upland species, as noted above, eighteen examples, upon a moist spot near a stream, three miles distant from Fenton's on the wagon-road to Lowville. Two visits to this locality were afterward made, without obtaining additional examples.

Among the Phalænidæ, are a larger number of northern forms than might have been expected from the comparatively moderate elevation at which the collections were made. The following of the species are recorded by Dr. Packard in the paper before referred to, as circumpolar or subarctic species, "ranging between the isotherm of 32° and 44°, and also following the isothermals of 44° and 48° southward into Colorado and California — in Colorado ranging from an elevation of 8,000 feet to the limit of trees, 11,000 feet:" *Larentia cæsiata*, *Oporabia cambricaria*, *Spargania magnoliata* *Cidaria truncata*, *C. hersiliata*, *C. cunigerata*, *C. Packardata* (populata of Pack.), *Coremia ferrugaria*, *Melanippe fluctuata* and *M. hastata*.*

To these ten species may be added *Thamnonoma brunneata*, which, if the identification of our species with the European *brunneata* and *pinitaria* be correct, also occurs in elevated regions in Europe.

In consideration of their very interesting distribution, we transcribe for these species the localities ascribed to them in Dr. Packard's monograph, adding to his table the new locality given them in this paper.

	Colorado or Pacific Coast or both.	Mt. Washington.	Adirondack Region, New York.	Labrador.	Iceland.	Lapland.	Alps of Central Europe.	Ural or Altai Mts., or both.
Thamnonoma brunneata	..	*	*	..	..	..	*	..
Melanippe hastata	*	*	*	*	*	*	*	*
Melanippe fluctuata	..	*	*	..	..	..	*	*
Coremia ferrugaria	*	*	*	*	*	*	*	*
Cidaria Packardata	*	*	*	*	..	..	..	..
Cidaria cunigerata	..	*	*	..	..	..	..	..
Cidaria hersiliata	*	*	*	..	..	..	..	..
Cidaria truncata	*	*	*	*	*	*	*	*
Spargania magnoliata	*	*	*	..	..	..	..	..
Oporabia cambricaria	..	*	*	..	..	..	*	..
Larentia cæsiata	*	*	*	*	*	*	*	*

*It will be observed that in several citations in this paper from Dr. Packard, his late generic references have not been followed, as I cannot regard the lists of Hübner as of any authority in nomenclature. I fully concur in the opinions so unequivocally expressed by Guenée, Dr. Speyer, Wallace, Dr. Boisduval (see Canad. Entomol. 8, p. 117.), Dr. Hagen, Edwards and other leading entomologists, that catalogue names (as were Hübner's), have no just claim for precedence over those of properly defined generic and other groups, and that the attempt to introduce them in nomenclature can result only in confusion and other serious evils.

Among these catalogue names of Hübner, "still-born" (Guenée) more than three score and ten years ago, recently galvanized into life, and for which Dr. Packard stands sponsor, are the following from the Tentamen: Epirrita, Petrophora, Rheumaptera, Hydria, Cymatophora; and from the "Verzeichniss" a few years later:

Ochyria,	Operophtera,	Semiothisa,	Calothysanis,	Therina,
Philereme,	Perconia,	Deilinia,	Eois,	Epirranthis.

It is probable that *Cidaria albolineata* should be numbered among the above subarctic forms, although not so included by Packard. The localities ascribed to it are all of northern or high elevations, as Quebec, Brunswick, Me., and the White Mountains. It has never been observed by me in this portion of New York, except upon the summit of the Catskill Mountains.

Phasiane trifasciata of the list, may also prove to belong to the above, — its only other recorded locality being Berlin Falls, Me.

Of the species cited above, *Larentia cæsiata*, *Oporabia cambricaria*, *Spargania magnoliata*, *Cidaria Packardata* and *Melanippe fluctuata* have not been collected in the vicinity of Albany.

Larentia cæsiata from this locality, is an interesting acquisition, it having been previously reported only from high elevations or northern regions, as the White Mountains, at an elevation of between 8,000 and 9,000 feet in Colorado, Okak in northern Labrador, and Caribou Island (Packard). Guenée states that it has the widest distribution of any species of the [Glaucopteryx, Hübner] group, occurring in the mountainous regions of France, Germany, Italy, Piedmont, Scotland, north of England, Iceland, etc., in barren places resting on rocks and the trunks of trees. From its range of variation, different species have been made of it, as *infrequentata* Haw., *flavicinctata* Steph., etc.

Of the Noctuidæ, the following species have not been found in the vicinity of Albany :

[a]Agrotis Chardinyi,
[a]Agrotis conflua,
Agrotis perconflua,
Agrotis rufipectus,
[a]Agrotis astricta,
Dianthœcea lustralis,
Hadena Hillii,
[a]Plusia bimaculata,
[a]Plusia u-aureum,
Plusia mortuorum,
Plusia epigæa.

At the present, we can only venture to indicate as probable subarctic forms, the five species of the above marked *a*. Of these, two are particularly interesting, from their northern and extended distribution and their occurrence in the lower Adirondack region. *Agrotis Chardinyi*=(A. gilvipennis *Gr.*) has been collected in several examples on Anticosti Island, Gulf of St. Lawrence, N. Lat. 49°; it also, according to Staudinger, occurs in Siberia; Guenée cites it as common in Russia. *Agrotis conflua* was found by Mr. Couper to be common on Anticosti Island; it occurs also in Iceland (Staudinger), Scotland (Norman), Switzerland (Grote), Prussia? and France (Guenée).*

The geographical distribution of the Noctuidæ of the United States has not been studied. No attempt has been made to assign the species to distinct faunal provinces, although the material for such an arrangement is unquestionably far more ample than with the Phalænidæ. Much of it is unpublished, although accessible, if sought for, from the various collectors to be found in nearly every State of the Union. It would prove of essential service if every skilled collector would furnish for publication, authenticated lists of the Lepidoptera known to him to occur in his vicinity. These, for greater convenience, could be combined into State Lists, similar to that given by Prof. Snow, of the Lepidoptera of Eastern Kansas.†

The editor of the *Canadian Entomologist*, in appreciation of such work, has made special request for such lists of the Diurnal Lepidoptera,‡ and has arranged for their tabulation previous to publication in the pages of that valuable journal. We feel assurance that lists of the Heterocera would also be gladly received and promptly published.

A very excellent local list, is that given by Mr. Roland Thaxter, in *Psyche*, Vol. II, pp. 34–38, 80, of collections of Noctuidæ, recently made, mainly by himself, at Newton, Mass., and vicinity. It enumerates three hundred and forty-seven species, and is accompanied with statements, in an abbreviated form, of the comparative abundance of each, month of appearance, means of capture of all (whether at light, at sugar, or at rest), and the food-plant of the larva of a number.

In Vol. VII of the *Canadian Entomologist*, pp. 3 and 21, Mr. George Norman records his captures of Noctuidæ at St. Catherines, Ont., during the year 1874, citing one hundred and seventy-four species, with their dates of capture and compar-

*For the determination of these species, and of many others, I am indebted to the kind services of Mr. Grote, always most cordially rendered.

†*Transactions of the Kansas Academy of Science*, Vol. IV, pp. 29–59. 1875.

‡*Canadian Entomologist*, Vol. VII, p. 72. 1875.

ative abundance: the larger number of these were attracted to sugar. In Vol. VIII of the same journal, p. 12, Mr. O. S. Westcott publishes a list of eighty-three species of Noctuas, taken by him at sugar, at Maywood, Cook Co., Ill.

In a list (still in MS.) of collections at sugar made by me at Schenectady, N. Y., in 1876, two hundred and twenty-six species of Lepidoptera are recorded, embracing a few Bombycidæ, about twenty-five species of Phalænidæ, a few Pyralidæ and Microlepidoptera — the remainder Noctuidæ.

In 1873, lists were given in the 23d N. Y. State Museum Report, of the Butterflies and the Sphingidæ of the State of New York. Similar lists of the Bombycidæ and Noctuidæ are in preparation.

Mr. Henry Edwards, of San Francisco, who has contributed so largely to our knowledge of the Lepidoptera of our western coast, is engaged upon a Synopsis of the Butterflies of the Pacific coast, which is promised for the press during the present year. Mr. Grote has published several lists of the Noctuidæ of Texas. Will not the members of the Cambridge Entomological Club favor us with a list of Lepidoptera collected on Mount Washington, as a companion to Mr. Austin's list of Coleoptera of the same locality.*

*Catalogue of the Coleoptera of Mount Washington, N. H., with Descriptions of New Species, by E. P. Austin and J. L. Le Conte, M. D.; in *Proc. Bost. Soc. Nat. Hist.*, Vol. XVI, pp. 265-276.

V. COLLCTIONS OF NOCTUIDAE "AT SUGAR."

AT SCHENECTADY, N. Y., IN 1875.

The list below given is a record of collections made between the 7th of July and the close of the season on the 25th of October. It includes all the species of Noctuidæ (one hundred and thirty-one in number) that were captured or observed on the fifty-three evenings devoted to the work — three or four evenings of each week. A few species of Bombycidæ, Phalænidæ, Pyralidæ and some Microlepidoptera were also taken, which are not embraced in the list.

The attempts previously made by me at collecting by sugaring, had been attended with no success. The satisfactory results obtained at this time are to be ascribed to persistent and more extended sugaring than before employed. The locality was not an unusually favorable one, for, instead of choosing a place for the purpose " on the border of some wood," as has been usually recommended, where the proper number of trees of a certain diameter and character of bark could be found, the collections were entirely confined to my garden — not a large one — in the city of Schenectady. It was an unexpected revelation that collections of such a variety and extent could be made within city limits, and in a garden where the presence of flowers undoubtedly interfered with the attractions of the bait. But as the convenient locality of one's own home may not always prove equally productive in other cities, the statement should be made that my residence was within a block of the Mohawk river which forms the northern boundary of Schenectady,—a city of comparatively small size, numbering under 13,000 inhabitants.

The slats and posts of a grape trellis of sixty feet in extent, offered a convenient place upon which to spread the bait: the leaves extending over the slats had been removed, except at intervals, where they were permitted to remain to serve as a cover or a lure to the moths attracted thither. The odor diffused from the area of surface sugared — computed at sixteen square feet — was evidently sufficient to draw the moths from quite a distance beyond the boundary of the garden. Farther experience has shown that a larger quantity of the bait might have been advantageously used in extending it to the fences inclosing the garden, and that the results to be obtained by this method of collecting, are to be measured by the extent of the locality, the area sugared, and the frequency of its repetition, provided the region be one where the Noctuidæ occur in reasonable abundance.

As directions for sugaring have already been published,* they will not be repeated here, but will be given in a paper in preparation, in which the various methods employed in the attraction of Lepidoptera will be detailed.

A brief account of the collections herein noticed was presented at the meeting of the American Association for the Advancement of Science, held at Detroit in 1875. Several of the members present were stimulated to test for themselves the recommendation made of this simple and most valuable means of enlarging their cabinets and extending our knowledge of an interesting and important class of insects. In every instance, it is believed, the results were highly gratifying.

Pseudothyatira cymatophoroides, July 17, 23, 27, 30, 31.
P. expultrix, July 12, 15, 17, 20, 21, 23, 24, 27, 28, 30, 31; Aug. 2, 5, 6, 7, 16.
Habrosyne scripta, July 24, 28, 31; August 7.
Acronycta occidentalis, July 27, 31.
A. Americana, July 14, 20, 21.
A. dissecta, July 24.
A. vinnula, July 7.
Agrotis sigmoides, July 24, 30; August 7.
A. badicollis, August 6.
A. baja, Aug. 16, 18, 19, 21, 25, 26, 28, 30; Sept. 2, 4, 7.
A. haruspica, July 7, 13, 17, 20, 23, 24, 27, 28, 30, 31; August 2, 5, 6, 7, 16, 19.
A. c-nigrum, July 7, 23, 24, 30; August 2, 25, 26, 28, 30; September 2, 4, 7, 9, 11, 13, 15, 18, 30; October 2, 4, 6, 10, 16, 21, 22.
A. bicarnea, July 30, 31; August 2, 6, 7, 21.
A. subgothica, August 2, 19, 21, 26.

* By Mr. George Norman, in the *Canadian Entomologist* for April, 1875, and by Prof. O. S. Westcott in the same journal for January, 1876.

Agrotis tricosa, August 19, 21, 28.
A. herilis, July 7, 20,
A. plecta July 7, 24; August 2, 5, 6, 7, 16, 18, 19, 21, 25, 26, 28, 30; September 2.
A. sexatilis, August 7.
A. pitychrous, August 30.
A. muræenula, August 21, 25.
A. messoria, August 19, 21, 25, 26, 28, 30; September 2, 4, 7, 9, 11, 13, 15, 18.
A. velleripennis, September 2.
A. venerabilis, September 15.
A. annexa, July 31; August 7.
A. ypsilon, July 7, 21, 24, 27, 31; August 2, 5, 6, 7, 16, 19, 21, 25, 26, 28, 30; September 7, 9, 11, 13, 15, 18, 25, 29, 30; October 2, 4, 6, 10, 19, 24.
A. saucia, September 7, 9, 11, 13, 15, 20, 29, 30; October 2, 4, 10, 19, 22.
A. alternata, July 7, 17.
A. herbida, July 7, 8, 10, 12, 14, 17, 24, 28, 30, 31; August 2, 5, 6, 7, 26, 28; September 4.
A. occulta, August 30.
Mamestra nimbosa, July 27, 31.
M. latex, July 7.
M. adjuncta, August 2, 6, 7.
M. subjuncta, July 31; Aug. 5, 7, 19, 21, 26; Sept. 15.
M. chenopodii, July 24, 30, 31; Aug. 2, 5, 6, 7, 19; September 9.
M. albifusa, July 28; August 6.
M. Godelli, June 8.
M. lorea, July 14.
M. renigera, July 7, 21, 23, 24, 27, 28, 30; August 2, 5, 6, 7, 16, 19, 21, 25, 28, 30; September 2, 4, 7, 9, 11, 13, 15, 27; October 2.
M. olivacea, August 25.
Hadena loculata, July 14, 15, 23.
H. devastatrix' July 8, 10, 13, 15, 17, 20, 21, 23, 24, 27, 28, 30, 31; August 2, 5, 6, 7, 16, 18, 19, 21, 25, 26, 28, 30; September 2.
H. sputatrix, July 8, 10, 12, 13, 14, 15, 17, 20, 21, 23, 24, 27, 28, 30, 31; August 2, 5, 6, 7, 16, 18, 19, 21, 25, 26, 28, 30; September. 2.
H. impulsa, July 8, 15.

Hadena apamiformis, July 24; August 6.
H. suffusca, September 7.
H. arctica, July 7, 8, 10, 12, 13, 14, 15, 17, 20, 21, 23, 24, 27, 28, 30, 31; Aug. 2, 5, 6, 7, 16, 19, 21, 25, 26.
H. lignicolor, July 7, 8, 10, 12, 13, 14, 15, 17, 20, 21, 23, 24, 27, 28, 30, 31; August 2, 5, 6.
H. verbascoides, July 12, 13, 14, 17, 20, 21, 23, 24.
H. modica, July 31; August 7.
H. sectilis, September 7.
Perigea luxa, August 7.
Dipterygia pinastri, July 28; August 5, 19, 30.
Hyppa xylinoides, July 28, 30, 31; August 2, 5, 7, 16, 18, 19, 21, 25, 26, 28, 30; September 2, 4, 7, 9, 11, 13, 15, 18, 20.
Laphygma frugiperda, September 15.
Euplexia lucipara, July 10, 21, 31.
Phlogophora iris, June 30.
Nephelodes violans, September 9.
Luperina reniformis, August 2, 5, 6, 7, 18, 19, 28; Sept. 2, 4, 7, 9, 11, 13, 15, 18, 20, 22, 25, 30; October 19, 25.
L. reniformis, v atra, August 19; September 4, 7.
Gortyna sera, July 7, 8, 10, 12, 13, 14, 15 17, 20, 21, 23, 24, 27 28, 30, 31; August 2, 5, 6, 7.
G. nictitans, July 17, 20, 21, 23, 24, 27, 28, 30, 31; August 2, 5, 6, 7, 16, 18, 19, 21, 25, 26.
Leucania pallens, July 31; August 18, 19, 21, 25, 26.
L. phragmitidicola, August 16, 19, 21; September 2.
L. lapidaria, August 7, 16, 19, 28.
L. adonea, July 7, 8, 15, 17, 24.
L. commoides, August 19.
L. unipuncta, July 7, 21; Aug. 7, 16, 18, 19, 21, 25, 26, 28, 30; September 2, 4, 7, 9, 11, 13, 15, 18, 20, 22, 25, 27, 29, 30; October 2, 4, 6.
L. pseudargyria, July 12, 15, 17, 20, 21, 23, 24, 27, 28, 30, 31; August 2.
Amphipyra pyramidoides, July 20, 24, 28, 30, 31; August 2, 5, 6, 7, 16, 19, 21, 26, 28, 30; September 2, 7, 13, 15, 25, 30; October 2, 4, 24.
A. tragopogonis, July 31.
Orthodes infirma, July 7, 8, 10, 12, 17, 20, 21, 23, 24, 27, 28, 30, 31; August 2, 5.
O. candens, July 7.

Tæniocampa oviduca, July 21.
Cosmia infumata, September 4, 9.
Plastenis pleonectusa, July 31 ; August 7.
Orthosia ferrugineoides, September 7, 9, 11, 13, 15, 18, 20, 22, 25, 27, 29, 30; October 2, 4, 6, 10, 16, 19, 21, 22, 23, 24, 25.
O. helva, August 5, 6, 7, 16, 18, 19, 21, 25, 26, 28, 30.
O. sp. (no. 3885), July 12, 17, 20, 21, 24.
Eucirrœdia pampina, September 13 ; October 10.
Scoliopteryx libatrix, July 8, 10, 12, 13, 14, 15, 20, 24, 30.
Xylina disposita, September 9, 11, 13, 25, 27, 29, 30 ; October 4, 10, 16, 21, 22, 23, 24, 25.
X. petulca, September 11, 13, 27, 29 ; October 21.
X. ferrealis, September 7, 11, 13.
X. Bethunei, September 4, 9, 13, 15, 18, 26, 29, 30; October 2, 10, 23, 24.
X. laticinerea, September 29, 30; October 4, 6, 10, 19, 21, 22, 23, 24, 25.
X. cinerea, September 15, 30.
X. pexata, September 26.
Anytus sculptus, September 7, 15.
Crambodes talidiformis, July 21.
Placodes cinereola, August 16, 19, 21, 25.
Plagiomimicus pityochromus, August 19.
Pyrrhia angulata, July 23 ; August 5.
Chamyris cerintha, July 23, 24.
Erastria carneola, July 7, 8, 10, 12, 15, 17, 20, 21, 23, 24, 27, 28, 30, 31 ; August 2, 5, 6, 7, 16, 19, 21, 25, 26, 28, 30 ; September 2, 4, 13, 15, 27.
E. synochites, July 15, 24, 27, 28 ; August 7.
E. nigritula, July 24, 27, 28, 30, 31 ; August 2, 5, 6, 7, 16, 18, 19, 21, 26 ; September 13.
Drasteria erechtea, July 27, 28, 30, 31 ; August 5, 7.
Ophiusa bistriaris, July 14, 28.
Euparthenos nubilis, July 14, 15.
Catocala Meskei, July 24.
C. Briseis, July 24, 28.
C. Ilia, July 27.
C. parta, July 24, 27, 28, 30, 31 ; August 5, 6, 7, 21, 26, 30 ; September 2, 13, 15, 25.
C. ultronia, July 8, 10, 13, 14, 17, 20, 21 23, 24, 27, 28, 30, 31; August 2, 5, 6, 7, 25.

Catocala concumbens, August 26, 28; September 15.
C. amatrix, August 21, 28.
C. cara, August 18, 21; September 2, 15, 26, 27.
C. cerogama, August 19, 26.
C. neogama, August 25.
C. habilis, September 15.
C. antinympha, August 26.
C. serena, July 24.
C. Clintonii, July 17.
C. polygama, July 7.
C. pretiosa, July 8, 10, 17.
C. nuptula, July 15, 20, 21, 23, 24, 27, 28, 30; August 2, 5, 6, 7.
C gracilis, July 21.
Homoptera lunata, August 26, 30; Sept. 2.
Homopyralis tactus, July 7, 8, 10, 12, 13, 14, 15, 17, 20, 21, 23, 24, 28, 30, 31; Aug. 2, 5, 6, 7, 16, 18, 19, 25.
Pseudaglossa lubricalis, July 17, 20, 21, 23, 27, 28, 30; Aug. 5, 7, 16, 18, 19.
Epizeuxis æmulalis, July 28, 31; Aug. 5, 28; Sept. 9, 25.
Xanclognatha marcidilinea, Sept. 9.
Clanyma angulalis, July 17, 31; Aug. 2, 16.
Renia Belfragei, Aug. 26.
Renia centralis, Aug. 30.
Renia lævigata, July 2.
Bomolocha abalienalis, July 21.
Hypena humuli, Aug. 25, 30; Sept. 2, 7, 13, 15, 27, 29; Oct. 22,
Plathypena scabra, Aug. 21; Sept. 9, 13, 26, 27; Oct. 22, 24.
Tortricodes bifidalis, July 28; Aug. 6, 7, 19.
Philometra serraticornis, July 20.

It will be observed from the above memoranda that a large number of the species (no less than forty, or nearly one-third of the whole) were quite rare, appearing on but a single evening, and usually in a single example. This, however, may not be taken as a measure of the actual rarity of the species in this portion of the State of New York. For some of the species other attractions would undoubtedly offer greater inducements. Many species are extremely local in their occurrence, perhaps abounding in a limited locality, and hardly to be found a mile or two distant. And again, the fact is well known to collectors that with nearly all the Lepidoptera, a year occasionally occurs when a species will appear in remarkable abundance. The two examples of the beautiful Noctua, *Chamyris cerintha*, taken as above stated on the 23rd and 24th of July, were very highly prized by their captor from their rarity up to that time. The following year, by the same method of sugaring and from the same grape trellis, between the 10th of June and 17th of August, nearly a hundred examples of it were taken — sixteen in a single evening. Such rarities in 1875 at Schenectady, as *Agrotis pitychrous*, *Agrotis alternata*, *Nephelodes violans*, *Cosmia infumata*, *Xylina ferrealis*, *Xylina pexata*, *Anytus sculptus*, *Catocala Briseis*, *Catocala antinympha*, *Catocala gracilis*, etc., — in 1877 at Center, by the captures there made, were consigned to the rank of common species.

In addition to a knowledge of the abundance of a species, the above and similar records may be serviceable in showing the duration of the period of apparition of the more common species, and also the succession of broods, when they occur.

On the first evening of collecting, July 7th, twenty-one species were taken, of which number one-third were species of AGROTIS. Of those present at this time, three, viz., *Agrotis c-nigrum*, *Agrotis ypsilon* and *Mamestra renigera*, continued into the month of October. In the record of *A. c-nigrum*, three intervals are shown of respectively sixteen days in July, twenty-three days in August and twelve days in September: may not three successive broods be inferred from this? *A. ypsilon* was not observed for the two weeks following July 7th, but continuously thereafter to October 19th, with the exception of five indicated absences of one and two evenings each.

Agrotis baja was captured in several examples on the 16th of August, and was observed each evening until its disappearance on Sept. 7th. The period of duration was probably a month (no collections were made between the 7th and 16th of August), and the same also of *Agrotis messoria*, from August 19th to September 18th.

In *Mamestra renigera*, two intervals appear of sixteen and twelve days each, in July and September.

Hadena devastatrix, *H. sputatrix*, *H. arctica* and *H. lignicolor* were among the most common species, and probably appeared in successive broods, as they were each present when the collections commenced, and two of the species continued into September.

Hyppa xylinoides, commencing the latter part of July, continued through all of August and September into the early part of October.

Luperina reniformis continued throughout August and September, and reappeared in the last half of October.

Leucania unipuncta, after the appearance of a single individual on two evenings in July, was present each evening for the space of two months. It was one of the most common moths at sugar, and was nearly always in remarkably good condition.

Amphipyra pyramidoides, appearing first on July 17th, continued, not every evening, however, for more than two and a half months.

Orthosia ferrugineoides was constant in its presence, and also an abundant species from its first appearance until the close of the season.

Orthosia helva was confined to the month of August, but was uniformly present after the 5th.

Of the Xylinas, *laticinerea* was the last to appear. It was in abundance and in perfect condition at the cessation of sugaring. It was the first to appear the ensuing spring (1876), on the 11th of April, followed a few days thereafter (April 25) by *X. Bethunei*.

Erastria carneola had a long duration, and was very seldom absent, although never appearing in large numbers. It continued until late in September. As it is known to occur in the month of May, its four months presence with us is evidence of a succession of broods, as is also shown in the freshness of examples collected at various times throughout the season.

Catocala parta continued for a very long time, having been taken on fifteen evenings in the months of July, August and September.

C. ultronia was the most abundant of the Catocalas, and, although not seen in the month of September, was observed on eighteen evenings during the preceding two months.

Homopyralis tactus was a remarkably constant visitor, having been unobserved on one evening only for nearly two months.

Pseudothyatira expultrix was present each evening of the month commencing July 15th.

The following table shows the period of duration of several of the species and their comparative constancy of attendance at sugar.

SPECIES.	First noted appearance.	Last appearance.	Evenings absent.	No. of days' duration.
Hadena arctica	*July 7	Aug. 26	1	†51+
Hadena lignicolor	7	6	0	31+
Hadena devastatrix	8	Sept. 2	2	57+
Hadena sputatrix	8	9	2	64+
Hydrœcia sera	7	Aug. 7	0	32+
Leucania unipuncta	Aug. 7	Oct. 6	0	61
Erastria carneola	July 7	Sept. 27	10	83+
Homopyralis tactus	7	Aug. 26	1	51+
Gortyna nictitans	17	26	0	41
Hyppa xylinoides	28	Oct. 2	5	67
Orthosia ferrugineoides	Sept. 7	25	0	49+
Catocala ultronia	July 8	Aug. 25	6	49
Catocala nuptula	17	7	2	22
Catocala parta	24	Sept. 25	12	64
Catocala Briseis	24	20	..	59
Catocala cara	Aug. 18	27	11	41
Catocala neogama	25	30	15	37
Xylina Bethunei	Sept. 4	Oct. 24	7	51
Xylina disposita	9	25	6	47
Xylina petulca	11	21	12	41
Xylina laticinerea	29	25	2	27

A few Microlepidoptera were among the preceding collections at sugar: for most of the determinations, I am under obligations to Prof. C. H. Fernald, of the Maine State College, at Orono.

Tortrix rosaceana *Harr*. .... July 4.
Exartema permundanum *Clem*. .... July 21.
Tmetocera ocellana (*Fabr.*) ‡ .... July 7.
Condylopeza nigrinodis *Zell*. .... June 23.
Depressaria atrodorsella *Clem* .... Sept. 15.
Depressaria pulvipurella *Clem*. .... Aug. 21.
Depressaria Fernaldella *Chamb*. .... Sept. 15.

* The collections were commenced at this date.
† The annexed + indicates that the entire period of duration is not shown.
‡ The Bud-moth (*Grapholitha ocellana*) of the Canadian Entomologist, v. III, p. 13, f. 9.

VI. ON SOME LEPIDOPTERA COMMON TO THE UNITED STATES AND PATAGONIA.

In the *Bulletin de la Société Impériale des Naturalistes de Moscou*, for 1875 (Vol. 49, Pt. 2d, pp. 191–247), an interesting paper is published by Prof. C. Berg, Director of the Museum of Natural History of Buenos Ayres, on "Patagonian Lepidoptera." It is based on collections made by the writer, in Patagonia, during a short visit in the year 1874. The collections were confined to the coast region, extending from the Rio Negro to the Rio Santa Cruz, or between 41 and 50 degrees of south latitude. The insect-fauna was found to be quite limited, as might be expected from the scanty vegetation of the coast. Could the interior country have been explored, it would, no doubt, have yielded much more abundantly.

Previous to this visit, but four or five species of Patagonian Lepidoptera were known. Fifty-six species were collected by Prof. Berg, at this time, of which twenty are described in his paper as new to science. Of these fifty-six species, nineteen were observed only in Patagonia, — the others had also been collected in the countries adjacent. The interesting statement is made that *Agrotis ypsilon*, *Heliothis armiger* and *Asopia farinalis* — species of extensive distribution throughout Europe and America — were apparently confined to those sections of the coast to which cultivation had extended, and, therefore, it was inferred that they had, in all probability, been introduced through commercial intercourse with other countries.

The collections were of the following groups: Of Rhopalocera, 14 species. Of Heterocera—Sphingidæ 3 sp.; Bombycidæ 5 sp.; Noctuidæ 11 sp.; Geometridæ 1 sp.; Pyralidæ 6 sp.; Chilonidæ 1 sp.; Phycidæ 4 sp.; Tortricidæ 1 sp.; Tineidæ 8 sp.; Pterophoridæ 1 sp.

A special interest attaches to the record of the above collection from the occurrence among them, in this remote region, of so large a number of species belonging to the United States — no less than seventeen species, or over thirty per cent of the entire number.

As the geographical distribution of our insects is at the present time receiving much attention, this list of Prof. Berg will be welcomed, from the care, apparently, with which he determinations have been made, and the extension of observations to a country of whose Lepidoptera scarcely any thing was previously known.

The species recorded which also occur in the United States are as follows:

Callidryas Eubule *Linn.*	[1]Leucania extranea *Guen.*
Danais Archippus *Fabr.*	Heliothis armiger *Hb.*
Pyrameis huntera v. Iole *Cram.*	Erebus odora *Linn.*
Pyrameis Carye *Hb.*	Asopia farinalis *Linn.*
Pamphila Phylæus *Drury.*	[2]Ephestia interpunctella *Hb.*
Philampelus labruscæ *Linn.*	Nomophila hybridalis *Hb.*
Philampelus vitis *Linn.*	[3]Plutella xylostella *Linn.*
Agrotis saucia *Hb.*	[4]Pterophorus leucodactylus *Fabr.*
Agrotis ypsilon *Rott.*	

Collections of the larvæ were also made, and a number of them described: their food-plants and transformations were also observed. A peculiarity of the caterpillars noticed by Prof. Berg presents so wide a departure from normal habits resulting from the modifying influence of surrounding conditions, that we are led to give the following translation, in full, of his statement:

"It still remains for me to note a peculiarity of the caterpillars, viz., their extreme ferocity — their cannibalistic propensities. All of them, irrespective of family or group, manifest the liveliest desire to kill their fellows. While confined they ate only one another, seldom, if ever, touching the food-plants. The caterpillars of the Bombycidæ completely devoured others of the same family, leaving absolutely no fragments of them. They even tore open the cocoons, from which they dragged out the pupæ and ate them — to which fact I called the attention of my traveling companions.

In like manner, the larvæ of the Noctuidæ acted among themselves and toward the Bombycidæ, and the latter toward the former. Among these last, *Heliothis armiger* was gluttonous beyond all measure,— one of them devouring in twenty-four hours, from six to seven others. The caterpillar of

[1] L. unipuncta *Haw.*
[2] Tinea Zeæ *Fitch.*
[3] P. cruciferarum *Zeller.*
[4] Aciptilia alternaria *Zell.*

Pyrameis Carye was also carnivorous, but to a very moderate extent, preferring at all times fresh plant-food to flesh, while others, as the Noctuidæ, *would not touch the plants after having once tasted flesh.*

This peculiarity of the Patagonian caterpillar is easily explained. During the summer, excessive heat and drought prevail, and these, coupled with dry winds, tend to wither and destroy all vegetation. As the caterpillar is then deprived of its proper nourishment it is compelled by the law of self-preservation to seek elsewhere for food, and so it comes that they eat one another. This habit becomes hereditary, and the descendants frequently practice it, even when there is no lack of vegetable food."

VII. ON LYCÆNA NEGLECTA EDW.

[From the Canadian Entomologist, for May 1, 1875.]

In the very interesting paper of Mr. W. H. Edwards, published in the May number of the Canadian Entomologist, in which another valuable addition is made to the knowledge of our Lepidoptera, by the identity therein shown of the Lycænas *pseudargiolus* and *violacea* — autumnal and vernal forms of the same species — it is suggested that *neglecta* and *Lucia* may prove to bear the same relationship to one another. The possibility of this is inferred by Mr. Edwards from observations made by him, that *Lucia* is an early spring form (April and May in New York), and *neglecta* a later one, " occurring at intervals from June till September."

I cannot believe that *neglecta* and *Lucia* will ever be united as seasonal varieties of the same species. Several years of diligent collecting by Mr. von Meske and myself in this portion of the State, embracing a range of ten miles of territory, have failed to reveal a single example of *Lucia*, nor has it come under our observation in any of the collections made by others in this part of the State. We might, therefore, be almost justified in asserting that it does not occur here. We have it from Long Island collected by Mr. Graef and Mr. Tepper.

On the other hand, in that famous collecting ground, Center, on the "pine-barrens," midway between Albany and Schenectady, upon the line of the N. Y. Central and Hudson River R. R., than which, we believe, the northern United States can produce no superior locality for the Lepidoptera, *neglecta* usually swarms at its proper season. There have been times and seasons when, as we have traversed the roadways leading over the yellow sands of Center and among its pines, that the air about us has seemed blue from the myriads of *neglecta* driven up from the damp sands by our approach. Here, certainly, one might confidently look for *Lucia*, were it but a varietal form.

Our observations and records do not agree with those of Mr. Edwards, giving June as the earliest appearance of *neglecta*. From notes made by me, and from dates of capture appended to examples in my collection, I cite the following :

In the year 1869, on May 21st, *neglecta* occurred in great abundance, all of which noticed, with three exceptions, were males. The worn condition of some of the captures indicated that they had already been abroad for several days. The locality had not been explored since the 11th of May, when the species was not found. About the 9th of June it was observed at its greatest abundance; it was seen for the last time during this year on the 30th of July. In 1870, it was first observed on the 14th of May (none in a collecting trip on the 6th). The last recorded appearance was on the 16th of June. *L. comyntas* was seen from May 6th to September 14th, continuously. In 1871, *neglecta* is recorded from May 16th to June 16th. In the following year its first record is on May 21st.

The latest date of my capture of this species is August 20th, at Schoharie, N. Y.; the earliest is at Bath-on-the-Hudson near Albany, on May 14th (the year not stated)

The observations which I have given above, when coupled with those of Mr. Saunders appended to the paper above referred to, of the frequent occurrence of *neglecta* in his neighborhood (London, Ont.,) and non-occurrence of *Lucia*, would seem almost to establish beyond question their non-identity. That these statements may receive all the consideration to which they are entitled, it may be proper to accompany them with the mention made to me by Mr. Scudder, not to be construed to the disparagement of the valued labors of others, that, as the result of an elaborate tabulation of the numerous returns made to him or collated by him, of the Rhopalocerous fauna of the various portions of the United States and Canada the two most thoroughly worked up fields were found to be those of London, Ont., and Albany, N. Y.

As a part of the history of *neglecta*, it may deserve mention that Mr. von Meske reports the species as quite rare this year at Center, where in so many preceding years it has abounded.

[Since the publication of the above *L. Lucia* has made its appearance for the first time at Center. Examples of it were collected by Mr. W. W. Hill, on the 16th of May, 1876, at this locality, where it was also captured on the 13th, 20th and 26th of May (5 specimens). At West Point, N. Y., it was observed in abundance on the 30th of April, when 8 males and 4 females were taken by a collector, and three or four times as many in addition, it is believed, were seen.]

VIII. DESCRIPTIONS OF TWO NEW SPECIES OF CALIFORNIAN BUTTERFLIES.

Lycæna Lotis n. sp.

Male. — Wings glossy violet-blue; margins bordered with black, extending narrowly on the costa to near the base — the black of the costa edged outwardly with white; veins defined by black scales; fringes white with black basilar scales. Palpi black above, white laterally. Thorax and abdomen black with long whitish hairs.

Beneath: wings gray. Primaries: the discal, extradiscal* and submarginal black spots, in appearance and position much as in *Scudderii;* the two rows of the submarginal series are more contiguous than in that species, nearly equally well defined, and without space between them for the fuscous spots usually present (at least in the median portion of the range) in the ♂ *Scudderii*, and always in the ♀. Secondaries: three white-annulated, black extrabasilar spots; the extradiscal doubly-curved series of similar spots, nearly as in *Scudderii;* the black spots of the submarginal series are nearly covered with metallic scales giving a green reflection (*blue in Scudderii*), anterior to which and resting thereon, a connected (on the veins) series of fulvous crescents, tending to a sagittate

*Reference in this description, to the shape of the spots in this series, is purposely omitted, for the reason that, in this group they are subject to so great variation that it is impossible to draw from them any reliable specific characters — at least from the inspection of a few individuals. In the twenty-eight examples of *Scudderii* before me, the following variations are noticeable: In one, all the spots are round (or nearly so) except the last interior one, which is geminate on the submedian fold: in another, not a single spot is round or even approximating that form: in one, the fifth spot, which, in the original description of the species is said to be "twice as long as the others," is in this, the shortest of all. In four examples, the second and third spots are prolonged inwardly toward the discal spot in a tail-like projection, while in others they are regularly rounded, and again in others, quite extended toward the outer margin. In one example, the first five spots are distinctly semi-oval in form; in others, the spots assume ovate, elliptical, triangular, crescentic and irregular forms. In five examples, there is an additional smaller spot between veins 8 and 9, preceding the one commonly called the first spot. A difference is frequently to be seen between the corresponding spots of the opposite wings form, narrowly edged before with black; at the tips of the veins, a row of subtriangular black spots. Legs, thorax and abdomen clothed with long, white hairs.

Female.—Above uniformly brown, being without the interior violet-blue shade characterizing *Scudderii;* a few (perhaps twenty on each side) purple scales are to be seen beneath the basilar portion of the median of the primaries and at the base of the secondaries. On the primaries, a submarginal crescentiform fulvous band, which is more distinct between the nervules of the median; on the secondaries a submarginal row of six semi-elliptical black spots, preceded by fulvous crescents, and followed by a few pale scales.

Beneath: on the primaries, the extradiscal row of black spots rather weaker than in *Scudderii* (as also on the secondaries), while the outer row of the submarginal series, which, in that species is often obsolescent, is in this, well defined and of nearly equal strength to the interior row. The secondaries show but two of the usual four extrabasilar black spots; the remaining ornamentation much as in the other sex.

Expanse of wings, ♂, 1.30 inch—♀, 1.25 inch. Length of body, ♂, .5 inch—♀, .44 inch.

This species differs principally from the allied species with which it is compared, in the black veins and brighter and more glossy wings of the ♂, and the uniform brown wings of the ♀, with its submarginal fulvous band on the primaries; in the stronger submarginal spots of the lower surface, and the weaker interior spots; the more numerous metallic scales and their peculiar hue; the shape of the fulvous crescents and the narrowness of the black lines bordering them; the heavy black termination of the veins, etc.

The secondaries are more prolonged on the submedian nervure, giving to the anal angle a greater prominence.

Hab., etc.—Mendocino, California. Two examples. Collection of W. H. Edwards.

Pamphila Osceola n. sp.

Wings above dark glossy brown as in *P. Metacomet;* outer margin blackish-brown; fringes, dark brown.

Male: primaries with some dull yellowish scales on the inner half of the costa, on the outer side of the discoidal stigma, and within it between the median and submedian nervures. Discoidal stigma velvety-black, consisting of two acutely ellipsoidal spots, which join on the 1st median nervule, and have their other extremities resting on the submedian and 2d median nervule — the inner spot with distinct black scales near the submedian. Beneath brown, blackish over the discoida stigma, with obscure yellow shades exterior to it as the only markings. Abdomen above, unicolorous with the wings, with yellowish scales laterally. Thorax beneath and abdomen contiguous, brown with some longer clay-colored hairs. Palpi clothed with bristling yellow scales, from which the tip of the last joint barely projects.

Female: primaries with dull yellow scales and hairs, more numerous on the inner half of the interior margin, and nearly absent from the outer margin; two yellow spots between the median nervules—the outer one scarcely more than a spot, the inner subquadrangular; no anteapical spots, but in their place some clustering yellow scales. Beneath, dark brown, the primaries reddish-brown basally, and the secondaries of the same shade throughout except toward their inner margin. The two spots of the upper surface of the primaries are reproduced beneath somewhat more obscurely. Thorax and front of head yellowish scaled; palpi with black scales above, and beneath with some clay-colored scales.

Expanse of wings, 1.28 inch; length of body, .62 inch.

The species is allied to *P. Metacomet;* the geminate character of its discoidal stigma is better defined; its fringes are darker; its median spots are yellow instead of whitish; it is without the anteapical spots, and lacks the band of pale spots on the secondaries beneath; the lower side of the abdomen is without the conspicuous mesial line of pale scales of *P. Metacomet.*

Hab., etc.—Mendocino, California. 1 ♂ and 1 ♀ in the collection of W. H. Edwards.

The references of these two species to the *28th Rep. N. Y. State Mus. N. H.*, made on pages 53 and 62 of *Edwards' Catalogue of Diurnal Lepidoptera*, require correction for the reason stated on page 70.

IX. ON SOME SPECIES OF NISONIADES.

Nisoniades Pacuvius n. sp.

Head and palpi thickly clothed with bristling brown and gray hairs, the obtuse tip of the third joint of the palpi only visible; antennæ brown above, the joints bordered with white beneath and within. Thorax and abdomen beneath with long brownish hairs; legs brown with pale hairs at their joints.

Wings approaching those of *N. Persius* in shape, but the primaries somewhat narrower.

Primaries umber-brown, mottled with black as in *Martialis;* near each extremity of the cell, conspicuously marked with a large black spot, the outer one having the hyaline white cellular spot on its outer margin. A row of black spots cross the nervules, upon which are the following white hyaline spots: four costo-apical ones, of which the costal one is scarce more than a dot, the second, the largest and quadrate, the third and fourth quite small, with their longest diameter in the direction of the breadth of the wing; in cells 2 and 3 each, a triangular spot with the apex directed toward the outer margin of the wing — that in cell 2 but partially hyaline; in cell 1 b, two triangular spots (not hyaline), marked with white scales so obscurely in the somewhat imperfect specimen, that possibly they may not prove a constant feature. Some white hairs and scales separate this row of black spots from a subterminal row of rounded black spots, which is again separated by a few similar white scales from the black terminal margin. Fringes umber-brown, their base cut by some white scales projected from the black margin.

Secondaries fuscous, faintly marked by some brown spots and an indistinct subterminal row of brown dots. Fringes snow-white with some brown scales of the terminal margin cutting their base, and at the apical angle of the wing, extending nearly to their outer edge.

Beneath, primaries pale brown, the hyaline spot in cell 3 showing conspicuously, and with white scales covering the extreme apical portion of the wing. Secondaries reddish brown basally and medially, and with a double row of pale brown spots before the outer margin between veins 1 b and 4. Fringes as above.

Expanse of wings 1.38 inch. Length of body, .58 inch.

Habitat.—New Mexico. Described from 1 ♂ in the collection of Mr. W. H. Edwards.

This species may be recognized among all those of the genus known at present, by the white fringes of the secondaries less sharply defined at their base than in *N. tristis*, by its smaller size, less pointed primaries and a less projected anal angle of the secondaries than in that species.

Nisoniades funeralis Scudd.-Burg.

Wings black, approaching *Pholisora Catullus* in shade; in the ♂, a few white scales on the outer half of primaries which cluster in a crescentiform subterminal line; an umber-brown spot resting on the discal cross-vein and another midway on the submedian fold; in the ♀, some white scales occur also on the basal half of the wing, the two umber-brown spots more conspicuous than in the ♂, and, in addition, a line of the same shade associated with the white scales of the subterminal line: in each sex four small (the two inner ones linear) anteapical white spots, and a larger one in cell 3; in the ♀, a discal spot in addition. Cilia brown, with some basilar white scales, more numerous in the ♀. Secondaries of ♂, prolonged at inner angle, nearly unicolorous; of the ♀, showing indistinctly two rows of umber-brown spots before the margin. Cilia snow-white, in the ♂ with black basal scales at and near the apical and inner angles, intermediately contrasting sharply with the black margin of the wing; in the ♀, the cilia longer, with some anteapical black basal scales, but none before the anal angle, where the white scales run over on the inner margin for a short space and then become dusky.

Head, thorax and abdomen above, black; terminal joint of palpi moderately projecting beyond the squarely-cut scales; antennal hook red.

Beneath, wings fuscous, the ♂ with the white discal spot indicated, and with two obscure rows of paler brown spots before the margin; in the ♀ the spots are much more conspicuous, of a much paler shade—the outer row of each wing consisting of whitish intranervular lines cutting the pale spots (the corresponding spots in *N. Brizo* and *N. Icelus* present this character in a degree). Antennal joints narrowly marked with white.

Expanse of wings; 1.75 inch.

Habitat. — Texas and California.

Described from 1 ♂ and 1 ♀ in perfect condition, received from Mr. Heiligbrodt, of Bastrop, Texas, and from 3 ♂'s in inferior condition from the Collection of W. H. Edwards (two from Texas and one from San Diego.)

This species is believed to be the *N. funeralis* of Scudd.-Burg. (*Proc. Bost. Soc. N. H.*, xiii, p. 293. 1870), it having been received from Mr. Edwards under that name. The marked contrast between the snow-white fringes of the secondaries and the black of the wings, in fresh examples, makes this the most beautiful species of the genus.

Nisoniades tristis (Boisd.).

Nearly allied to *N. funeralis*, and may best be separated from it by comparison. It is a little smaller in size; the secondaries of the ♂ are apparently less prolonged at the anal angle; the white spots of the primaries are larger, and the ♂ has also a white cellular spot; the brown basal scales of the cilia of the secondaries are not confined to the angles, but cut the white scales throughout the entire margin although less numerously intermediately. The wings on the under side lack the whitish intranervular lines upon the submarginal spots seen in *N. funeralis.*

Expanse of wings; 1.55 inch.

Habitat. — California. Material under observation, 2 ♂'s and 1 ♀, in imperfect condition.

The diagnosis of this species as given by Dr. Boisduval (*Lépidoptères de la Californie*, p. 22; 1869) is as follows: "*Alæ nigro-fuscæ; anticæ punctulo medio strigaque e punctulis sex similibus transversis albidis; posticæ fimbria alba.*

Elle a le port et la taille de *T. Juvenalis.*"

The description given in Morris' Synopsis, in addition to the above features, represents the line of white points as "separated into two groups — the one of four near the upper edge, the other of two, beyond the median nerve."

The spots "beyond the median nerve" vary in size and in number in the same species, as will readily be seen by an inspection of a number of individuals. In two of the above examples of *N. tristis*, there is a white spot in each of the cells 2 and 3; in the third, in cell 3 only. When but one spot is present, it is always that in cell 3, — the smaller of the two having disappeared. Nor does the number of anteapical white spots afford a specific character, for while the normal number is four, some examples of *N. Martialis* in my collection show but three, and others (more rarely) five.

Nisoniades Afranius n. sp.

Thorax and abdomen above, black; beneath, with brown hairs. Palpi clothed with long brown hairs. Legs fuscous.

Primaries with the costal margin nearly as straight as in *N. Persius*, but rounded toward the apex; moderately bent basally. Outer margin more rounded than in any ♂ Nisoniades known to me (the ♀'s, as a rule, having more rounded wings), as much so as in *N. Brizo* ♀. Inner angle rounded, with internal margin short.

The usual black markings in the basal region of the wing; the remainder clouded with brown, distinctly relieving the transverse line of elongated black spots, and the row of rounded submarginal black spots; a few gray scales are sprinkled over the brown ground. The black spots of the transverse band above vein 2 are more elongated in proportion to their width, more acute toward the outer margin, and more sharply defined than in any other known species — even than in *N. Ausonius*. The line of four small, anteapical, white, hyaline spots is sensibly drawn inward toward the base, so that an imaginary line traversing these spots will cut the outer margin within its apical half. A white hyaline spot rests on the black spot in cell 3, and the three black spots in cells 2 and 1 b, have some gray scales centrally. There is a trace of a small, whitish, hyaline, discal spot. The terminal margin is without the black line seen in *N. Martialis*.

Secondaries, dark umber-brown, with the two rows of pale brown spots, similar to those of *N. Persius* ♀.

Wings beneath, a rich umber-brown, showing on the primaries the discal and anteapical spots more plainly than above, and a white spot each in cells 3 and 2. The two rows of pale brown spots on the secondaries are strongly relieved by the dark ground. The margins of the wings bear a black marginal line, obsolete toward the apex of the primaries.

Expanse of wings, 1.20 inch : length of body, .48 inch.

Habitat.— Colorado.

I venture, from a single example, to designate this as a distinct species, in consideration of the entirely different aspect it presents from the other forms. It is one of the smallest of our species, about equal to *N. Ausonius;* has unusually rounded wings, and is more distinctly marked than any other species, except *N. Martialis*, from which it differs materially in the shape of its wings and its transverse band of spots less inflected at its last fourth toward the outer margin.

I have no opportunity of determining at the present, if the above may not be one of the two species from Colorado, to which Mr. Scudder has given the MS. names of *N. Petronius* and *N. Rutilius*, in Lieut. Wheeler's *Report upon Geographical and Geological Explorations and Surveys West of the One Hundredth Meridian*, 1875, pp. 786, 787.*

Through the kindness of Mr. W. H. Edwards, I have been permitted to examine a number of examples of NISONIADES collected in 1877 by Mr. H. M. Morrison, in Colorado. They were all perfectly fresh, in fine condition for examination, and were as follows:

Nisoniades Icelus Lintn.

Several examples did not differ, apparently, in the slightest particular from New York specimens, except in one small individual, of less than an inch expanse of wings, in which the pale color, indistinct ornamentation, and small size, are, in all probability, the result of imperfect development in the larval stage. One specimen of this species is reported by Mr. Mead, loc. cit., as having been taken in Central Colorado, but in Edwards' *Catalogue of Lepidoptera* (1877), its greatest western distribution is given as Illinois.

Nisoniades Brizo Boisd.-Lec.

The examples of this species in their bright coloring and distinct ornamentation were more beautiful than any which

*Chapter VIII Report upon the Collections of Diurnal Lepidoptera made in Colorado, Utah, New Mexico and Arizona, during the years 1871-1874. By Theodore L. Mead, pp. 738, 794; plates xxxv-xxxix.

had previously come under my observation. The series of six gray-centered and black-bordered spots crossing the cell parallel to the outer transverse row, which usually forms an interrupted series, in these, blend in a connected band, nearly as conspicuous as the outer row. This species, I believe, has not been previously reported from Colorado.

Nisoniades Martialis Scudd.

One of the two examples of this species corresponds with our usual New York forms, and the other, in the more subdued tone of its ornamentation, is similar to the individuals of our second brood, appearing in July and August. It is unfortunate that no dates of capture are appended to these specimens.

Nisoniades Persius Scudd.

The examples which I refer to this species, present some differences as compared with our eastern forms. In Mr. Mead's Report, ut cit., Mr. Scudder is quoted as having noticed some points of difference. Although Mr. Mead represents this species as the most common of its genus in Colorado, I have but three examples before me; and upon so small a number, I am unable to form a decided opinion.

Nisoniades Juvenalis Fabr.

I have, with some hesitation, labeled several examples agreeing among themselves, with this specific name, as I am unable to trace any constant features in which they differ from some of our New York forms. I am, however, of the opinion, that in the collections made in the vicinity of Albany, two species are included in our *N. Juvenalis*. Marked differences are noticeable in size, shape of wings, and markings, which are hardly consistent with a single species. The smaller form is that in which are seen more pointed wings and narrower, less rounded outer margins, and plainer ornamentation. Still, I have not been able to discover any marked features by which a separation can be made. The larger form with broader wings and conspicuous markings is of less frequent occurrence than the other. A large series from Center, N. Y., submitted some years ago to Messrs. Scudder and Burgess for the examination of the genitalia, contained both of these forms, but were all returned to me labeled as *N. Juvenalis*.

It is very desirable that large collections of these forms should be made for study, and that broods of them be reared from captured females imprisoned over their food-plant, upon the plan practised by Mr. Edwards with such signal success, and extremely valuable results.

X. TRANSFORMATIONS OF NISONIADES LUCILIUS LINTN.

The egg measures .03 of an inch in diameter. Its shape was not noted. An example examined was marked with fourteen ribs and twenty-five transverse striæ.

The larva before its second molting measured .30 of an inch in length, and previous to its third molting .55 of an inch. After the molting, its length was .70 of an inch. Its body bears numerous short, white, downy hairs, and is marked with white dots. Its color is yellowish-green, especially on the incisures, with a blue-green vascular line. The legs are tipped with fuscous, particularly the anterior pair; the prolegs are green. The segments show four annulations on the posterior half.

The body of the larva is translucent, allowing the internal organs to be seen. On the eighth segment an oblong yellow spot on each side of the vascular line, as in *Pieris oleracea*, marks the position of some of the viscera, and on the second segment is a similar mesial mark. The pulsations of the dorsal vessel are quite conspicuous. With a magnifier, ramifications of the branchiæ are to be seen, surrounding the stigmata.

After its third molting, the two brown spots on the head of the larva appear, which thenceforth are so marked a feature. At maturity the larva has attained a length of .8 of an inch, with a diameter in its broadest part of .16 of an inch; diameter of head .10 of an inch.

The last molting was on August 3d, and on the 6th the chrysalis was formed.

The chrysalis is cylindro-conical in form, not angulated; thorax slightly elevated; head-case rounded in front, depressed below a line drawn from the anal spine across the bases of the wings to the humeral tubercle — this tubercle dark brown in color, cylindrical, truncated at the apex, and located a little before the base of the anterior wings. The stigmata are white.

At this stage the transparency of the chrysalis permits the rapid pulsations within to be clearly seen. The nervulation of the anterior wings is perfectly visible, and that of the posterior pair, indistinctly.

Five days after pupation (August 11th), the following changes were noticed. The eye-cases had become purple; the wing-cases were whitish, perfectly relieving the nervulation; the abdomen green except at its tip where it was brown. The antennæ folded over the eyes, cutting off a small section of their upper portion, have the club brown, and showing the joints; the posterior leg-cases show numerous brown spinules on the inclosed legs.

On August 12th, a few hours before the escape of the butterfly, the chrysalis was brown, except at the abdominal incisures, where it was green and of a transparency disclosing some of the internal organs. The white annulations of the antennal joints were visible, and through the wing-cases could be seen the gray scales of the margin, the disk and the cilia of the wings. The butterfly emerged in the afternoon of the 12th.

From two other larvæ which had been reared on *Aquilegia canadensis** and changed to chrysalis on the 8th and 9th of August, butterflies were obtained on the 15th, giving for the length of pupation of the three examples, six, seven and six days respectively.

The following captures in the field of *N. Lucilius* were made during the year (1870) when the above notes were taken: May 16th, at Bethlehem, Albany county, 3 ♂'s; May 21st and 31st, one ♂ each, at Center; July 6th, 9 ♂'s at Bethlehem, and another at same locality on the 28th; and others again on August 26th, and September 9th and 14th at the same place.

On August 25th and 28th, five butterflies were obtained from larvæ which had been collected at Bethlehem. So late as September 9th, larvæ just emerged from the egg were taken, associated with others about half-grown.

There are two annual broods of this butterfly, and possibly a third.

---

* See Twenty-fourth Report on the N. Y. State Museum of Nat. Hist., p. 164.

XI. DESCRIPTION OF EUDAMUS EPIGENA BUTL.

Eudamus Epigena BUTLER. Lepidop. Exot., p. 65, pl. 25, f. 6. 1871.
Thymele " " KIRBY. Syn. Cat. Diurn. Lep., p. 655. 1871.
Eudamus " " EDWARDS. Cat. Diurn. Lep. N. A., p. 58. 1877.
Eudamus Orestes LINTNER MS.: non 28th Rep. N. Y. St. Mus. N. H.

Thorax, abdomen and wings dark brown, nearly unicolored, but rather deeper toward the terminal margin.

Primaries: costa moderately curved, outer margin nearly straight; in general shape in the ♂ resembling *E. Bathyllus* of same sex, but in the female with its prolonged secondaries, approaching *E. Tityrus* ♂. Cilia, fuscous on primaries merging into white toward the inner angle; on secondaries, white with black basilar scales opposite the veins, until to the angle on the internal vein, thence black. Eight transparent white spots on each wing, viz.: three small disconnected anteapical ones; one triangular cellular spot; a small one in cell 1a, touching vein 2; a larger double-concave one reaching from vein 2 to vein 3; a subtriangular one extending from vein 3 to vein 4; a minute one just above vein 4, equidistant from the margin with that in cell 1a.

Beneath: primaries black costally and above the 1st median nervule (vein 2) outwardly to the white spots — remainder, brown; spots same as above, margined with black. Secondaries, with the bands much as in *Lycidas*, except that they do not contrast so strongly with the ground, producing less of a mottled effect; the outer fourth (third in *Lycidas*) bordered with white (except at anal angle), traversed by numerous short, wavy, brown lines.

Expanse of wings: male, 2 inches, female 2.15 inches.

Habitat.—Texas.

From a pair in the collection of Mr. Otto von Meske, received from Mr. Heiligbrodt, of Bastrop, Texas, to whose faithful labors science is indebted for the discovery of a number of new and peculiarly interesting species of Lepidoptera.

The above species is of special interest from its uniting the principal features of *Bathyllus* and *Pylades*, and the consequent argument which it furnishes against the adoption of proposed genera, resting on microscopic detection of some slight variation in form or proportion.

In the belief that the insect was new to science, it was described by me as *Eudamus Orestes*, for publication in the 28*th N. Y. St. Mus. Report*, then passing through the press; but in the necessitated printing of the Report at an earlier day than was anticipated, the description could not (together with other papers in readiness) be given place. Hence, the erroneous reference made to *Orestes* on page 58 of Edwards' *Catalogue of the Diurnal Lepidoptera of North America.*

Subsequently, Mr. W. H. Edwards identified the species with a figure of Butler in his *Lepidoptera Exotica.* As the figure is accompanied by only a brief diagnosis, and but a few copies of the work are to be found in this country, it is thought that the above description may be of service.

XII. A SYSTEMATIC ARRANGEMENT OF THE EUROPEAN AND SOME AMERICAN HESPERIDÆ.

During the preparation of the Edwards' *Catalogue of the Diurnal Lepidoptera of North America*, the aid of Dr. Speyer, of Rhoden, Prussia, was solicited in the rearrangement of the difficult group of Hesperidæ. The revision kindly undertaken by him embraced only those of the North American species — forty in number — which were represented in his cabinet, together with the European species, of which twenty-three are enumerated. These latter could not conveniently be given in the pages of the catalogue, but as it is the first satisfactory arrangement of the European forms — the more valuable to us from its incorporation with our more numerous species — the present opportunity is taken to present the arrangement in full as furnished by Dr. Speyer.

It is proper to state that the free use which was made in the catalogue of the MS. of Dr. Speyer, especially in the publication of the generic definitions, was not in accordance with his intention, and has called from him an expression of regret. We hope that this further use of the MS. may prove less objectionable.

A. ASTYCI Scudd.

1. Carterocephalus Led.
Palæmon *Pall.*
=*paniscus* Fabr.
2. Cyclopides Hübn.
Silvius *Knoch.*
Morpheus *Pall.*
=*Steropes* W.-V.
3. Ancyloxypha Feld.
Numitor *Fabr.*
4. Copaeodes n. g.
Waco *Edw.*
minima *Edw.*
5. Thymelicus Hübn.
Thaumas *Hufn.*
lineola *Ochs.*
Actæon. *Rott*
6. Pamphila Fabr.
a Massasoit *Scudd.*
Zabulon *Bd. Lec.*
Hobomok Harr.
b **Sylvanus** *Esp.*
comma *Linn.*
Sassacus *Harr.*
Metea *Scudd.*
Leonardus *Harr.*
Huron *Edw.*
Phylæus *Drury.*
Brettus *Bd.-Lec.*

conspicua *Edw.*
Ætna *Boisd.*
Peckius *Kirby.*
Mystic *Edw.*
Manataaqua *Scudd.*
Cernes *Bd.-Lec.*
=*Ahaton* Harr.
Metacomet *Harr.*
bimacula *Gr.-Rob.*
Vitellius *Sm.-Abb.*
=*Iowa* Scudd.
Osyka *Edw.*
verna *Edw.*
Hianna *Scudd.*
7. Amblyscirtes Scudd.
vialis *Edw.*

B. HESPERIDES Scudd.

8. Pyrgus Hübn.
a **Lavateræ** *Esp.*
alceæ *Esp.*
=*malvarum* O.
althææ *Hübn.*
Proto *Esp.*
b **Sao** *Hübn.*
=*Sertorius* O.
orbifer *Hübn.*
c **malvæ** *Linn.*
=*alveolus* Hübn.
alveus *Hübn.*
serratulæ *Ramb.*
cacaliæ *Ramb.*
andromedæ *Wall.*
centaureæ *Ramb.*
carthami *Hübn.*
sidæ *Esp.*
d tessellata *Scudd.*

9. Nisoniades Hübn.
Tages *Linn.*
Persius *Scudd.*
Lucilius *Lintn.*
Icelus *Lintn.*
Brizo *Bd.-Lec.*
Martialis *Scudd.*
Juvenalis *Sm.-Abb.*
tristis *Boisd.*
10. Pholisora Scudd.
Catullus *Fabr.*
Hayhurstii *Edw.*
11. Eudamus Swains.
a Pylades *Scudd.*
Bathyllus *Sm.-Abb.*
Lycidas *Sm.-Abb.*
Cellus *Boisd.*
b Tityrus *Fabr.*
c Proteus *Linn.*

Since the above was in type, a copy has been received of a paper on the Hesperidæ of the European Fauna (*Die Hesperiden-Gattungen des europäischen Faunengebiets*), by Dr. A. Speyer. The author was not satisfied with the arrangement above presented, which had been drawn up at the request of some of his American friends, and which, from the limited time that he was able to devote to it, and the partial examination of species upon which it was based, was contributed only for private use — not for publication. Since then, he has undertaken a more thorough study of the species pertaining to the European Fauna, and the result, published in the *Stettiner Entomologische Zeitung* for 1877, pp. 167–193, is the exceedingly valuable contribution to the knowledge of this interesting group, which is cited above.

At the present time, as these pages are passing through the press, there is only the opportunity of presenting, in justice to Dr. Speyer, in company with his provisional arrangement, the following carefully prepared one recently given to the public. It embraces not only the Hesperidæ of Europe proper, but all those occurring within the European Faunal Division, which includes some of the northern and eastern portions of Asia, as defined in the author's "*Geographischen Verbreitung der Schmetterlinge.*" The Asiatic species are indicated by an asterisk. Forty-one species are recorded in the list, of which twenty-nine belong to Europe. The following is the list, which, in consideration of its careful arrangement, we transcribe literally:

HESPERIDES Latr.

1. **Cyclopides** H. (p.)

1. Morpheus (Pap. m.) Pall. = Steropes WV.
*2. Ornatus Brem.

2. **Carterocephalus** Led.

1. Palaemon (Pap. p.) Pall. = Paniscus F.
2. Silvius (Pap. s.) Knoch.
*3. Argyrostigma (Steropes a.) Ev.

3. **Thymelicus** H. (p.)

1. Lineola (Pap. l.) O.
2. Thaumas (Pap. th.) Hufn. = Linea WV.
*3. Hyrax (Hesp. h.) Led.
4. Actaeon (Pap. acteon) Rott.

4. **Pamphila** F. (p.)

A.

1. Comma (Pap. c.) L.
2. Sylvanus (Pap. s.) Esp.
*3. Ochracea Brem.
(Ætna Bdv. spec. americana ?)

B. (Goniloba HS.).

*4. Alcides (Hesp. a.) HS.

C. (Goniloba HS.).

*5. Mathias (Hesp. m.) Fabr. = Thrax Led., non Lin.
*6. Zelleri (Hesp. z.) Led.
7. Nostrodamus (Hesp. n.) F. = Pumilio O.

D.

*8. Inachus (Pyrgus i.) Mén.

5. **Catodaulis** n. gen.

*1. Tethys (Pyrgus t.) Mén.

6. **Pyrgus** H. (p.)

A. a. (Carcharodus H., Spilothyrus Bdv.).

1. Lavaterae (Pap. lavatherae) Esp.
2. Althaeae (Pap. altheae) H.
   Var. b. Baeticus (Spil. b.) Ramb. = Floccifera Zell.
3. Alceae (Pap. a.) Esp. = Malvarum O.
4. Proto (Pap. p.) Esp.
5. Tessellum (Pap. t) H.
   *Var. b. Nomas (Hesp. n.) Led.
6. Cribrellum (Hesp. c.) Ev.

B. a.

*7. Poggei (Hesp. p.) Led.

B. b.

8. Phlomidis (Hesp. phl.) HS.
9. Sao (Pap. s.) H. = Sertorius O.
10. Orbifer (Pap. o.) H.

7. **Scelothrix** Ramb.

*1. Maculata (Syricht. maculatus) Brem. et Grey.
2. Sidae (Pap. s.) Esp.
3. Cynarae (Hesp. c.) Ramb.
4. Carthami (Pap. c.) H.
5. Alveus (Pap. a.) H.
   Var. b. Fritillum (Pap. fr.) H.
   Var c. ? Cirsii (Hesp. c.) Ramb.
   Var d. ? Carlinae (Hesp. c.) Ramb.
6. Serratulæ (Hesp. s.) Ramb. HS. An praeced. var. ?
   Var. b. Caeca (Hesp. caecus) Fr.
7. Cacaliae (Hesp. c.) Ramb. HS.
8. Andromedae (Syrichth. a.) Wallengr.
9. Centaureae (Hesp. c.) Ramb.
10. Malvae (Pap. m.) L. = Alveolus H.
    Ab. Taras (Hesp. t.) Meig.
    *Var.b. Melotis (Hesp. m.) Dup. = Hypoleucos Led.

8. **Nisoniades** H. (p.)

*1. Montanus (Pyrgus m.) Brem.
2. Tages (Pap. t.) L.

9. **Thanaos** Bdv. (p.)

1. Marloyi Bdv. = Sericea Fr.

The List is followed by a Diagnostic Table of the Genera, after which, twelve pages are devoted to descriptions of, and remarks upon, the several genera.

A translation of the entire paper of 27 pages is contemplated, that American students of Lepidoptera may have the benefit of the highly valuable observations and criticisms which it contains.

XIII. NOTES ON NOTODONTA DICTÆA LINN.

? *Phalæna tremula* CLERCK. Icon. pl. ix, f. 13. 1759.
? *Phalæna Bombyx tremula* LINN. Faun. Suec., Ed. ii, p. 298, no. 1121. 1761.
? " " " LINN. Syst. Nat., Ed. xii, p. 826, no. 58. 1767.
" " *dictæa* LINN. Syst. Nat., Ed. xii, p. 826, no. 60. 1767.
Leiocampa dictæa STEPHENS. Ill. Brit. Ent., Haust. ii, p. 25. 1829.
Pheosia rimosa PACKARD; in Proc. Ent. Soc. Phil., iii, p. 358. 1864.
Notodonta Californica STRETCH. Zygænidæ-Bombycidæ N. A., i., pp. 116, 240, pl. 4, f. 5; pl. 10, f. 9, 1872–73.
Notodonta tremula STAUD. Cat. Lep. Eur. Faun., Ed. ii, p. 72, no. 975. 1871.—p. 72. 1877.

A larva of the above species was taken at Bath-on-the-Hudson, Sept. 9, 1869, on willow. It molted during the night, and on the following day it measured, when at rest, .95 of an inch. The following were its features: Head light yellow-green, subquadrangular, with an impressed median line; mandibles yellow, tipped with black. Body white dorsally, with a bright yellow stigmatal stripe bordered above with green. Caudal horn conical, white, tipped with glossy black and with a black stripe laterally. Caudal shield granulated, broadly elliptical in outline—its largest diameter transverse to the body. Stigmata broadly oval, velvety-black on a white ground. Legs ferruginous, with a black spot above them: prolegs with a glossy black spot laterally, and a dull black larger one above them, extending upward to the stigmatal line.

The larva was of remarkable transparency, exceeding that of any other which had come under my observation. The lateral and ventral regions had almost the transparency of glass.

It matured on the 19th, when it measured 1.1 inch long and .17 inch broad. It was not suspected at this time of having reached maturity, but was thought to be a young Sphinx, with probably one or two additional moltings to undergo before its pupation. The diminished activity shown by it, and its refusal of food, was ascribed to its change to poplar soon after its capture. On the 21st, at the suggestion of a friend that it had possibly matured, it was placed on some ground, when, much to my surprise, notwithstanding its weak condition, it speedily buried itself beneath the surface for pupation.

The moth was not obtained from it.

On Sept. 14, 1869, a second larva was found at Bethlehem, Albany county, feeding on the aspen (*Populus tremuloides*), in an earlier stage of its growth, and just after a molting, judging from the comparative size of its head, which was twice the breadth of its body. Its length was .56 inch, and diameter .05 inch.

It was fed on aspen leaves, and on the 19th it again molted. The following day it resumed its feeding, and the day thereafter its dimensions were, length — , diameter .08 inch, diameter of head .12 inch. It was of a yellow-brown color dorsally, with transverse slate colored markings centrally on the segments. (No further record of the larva: it probably died before its maturity).

On Sept. 5, 1872, another larva, 1.65 inch long, was taken on poplar. Body greenish-white dorsally, shading on the side into green; substigmatal stripe bright yellow, interrupted below the stigmata by the extension of the oval white spot encircling the stigma. Caudal horn black. Caudal shield broadly crescentic, granulated, with a glassy tubercle centrally and margined with brownish-red. Legs and prolegs having the portions of the body above them of a violet color — the prolegs with an acutely elliptical ferruginous spot upon them outwardly, crossed on their anterior part by a quadrilateral black spot.

Sept. 14, 187–, larva feeding on *Populus tremuloides*, at Bethlehem. Length at rest, 1.3 inch; diameter .18 inch; the head and first pair of legs extended in line with the body. Head of the diameter of the thoracic segments, subquadrangular, deeply impressed medially, smooth, of a bluish-gray color, showing reticulations under a magnifier; mandibles and a crescentiform spot bearing the eyes dull yellow. Body with a marked degree of transparency in its lower portion, shining, without the usual annulations of the segments, nearly cylindrical to the tenth segment, the eleventh broad, elevated in a prominent cone; the thoracic segments contracted when at rest, forming each three distinct wrinkles, making these segments broader than the succeeding ones; incisures deep; color bluish-gray, a yellow ventral line, and a bright yellow substigmatal one indicated by obscure yellowish markings at the incisures; a dark, bluish-gray transverse line on the side of each segment — the same shade surrounding the stigma and extending to the proleg; on the eleventh segment a blackish transverse line running behind the stigma upon the caudal horn — the latter .09 inch long, glossy black. Caudal shield ferruginous, rugose, elevated marginally and in a small tubercle centrally, in outline a broad ellipse having a lenticular portion excised from its upper fourth; anal plates subtriangular, and of a similar color and surface. Stigmata depressed, elliptical, surrounded (except the first) with a well defined white ring which is more broadly elliptical than the stigma.

Entered the ground for its pupation on Sept. 16th.

The larvæ briefly described by me in the *Proceedings of the Entomological Society of Philadelphia*, vol. III, p. 670, were in all probability this same species. Their color is given as bluish-slate, of about the shade of the branches of the poplar, on which they were feeding (*Populus nigra*). The body is described as gradually increasing in size to the tenth segment; the eleventh segment elevated in a hump, bearing the black caudal horn, one-tenth of an inch long; the stigmata broadly oval, black, white annulated. It is probably identical with the form occurring in Europe, and mentioned as a variety of the *dictæa* larva.

All my efforts to obtain the imago from the above larvæ, have resulted in failure. Examples of the larvæ collected by Mr. von Meske, have also failed to give the imago; after having safely reached the pupal state, they have uniformly died while in that stage. A figure of the larva, taken from a drawing made by me, is given in Glover's Plates of Lepidoptera, XCIX, fig. 16.

No examples of the pupæ, unfortunately, have been retained, and I am only able to recall their smooth and shining surface, and the tapering form of their abdominal segments tipped with a rather long, bifid anal spine.

Mr. Græf, of Brooklyn, has succeeded in rearing the moth from larvæ collected by him, and to him I owe the privilege of being able at the present to refer descriptions made by me several years ago of larvæ which have meanwhile proved an enigma to me, to the imago which they produce, which is apparently identical with the *Notodonta dictæa* of Europe.

I regret that I have not at hand a detailed description of the European larva, to compare with our own. Stephens (*Illustrations of British Entomology, Haust. II., p.* 25), says of it: "Larva naked, with a small conical protruberance on the anal segment; reddish-brown, green on the sides and glossed with violet above, with a black dorsal streak: it feeds on poplar, willow and birch, and is found in July and September."

Newman, in his *History of British Moths*, page 228, gives the following description: "The caterpillar has rather a large head, which is very slightly notched on the crown and shining, and is of a pale green color. The body is almost uniformly cylindrical until the twelfth segment, which is humped, and the hump terminating in a moderately sharp point; the color of the body is whitish or glaucous-green on the back, with a broad paler green stripe on each side—and adjoining this there is a narrow raised yellow-green stripe, just below the spiracles, and touching all of them except that on the twelfth segment; it extends the entire length of the caterpillar, terminating in the anal claspers; on the summit of the twelfth or hump segment, is a black transverse line. It feeds on the sallow (*Populus nigra*), etc. There is a common variety of this caterpillar which is plain brown, without the slightest appearance of the lateral stripe; this occurs after the last change of skin."

Dr. Speyer writes of the larva of the European *N. dictæa*: It has a tubercle of pyramidal shape on the 11th segment, which is much more pointed in the younger stages of the larva, so as closely to resemble the horn of a Sphinx. The mature larva has a strong porcelain lustre, and occurs in two varieties: one is of a green color with a yellow stripe on its sides, and the other is brown without the stripe. It lives on poplar.

Several examples of the American *N. dictæa* were sent by Mr. von Meske to Dr. Speyer to compare with the European forms. Having made the comparison, he does not doubt that they are identical, although the following differences are noticeable. "The American form has a bent white cross-line on the inner part of the brown portion of the anal angle of the secondaries, which is not found in the European. In the former the interior branch of the median nervure [vein 2] has the white streak with which it is marked, shorter and narrower than in the European, and entirely wanting from the middle branch [vein 3]. These are the only differences observable, and they are too slight to afford grounds for their separation."

In addition to the two examples in my own collection, I am indebted to the kindness of Mr. von Meske for the opportunity of comparing two examples of *N. dictæa* from Germany, and two from Racine, Wis.

In the European, the white stripe which traverses the brown anal patch very near the margin, commences in cell 1, within the internal vein, and is continued until near or just beyond vein 2. In the American, this line is not so distinct, and in an ex-larva example, from Albany, it is obsolete, being represented only by a few white scales; in another example, it commences on the fold and continues to vein 3.

The bent white cross-line pointed out by Dr. Speyer as characterizing the American form, is not equally well marked in all. In the Racine specimens, the line commences on the internal margin, runs for a short distance parallel with the general direction of the outer margin of the wing (not of the anal angle portion), and curves inward toward, and is lost in, the submedian fold. An exserted portion of the brown patch lies inside of this white line on the submedian vein (1 *b*). In the Albany example, the line is less conspicuous, and the brown portion inside of it is barely indicated.

The American examples, besides having the brown patch larger, have also the brown border of the secondaries heavier than the European, and continued to the apex.

The comparison of Dr. Speyer of the length of the white lines on the veins of the primaries is not sustained by the examination of other examples, as they vary in length and distinctness. In one before me, the white lines are of the same length on veins 2 and 3 — in another, shorter on 3. The more conspicuous bifurcating white line on the submedian fold, also varies in length. All the above lines also vary in their breadth; those on veins 2 and 3 being nearly as heavy as in the European, while in the Albany example, they are much more delicate.

In the other markings of the wings I find no differences of sufficient constancy to aid in the separation of the forms of the two continents.

So far as we are able to judge from the descriptions at hand, the larval forms also agree; and it is quite an interesting fact that the European variety which is destitute of the yellow lateral stripe, has also its counterpart in the example found at Schoharie, of which the description has been given, and in another taken at Sharon Springs by Mr. von Meske.

From an example of our eastern form sent by me to Mr. Stretch, of San Francisco, he has identified it as his *N. Californica.* If, however, the sketch of the larva made by Dr. Behr and the information which he gives of the larva, be correct, then there is a possibility that a comparison of additional examples of "*N. Californica*" may show it to be distinct. The figure of the larva, as reproduced by Mr. Stretch, is certainly quite different in its appearance from any of those which have come under my observation.

It will be seen that I have included in the synonymy of this species, references to the *Ph. Bomb. tremula* of Linnæus and Clerck, as probably identical with it. This opinion is held by several of the best European Lepidopterists, who claim that the same species was twice described by Linnæus. Others, as Staudinger, entertain the belief that the *dictæa* of Linnæus is a distinct species,—the one ordinarily occurring in Europe being the *N. tremula*, to which the name of *dictæa* has been improperly applied. Staudinger, in his citation of *N. tremula* in his catalogue, includes as a synonym "*Dictæa* (L. S. N. xii, 826, ex Barbaria, alia species esse videtur)," adding the references to the following authors who give *tremula* under the name of *dictæa:* "Esper, 58, 5; 84, 2; Hübner Beitr., 22; Ochsenh., iii, 63; Godart, iv, 19, 1; Freyer, 579."

For the present I think it proper to retain the familiar name of *dictæa* for the species, as the necessity for the proposed change does not appear to be clearly shown.

XIV. ON SOME NEW SPECIES OF CERURA.

Cerura occidentalis n. sp.

Head white. Palpi white, blackish laterally. Antennæ white with black pectinations.

Collar pale cinereous, traversed by a darker band and edged behind by a black band. Tegulæ pale cinereous, darker posteriorly; the narrow black band crossing their front, followed by a patch of orange scales, and a few black scales on their inner side. Thorax marked with black and orange bands of raised scales (apparently three orange bands).*

Abdomen above cinereous, the segments bordered behind with pale cinereous; beneath whitish; sides tufted with a lateral row of small black spots.

Primaries whitish basally, sprinkled with some black hairs; medially and terminally pale cinereous with more numerous black hairs. A black basal dot on the subcostal; an extra-basilar row of five black spots on the nervures, usually, in the males, in a straight line (5 examples), but sometimes the two superior are nearer the base (2 examples)†; in the ♀'s (5 examples) the two superior spots are considerably drawn in toward the base, the line presenting quite a curve costally. The median band of black and a few orange scales, paler than in *borealis* and *aquilonaris*, broadest on the costa, elsewhere of nearly uniform width; its black borders subparallel; the inner border more distinctly marked; its general course in the male, direct or slightly excavating the band below the median, while in the female it is conspicuously bent, on or below the same nervure; the outer border usually not well defined below the submedian fold. Behind the median band, a black transverse line, interrupted on the cell and indistinct over the submedian fold. On the discal cross-vein, an elongated black spot. Beyond this, two or three subparallel crescentiform

*A cabinet specimen of this species is rarely seen, in which the thoracic scales have not been so affected by greasing, that the bands can with difficulty be traced.

†In five examples of the European *bifida*, this line curves outwardly at the costal or on the inner margin; in one example (female) it is straight.

black lines (the inner of the three sometimes obsolete), preceding the abbreviated blackish subterminal band — the band usually terminating at the second median nervule (vein 3). The nine marginal intra nervular black spots smaller than in *borealis*, but larger than in *aquilonaris*.

Secondaries white, with traces of the inner margin of an outer border, mainly seen on the nervules and at the anal angle, and sometimes with indications of a mesial band behind the obscure discal spot.

Beneath: primaries as above, but less distinctly marked; secondaries with a large discal spot.

Described from 7 ♂'s and 6 ♀'s from the Collections of the Buffalo Society of Natural Sciences, Messrs. von Meske, Hill, Riemann, Tepper, Strecker, Kuetsing and Lintner.

Habitat.—From New York, Pennsylvania, Wisconsin and Canada (Montreal). It will probably be found to extend throughout most of the eastern portion of the United States. It has not, to my knowledge, been observed west of the Mississippi.

The above insect may be found in nearly all the principal collections of the country, under the name of *Cerura borealis*, it being the one which was described by Dr. Harris in his *Report on the Insects of Massachusetts*, 1841, p. 306, and referred by him to the *borealis* of Dr. Boisduval. The description is as follows:

"The ground-color of our moth is dirty white; the fore-wings are crossed by two broad, blackish bands, the outer one of which is traversed and interrupted by an irregular, wavy, whitish line; the hinder margins of all the wings are dotted with black, and there are several black dots at the base and one near the middle of the fore-wings; the top of the thorax is blackish, and the collar is edged with black. In some individuals the dusky bands of the fore-wings are edged or dotted with tawny yellow; in others [*Cerura cinerea*] these wings are dusky, and the bands are indistinct. They expand from one inch and three-eighths to one inch and three-quarters."

The extrabasilar straight row of five spots readily distinguishes this species from *borealis* and *cinerea* (but not from *aquilonaris*) and ally it with *furcula* and *bifida* of Europe. It is usually of a smaller size than our other species.

Cerura borealis (Boisd).

This is quite a different insect from the preceding, and need not be mistaken for it, or any other species. It is figured in *Cuvier's Animal Kingdom*, London, 1836, vol. IV, pl. 98, fig. 5, as *Dicranura borealis* Bdv. The figure leaves no doubt of the species intended. On the right wing of the illustration, the five inferior nervular spots (see description below) are faithfully depicted in proper position on the median nervules (the two superior ones not shown). On the left wing the two inner sinuses and the three outer of the mesial band are correctly represented in form, size and position. The absence of the two costal black spots between the bands, and of some of the extrabasilar ones, would indicate some imperfection in the example figured. It is also well figured in Smith and Abbot's *Lepidopterous Insects of Georgia*, London, 1797, p. 141, pl. 71, as *Phalæna furcula* — believed by Smith to be identical with the *furcula* of Europe; but to this insect it bears no greater resemblance than to *C. occidentalis*.

It is characterized by its white head and collar, thorax in front marked with a conspicuous transverse black line, abdominal segments broadly banded with fuscous dorsally. Primaries of a snow white ground color, a basilar black dot, followed by four other nervular ones, forming an angular line; a broad, centrally constricted, well-defined, mesial band, of about the same width on the two margins; between this and the subterminal band are two distinct costal spots, and below these on the nervules, seven black spots arranged in an oval, as follows: the two inferior ones on the first median nervule (vein 2), the two medial spots of the exterior four, at about the inner third of the second and third median nervules — veins 3 and 4; of the two medial spots of the interior ones, the lower is at or just before the bifurcation of veins 3 and 4, and the upper is on the discal cross-vein; the superior spot of the oval is on vein 5, equidistant from the two superior medial spots below it. The subterminal line is distinctly marked, and followed by the subterminal band, reduced to a line as it crosses veins 3 and 4, but again expanding on the inner margin.

Secondaries with a well marked discal spot, and with a broad marginal band; the latter in some examples is obsolete.

Expanse of wings: from 1.50 to 1.70 inch.

Habitat. — New York, Pennsylvania, Virginia, Georgia, Missouri (Aug. 26, at light, Riley).

From 3 ♂'s and 6 ♀'s, in the Collections of Messrs. von Meske, Tepper, Riley, Strecker and Lintner.

The seven black spots on the white ground intermediate to the bands, arranged in an ellipse as above described, readily distinguish this species from any other of the genus.

It is closely allied to the *C. bicuspis* of Europe. Mr. A. G. Butler, of the British Museum, to whom I communicated an excellent photograph of it, writes: "it precisely agrees with some of our European examples of *bicuspis*." Dr. Speyer who has received an example from Mr. von Meske, remarks of it: "it is very near to *bicuspis*."

Although quite dissimilar in color from *C. cinerea* Walker, yet it is closely related to that species in the form of the bands (often imperfectly defined in *cinerea*) and in the arrangement of the intermediate nervular spots.

The moth has been reared from larvæ found by Mr. F. Tepper of Flatbush, L. I., feeding on wild cherry, when near their maturity. I am indebted to him for the following note in regard to them. "They are of the same shape as *borealis* [*occidentalis*] but differently colored. Instead of the green of that species, the color is greenish-yellow, the dorsal patch is rather smaller and of a brighter shade, and the minute spots on the sides are more delicate and brighter. The examples met with have been larger than *borealis* and somewhat heavier in appearance. Three mature larvæ were taken by me between the 10th and 15th of July, 1875; one was ichneumonized; the other two spun up in the same manner as *borealis*, within a few days after their capture, and the moths emerged in from two to three weeks."

In Smith and Abbot's *Insects of Georgia* it is said of it: "The caterpillar was taken the latter end of July, feeding on that kind of poplar vulgarly called the cotton-tree. When disturbed, it shoots out of the ends of its forked tail two soft orange-colored threads. Early in August it shed its skin, and on the 10th of that month it inclosed itself in a case made of chips of wood and affixed to a branch. The moth came out April 24th. It likewise feeds on the wild cherry and willow, and is found also in Virginia, but it is a very rare species."

Cerura aquilonaris n. sp.

Head, collar and tegulæ white, the latter crossed anteriorly by a row of glossy purple-black scales; the collar bordered behind by a similar row. Thorax with elevated purple-black and orange scales, which are probably, when in perfect condition, arranged in transverse rows. Abdomen above black, with white borders to the segments, which increase in width as they recede from the thorax; a lateral row of black spots; beneath white.

Wings white as in *borealis*. Primaries with a black basilar spot on the subcostal; beyond, four black spots in a straight line, of which the superior one, under a magnifier, is shown to be triple and the remaining three, in perfect examples, double. Median band consisting of black, orange and pale ash scales and prominently bordered with shining purplish-black scales. The band, in the female, is twice as broad on the costa as on the internal margin, and in the male but one-fourth broader costally; twice equally constricted between the median and submedian; its inner border projected on the subcostal, median and submedian nervures, and slightly on the median fold, on either side of which the excavations are deeper than elsewhere, making the general course of the line a little indirect; the outer border projected on the median, median fold and submedian. Subterminal band less sharply excavated on vein 7 than in *C. occidentalis*; inner border less enlarged at internal angle than in *C. borealis*; outwardly with a black dash before the apex on veins 6, 7, 8 and 9 (the last, costal). Of the usual three lines crossing the wing between the median and subterminal bands, the anterior one is broken and the other two faintly continuous. The terminal intranervular spots are smaller than in *cinerea*, *occidentalis*, *borealis* or *multiscripta*.

Secondaries, without a border, but with traces of an obsolete outer margin near the apex and more distinctly at the inner angle; a few black scales on the discal cross-vein.

Beneath, the primaries have a distinct discal spot and are marked on the costa with black spots at the points where the lines of the opposite side commence, which lines are seen in transparency.

Expanse of wings: ♂ and ♀, 1.60 in.

Habitat. — Canada, Montreal.

From two examples in the Collections of Mr. C. W. Pearson and Mr. F. B. Caulfield, of Montreal, captured in that city. It is probably quite a rare species.

This species, which bears a general resemblance to *C. borealis*, may be distinguished from all others known to me by the black marking cf the veins sub-apically. The conspicuous black bands of the abdomen above, in one example (the ♀, the ♂ being without its abdomen) may also prove to be a good distinctive feature.

Cerura candida n. sp.

Antennæ white, with black pectinations of considerable length in the ♀. Palpi porrected, white, outwardly with black hairs. Front, patagiæ, thorax and collar white — the latter with a single fuscous band. Abdomen white, the segments on their anterior border with a few dusky hairs; beneath white; anal region, pale brown. Legs white, banded with black; tarsi black, banded with white.

Primaries, silvery white, with four interrupted black bands before the discal spot — the discal spot forming a small oval ringlet — followed by three interrupted black bands, and a fourth heavier one at the apical and anal region; the usual line of eight intranervular marginal blacks spots, extending on the fringe, less conspicuously marked than in *C. multiscripta*.

Secondaries wholly white, without marginal spots, showing faintly, by transparency, the discal dot of the under surface.

Beneath, primaries with the outer lines of the upper surface heavily marked on the costa, and a dusky cloud behind the cell. Secondaries with a fuscous spot on the outer third of the costal margin.

Expanse of wings, 1.75 inch; length of abdomen, .8 inch.

Habitat. — Kansas.

From a specimen in my Collection, received from Mr. H. Strecker.

This beautiful species is allied to *C. scitiscripta* Walk.,* and *C. multiscripta* Riley.† It cannot be the former, which it more nearly resembles, as that is described with three thoracic bands, the fore-wings with an ochraceous tinge, and the wings [the four] with black marginal dots. In a pen-and-ink sketch of the typical specimen in the British Museum, kindly sent me by Mr. Butler, the marginal dots of the secondaries are represented. *C. candida* is the only species of the genus which we have seen, in which these spots are absent from the secondaries, and in which the wings are entirely white.

* *List Lep. Ins. Br. Mus.*, Pt. xxxii, p. 408. 1865.
† *Trans. St. Louis Acad. Sc.*, vol. iii, p. 241. 1875.

Mr. Strecker, to whom I communicated the MS. name of this species, together with its distinctive features as observed by me, expresses, in letter, his opinion that it may prove to be but a form of *scitiscripta*, inasmuch as some of his examples "show no thoracic band at all, not even traces of it, and again, others (four examples ♂ ♀) have the usual black marginal spots of the secondaries very distinct, so much so as any *scitiscripta* I ever saw; the veins on secondaries in some are also dark, like *scitiscripta*."

Should this species be shown to vary to the above extent, it would be an anomaly in the genus, for nothing approaching so great variation has come under my observation in the course of my critical study of the several species. To the contrary, I have found the species to be remarkably constant in their ornamentation.

I would not hesitate to refer examples so differently marked as indicated by Mr. Strecker to distinct species, in the absence of sufficient evidence of their identity.

The above descriptions are published at the present time, in advance of an extended paper on the species of CERURA — American and European — which, as is known to many of my correspondents, was commenced some time ago. It has been delayed, from my inability to obtain a few species which it seemed desirable to embrace in it, and from not having been able to arrange for the satisfactory illustration of the paper.

I avail myself of the present opportunity to express my obligations to my friends — to Messrs. Bowles, Caulfield, Grote, Hill, Hoy, Kuetzing, von Meske, Pearson, Riemann, Riley, Strecker and Tepper — who have freely loaned me all the examples contained in their collections, and have most generously permitted me to retain them for an unusual length of time.

XV. ON CARADRINA FIDICULARIA MORR.

This species was described as *Segetia fidicularia* by Mr. Morrison, in *Proc. Bost. Soc. Nat. Hist.*, vol. xvii, p. 145. In the Grote Check List (No. 456) this name is recorded as a synonym of *Caradrina ?multifera* Walker. In vol. viii, p. 188 of the *Canadian Entomologist*, Mr. Grote refers the species to the European *C. cubicularis* S. V., from a comparison made by him, with a male example of the latter, noting as the only difference between the American and the European forms, that the latter has the common line beneath more extended, and the hind wings white.

With four examples of *C. fidicularia* before me (three in nearly perfect condition), and two of *cubicularis*, I am compelled to differ from the reference made by Mr. Grote. I note the following points of difference, as my reasons for regarding them as distinct.

While in *C. fidicularia* the primaries are of a clear gray, in *cubicularis* they are of a peculiar pale brownish shade, difficult to designate, bearing what might be called an amber tint. In the former, the intranervular marginal black dots are more conspicuous than in the latter, and its subterminal line is not followed by whitish ; the posterior transverse line is farther removed from the reniform than in *cubicularis*. *C. fidicularia* has a well marked discal dot, which the other has not. The primaries of the American species are the broader.

The secondaries of *fidicularia* are of a fuscous shade, increasing in depth toward the margin ; in the other they are white, with the end of the nervules and extreme margin tinted with ochraceous. In the latter, no discal spot is seen from above, while in the former it is quite conspicuous — the heavily marked dot of the lower surface showing in transparency.

C. cubicularis is known to be a variable form, differing greatly, according to Guenée, in size, depth of color, and intensity of designs, having been described under the several names of *quadripunctata* Fabr., *segetum* Esp., *callisto*

Engr., *blanda* Haw., *superstes* Steph., and *leucoptera* Beck. But the difference of color between our form and the European does not come within the range of variation thus far shown in any of the species accepted as common to the two countries.

These colorational differences have been critically studied by Dr. Speyer, during the comparisons in which he has been for some time engaged, of the identical and closely allied Noctuidæ of Europe and America, and some of the results of which have been given in a series of papers, four in number, communicated to the *Entomologische Zeitung zu Stettin* for the years 1870 and 1875. In his second paper on *Europäisch-amerikanische Verwandtschaften*, p. 102, these differences (probable climatic modifications) are so admirably presented, that a translation of that portion of the paper, kindly furnished for the purpose by Mr. Grote, cannot but prove most acceptable to those engaged in the study of the Noctuidæ who may not be able to avail themselves of the original.

"In the gray and brown colors usual to the Noctuidæ, arising from a mixture of black, white and red, the American specimens, as a rule, show less red than the European, and more black.

"This appears generally in the color of the abdomen and the hind wings; the brown-gray of the European forms becomes clear gray or blackish-gray in the American. The red cast which shows on the gray undersurface of the wings of many brown species, especially on the edges, becomes very faint in American varieties, and is even, at times, wanting. The red brown of the back and fore-wings becomes more gray, blackish or bluish. A stronger mixture of black darkens the colors of many American forms. The ornamentation (such as costal marks and sagittate points) is often made more distinct and coarser; the transverse lines and discal spots are thrown more into relief by their deeper black defining lines.

"But, in opposition to this rule, with regard to gray and brown colors, those arising from a mixture of yellow and red, show more red in American specimens; and where there is a mixture of black with these latter colors, producing a rust-color, the black is less perceptible, as for example, in the case of *Orthosia ferrugineoides*, *Hydrœcia nictitans*, *Plusia Putnami* and *Brephos infans*.

"If there is any change in the shape of the wings, it seems

to be more usual for the American specimens to have them broader and shorter than the European.

"Only to the first-named modification in the mixture of gray and brown colors, might some importance be attached, since it seems to pertain, although not without exception, to the majority of compared species, and may therefore be referred to a common cause, arising from the different climates of continental eastern North America and insular western Europe. Which of the many climatic factors influences this modification of color — whether it is due to the more intense heat and dryness, or to the severe winter of the transatlantic faunal territory, or to both combined and as opposed to the cooler and damper summers and milder winters of the cis-atlantic — cannot now be determined. It is well known that the coloration becomes generally darker and blacker as we approach the pole or ascend in elevation; but it is doubtful if the greater cold of the winter is the real cause of this effect."

C. fidicularia appears to be a rare species with us. Mr. von Meske has taken it, at sugar, at Sharon Springs, N. Y., on August 15th, in two examples. Mr. W. W. Hill has captured it in Lewis county, N. Y., on August 1st, also at sugar. I have taken it on but one occasion, at Schoharie, N. Y., on 5th of September. It has not made its appearance, at sugar, during my two years' collecting by that method at Schenectady. Mr. Morrison's example was from the Adirondack region. A species, believed to be the same, and referred to *C. multifera* Walker, in Vol. I, of the *Canadian Entomologist* (page 84) was captured at Coburg, Ontario. Mr. Walker's specimen was from Nova Scotia.

C. cubicularis is stated by Guenée to be very common in the months of June and September. Wood (*Index Entomologicus*, p. 44) refers to it as common in gardens and meadows during the middle of June. Dr. Speyer in his *Fauna of Waldeck*, represents it as common throughout the month of May, and with a second generation, less abundant, extending from the latter part of June to the last of August.

A delay in the printing of these papers enables me to add to the manuscript as above prepared, a comparison made by Dr. Speyer, at my request, of *C. fidicularia* with the European species. He writes as follows:

"Of *C. fidicularia* I have but a single specimen, and I should not dare to give an opinion as to its distinctness, were it not, fortunately, for a decided difference presented in the antennæ. In *C. cubicularis* the antennæ are short and evenly clothed with cilia which are only one-half so long as in *C. fidicularia*; in the latter species their length is about equal to the diameter of the antennal stem. Furthermore, my example lacks the rust-brown bordering of the subterminal line, which is so conspicuous a feature in *cubicularis*. The apex of the wing in *fidicularia* shows a lighter shade between the subterminal line and the margin, while in *cubicularis* this portion is equally dark with the rest of the subterminal region. With these exceptions, I find no difference, except in the very different colors of the two. The white spots of the reniform, are also more or less distinctly seen in *cubicularis*. Whether the differences above stated are reliable, could only be determined by an examination of a number of examples. But the decided difference in the antennæ cannot be questioned, and this feature will sufficiently establish the specific distinction of the two forms. Their comparative size is of little importance, as CARADRINA varies very much in that respect."

XVI. THE LARVA OF HOMOHADENA BADISTRIGA.

Hadena badistriga GROTE: in Trans. Amer. Ent. Soc., iv., p. 20. 1872.
Homohadena badistriga GROTE: in Bul. Buf. Soc. Nat. Sci., i. p. 180. 1873.

The young larvæ, three-eighths of an inch in length, were discovered on May 30th, at Schenectady, N. Y., feeding on the leaves of the honeysuckle.

The mature larva measures 1.12 in. long, by .18 in. diameter. Head small, about one-third the breadth of the central segment, flesh-colored, spotted with dull green and with the frontal triangle bordered within and without by black; ocelli black.

Body cylindrical on segments 3–9, the last three rapidly tapering, the last one being less than half as broad as the central ones; the second segment (head not counted) slightly smaller than these, and the first a little broader than the twelfth; surface smooth, without hairs, except the usual minute setæ of the setiferous spots, conspicuously striped as follows: a broad substigmatal band, traversed by longitudinal waved lines, limited above by a black line on the thoracic segments which becomes obsolete on the fourth segment; a somewhat narrower stigmatal band of dull green — a whitish line traversing the stigmata, dividing the band equally in different shades of green; a subdorsal pale band limited above by a black line which is more distinct on the central segments where also the band is paler; above this a black stripe commencing on the fourth segment, becoming more marked on the central segments, and terminating on the ninth — this line bordered above by a corresponding one of white; dorsal stripe geminate, whitish on the thoracic and terminal segments — intermediately, expanding between the middle of each segment and its posterior portion to inclose a mesial black spot or spots resting on the incisure — bordered outwardly by olive-green, which by being broken at the incisures gives the conspicuous dorsal feature of two oblique dashes traversing the segments and approximating anteriorly; the four trapezoidal spots of each segment which rest on these dashes anteriorly, are white — the front ones the larger and marked with black on nearly one-half of their outer portion; on the second segment a large white spot rests on each green line bordering the subdorsal.

Stigmata small, broadly oval, black ringed.

Legs spotted with brown.

Several of the young larvæ were found during the month of May, at Schenectady, feeding on the leaves of the trumpet honeysuckle — the woodbine of Europe (*Lonicera periclymenum*). They were secured in a tin box and supplied daily with fresh food. Toward the latter part of the month and early in June, larvæ nearly full grown, were taken from the vines. They were generally found extended at rest upon some portion of the stem which so closely resembled their markings, that, added to their tapering extremities, they were with difficulty detected — appearing rather as enlargements of the vine.

From about twenty larvæ collected, twelve matured and formed their cocoons between the 9th and 16th of June.

The cocoons are elongate-oval in form, and five-eighths of an inch long by nearly one-fourth inch broad. They are of rather a slight texture, inclosed by leaves drawn around them, and were attached by their flattened under surface to the bottom and sides of the paper box to which they had been transferred; one cocoon, only, was fastened to a twig.

The remains of the pupa-cases found within the cocoons are of a chestnut-brown color, but were too much broken to afford any special features for description.

The first imago emerged on June 29th, having been twenty days in its cocoon. June 30th, five of the moths emerged, followed by others on July 3d and 4th.

Prof. Grote, in the *Buffalo Bulletin*, ut cit., mentions the larvæ as occurring on the common honeysuckle — *Lonicera sempervivum.*

XVII. DESCRIPTIONS OF TWO NEW SPECIES OF XYLINA.

Xylina lepida n. sp.

Anterior wings plumbeous gray ; lines distinct, pale gray. Demi-line in two elongated teeth, bordered on each side with black. Anterior transverse line bordered behind by black, quite angulated, united with the orbicular, sharply toothed (more than in *X. Thaxteri*) on the submedian fold. Posterior transverse line bordered before with black, distinctly toothed on the veins, touching the reniform beneath in two of its inward inflections, and connected with the anterior transverse by a black line on the submedian fold. Subterminal line less sharply projected inward opposite the cell and on submedian fold than in *X. Thaxteri*, and bordered outwardly by a series of connected black lunules which are heavier than elsewhere against its more prominent inward inflections.

Discal spots distinct, gray, paler just within their black border, making almost a double annulation: orbicular large, oval, oblique, very near to reniform beneath: reniform elongate, nearly straight outwardly, quite exserted inwardly, with a black line within it on the median vein, distinctly bordered on all sides with black, but more heavily below; in *X. Thaxteri* the spot is more broadly edged beneath with black, and more deeply exserted outwardly; the two spots are connected at their extremities by a black line. A brown shade-band traverses the median space between the discal spots, terminating on the internal margin midway between the transverse lines.

Posterior wings pale fuscous, with a discal spot and an indistinct fuscous band before the outer margin.

Thorax, color of primaries, sprinkled with gray and prominently tufted. Frontal tuft and shoulder covers bordered with a black line followed by white. Abdomen untufted, reddish.

Beneath, wings reddish: primaries with a heavy discal spot and partly crossed by two extradiscal bands: secondaries with a distinct discal spot and median line.

* *Lithophane lepida* Lintner MS. GROTE. in Bull. U. S. Geolog. — Geograph. Surv. Terr., iv, no. i, p. 181. February, 1878.

Expanse of wings from 1.50 to 1.56 inch.

Habitat. — New York; Maine.

Described from 2 ♂'s and 3 ♀'s, taken, at sugar, at Center, N. Y., on October 1st, 8th, 9th 12th and 15th, by Mr. W. W. Hill. The types are in Mr. Hill's cabinet.

This species has more resemblance to *X. Thaxteri*, than to any other of our known species. It lacks, however, the marked contrasts presented in paler ground color, the heavier black markings and the red dashed reniform of that species. The black line on the submedian fold is not so long or so heavy, and its posterior wings are not so dark.

It resembles more closely the *X. conformis* of Europe, but it has not the distinct basilar line, the claviform spot, or the broad reniform of that species. Its transverse lines are also more sharply angulated, and it is apparently a shorter-winged species.

Xylina unimoda n. sp.

Head and thorax cinereous. Abdomen above brownish, darker than the posterior wings.

Primaries glossy, bluish-cinereous, slightly paler at the costo-basal space, and with a few white scales on costal and basal regions. Transverse lines inconspicuous, scarcely visible without a magnifier, pale gray, faintly bordered by a darker shade. Median band faintly visible, angulated on the lower portion of the reniform. Anterior transverse band prominently toothed; the cellular teeth separated from the orbicular; the two small teeth inclosing the submedian fold, with some whitish scales interiorly, and outwardly continued in an acute claviform mark. Posterior transverse line dentate, removed from the reniform. The obsolete subterminal line preceded by intranervular fuscous sagittate spots. Orbicular spot pale gray, darker scaled interiorly, either connected with or detached from the suborbicular, which is gray-bordered above. Reniform inconspicuous, cinereous, with a black border more heavily markéd below, and edged within by gray.

Secondaries pallid, slightly darker along the margin, with paler fringes, which are almost white near the anal angle, a distinct cellular spot, and faint median line.

Beneath, secondaries paler than in *X. laticinerea* and less thickly sprinkled with fuscous scales.

Size intermediate to *X. cinerea* and *laticinerea*.

Habitat.—Center, N. Y.

Described from seven examples — 3 ♂'s, 4 ♀'s — taken at sugar, by Mr. W. W. Hill on the 5th, 8th and 12th of October, 1877. Types in Mr. Hill's Collection.

This species can be separated from *X. laticinerea* to which it is closely allied, by its uniform cinereous shade, without black lines or spots; by the absence of the black basilar line on the submedian fold edged above by white, and of reddish-brown scales in the reniform. Its orbicular is nearly round, and lacks the extension and obliquity seen in *laticinerea*.

XVIII. NOTES ON CUCULLIA LÆTIFICA LINTN.

In the *Check List of the Noctuidæ of America*, by A. R. Grote, I had described (on page 24), the above species from a male specimen received by Mr. O. von Meske from a correspondent in Bastrop, Texas, as follows :

Closely allied to *C. Speyeri*. The anterior wings are narrower and less curved anteapically than in that species ; they are of a paler gray shade. The subobsolete reniform and orbicular spots are marked with ochraceous-yellow dashes ; a streak of the same color rests on the subcostal nervure at its base and another within the inferior tooth of the anterior transverse line. This line is more acutely toothed than in *Speyeri*. The oblique black streak in cell 1 b, is faintly bordered above with ochraceous-yellow ; the two small teeth of the posterior tranverse band, which are divided by the submedian fold are of nearly equal length, while in *Speyeri* the one below the fold is much the longer ; between these teeth and the opposed teeth of the anterior transverse line is a white spot, resting on the fold and reaching nearly half way to the nervure on each side. Terminal margin lined distinctly with black, interrupted by the nervules.

Posterior wings hyaline, with a very narrow lustrous brown border, and nervules covered with brown scales. Cilia white.

Expanse of wings 1.90 in. Length of body exclusive of anal tuft, .80 in.

Through the kindness of Mr. E. L. Graef, I have been permitted to examine three females of this species, received from Texas, one of which had been sent to him under the name of *Speyeri*. While closely allied to *Speyeri*, as above indicated, the distinctive features of narrower and less apically-rounded primaries, a lighter gray shade, ochraceous markings, et cet., are fully sustained by this additional material. The posterior wings of this sex have a narrower marginal brown border than in *Speyeri*, where it occupies, opposite the cell, nearly one-third of the wing, but in this, less than one-fourth. The wings in this species are more hyaline than those of *Speyeri*.

The nervular-interrupted black terminal line of the posterior margin of the primaries, is a good feature to distinguish this species from *Speyeri*.

In none of the four examples of *lætifica* before me, can the outlines of the orbicular and reniform spots be traced, or even approximately lined by comparison with congeneric examples. (It is proper to state that they are all in a somewhat imperfect condition.) In each, the lower portion of the reniform is so well defined, that, from its position and extent it might readily be mistaken for a portion of the posterior transverse line. It can be followed as a black line from vein 5, curving downward over 4 and 3, and then upward over the median at about the anterior third of the space between 3 and 2. In *C. intermedia* this line has the same extent but is less curved over 4 and 3.

In the type specimen of *lætifica*, traces of the orbicular are visible as two pale brown dashes, separated by a tooth of the anterior transverse line lying above and back of the point of bifurcation of the median and vein 2. It is believed that the form of the orbicular of *lætifica* (and also of *Speyeri*), should it hereafter be traceable in perfect examples, will prove to be that of a figure 8, or two contiguous ellipses of which the lower is the larger — quite unlike the quadrate form in *postera* and *asteroides*, and in *absynthii* and *asteris* of Europe.

Expanse of wings of ♀, from 1.58 to 1.88 inch. Length of body from .7 to .78 inch.

From 1 ♂ in the Collection of Mr. Otto von Meske, 2 ♀'s in the Collection of Mr. E. L. Graef, and 1 ♀ (through the favor of Mr. Graef) in my Collection.

Previous to the detection of this species, I had determined a Texan CUCULLIA of Mr. Morrison, as *Speyeri*. Since then, it has not been convenient for me to review the determination, but I think it probable that a re-examination would show it to be a *lætifica*, and consequently that *Speyeri* has not been received from Texas.

XIX. NOTES ON CATOCALA PRETIOSA LINTN.

This species is closely allied to *C. polygama* Guen. Its distinctive features may be more clearly appreciated by a differential comparison with that species. The basal region is conspicuously and broadly shaded with black, deepening toward the anterior transverse line; in *polygama*, shaded with ferruginous. The anterior transverse line is geminate, moderately oblique in its general direction, tending to the posterior third of the internal margin, distinctly separated by white below and slightly above the submedian: in *polygama* the line is quite oblique, tending to, or very near to, the internal angle, and is preceded below the submedian by gray and ferruginous scales.

The posterior transverse line has the extra-cellular teeth moderate, unequal, the lower one in cell 4 being improminent; moderately angulated outwardly (not toothed) on the submedian fold before the sinus; the sinus short, not extending to the middle of the wing, the line narrow with ferruginous and white below it; from the sinus, running direct and slightly oblique outwardly to the internal margin, followed by a white line. In *polygama* the two teeth are conspicuous and nearly equal; sharply toothed outwardly on the submedian fold, as in *C. cratægi;* sinus long, reaching the middle of the wing, the line broad, with ferruginous on each side and without white below; below the sinus, a long and sharp tooth bordering the internal margin.

The two transverse lines are separated on the submedian nervure by a space equal to the width of cell 2 on the terminal margin, whence they run parallel to the internal margin; in *polygama*, they are nearly or entirely united on the submedian, beyond which they widely diverge and again wholly or nearly unite on the internal margin.

The reniform is broadly surrounded by white; in *polygama* narrowly. The subreniform is round, its outline defined by black scales; it touches outwardly the median shade-line on vein 2; of the two transverse lines, it is nearer to the posterior, or midway between them: in *polygama* it is subquadrangular, defined by ferruginous scales, is quite removed from the median shade-line, and is nearer to the anterior transverse line, sometimes quite approximate to it.

The subterminal line is dark brown; in *polygama*, pale gray. The posterior wings have the marginal band slightly narrowed on the submedian fold: in *polygama*, it is separated or quite constricted; beneath, the cellular fold is shaded with black (not in *pretiosa*).

In size it is smaller than *polygama*, five examples of which before me measure in expanse of wings, males 1.80, 1.85 and 1.90 inch; females 2 and 2.1 inches. *Pretiosa* males 1.60 and 1.70 inch; females 1.80 inch. The wings are proportionally broader than in *polygama*; they are more clouded with black basally, with more white medially, and with less ferruginous in the terminal region.

Three examples of the species were captured by me at sugar, at Schenectady, N. Y., last year — the two males, in perfect condition, on July 8th and 10th, and the female somewhat worn, on July 16th.

A fine example of *C. cratægi* Saunders was also taken by me at sugar, on the 17th of July. I had recognized it as an undescribed species at the time of its capture, and had so indicated it in my Collection. With the larval state of nearly all of our Catocalas unknown, it is very gratifying that Mr. Saunders has been so fortunate as to be able to accompany the description of the imago with that of its larva

C. polygama was taken but once by me last season, viz., on the 7th of July, in perfect condition. The examples which I have seen of this species present very little variation. The variability which has been ascribed to it probably arises from the confounding with it of *cratægi*, *pretiosa*, and perhaps some other species. — *Canadian Entomologist* for July, 1876.

Since the publication of the above, through the favor of Mr. G. W. Peck, of New York, I have been able to see additional examples of the species, which show some variation from my type specimens. In those, the marginal black band of the secondaries is continuous, presenting only a constriction on the submedian fold, acute in the female, and approaching to a separation, but slighter in the male. In a pair received from Mr. Peck, the band is disconnected on the fold, to the same extent in each sex. They also show variation in the shape of the subreniform, — that of the female being more oval than in the type, and of the male, subquadrangular.

It is believed that the tawny band of the primaries beneath, separating the marginal and median black bands, may afford valuable differential features in closely allied species of CATOCALA. Thus, in the four examples above referred to, the band is sharply and almost rectangularly reflected on vein 4, thence running direct with regular contraction to the internal margin.

In *C. cratægi*, the form and course of this band are similar to that of *C. pretiosa*, but its breadth is less.

In one example of *C. præclara*, the band is less acutely bent between veins 3 and 4, and again sharply between veins 1 and 2, from which point the lower portion of the median black band is continued very narrowly to the inner margin.

In one *C. fratercula*, female, the tawny band is regularly curved, with the exception of a moderate outward bending on the submedian fold.

Mr. Peck informs me that he captured a number of examples (12) of *pretiosa* last year (1876) at Morristown, N. J. For several years the species had been in his collection under the name of *polygama*. This latter species had never, to his knowledge, been taken in the vicinity of New York, but he had received it (not identified at the time) from Canada. Two examples of *C. pretiosa* were also taken by Mr. F. Tepper, at Flatbush, L. I., during the season of 1876.

C. polygama, thus far, has proved to be comparatively rare.* Its name is probably misapplied in many collections. The reference of *pretiosa* as a variety of that species may be presumed to arise from erroneous identification of one or both species.

* Both *C. pretiosa* and *C. polygama* were collected, at sugar, at Center, in quite a number of examples, during July and August of 1877. The former species was not at all rare throughout the month of July. *C. polygama* was less common, but of longer continuance, extending into August.

XX. ON A NEW SPECIES OF HYPOCALA.

Among the unequaled collections in number, variety and rarity, made at the famous Center locality during the season of 1877, perhaps the most interesting capture is that which gives us, for the first time, representation of a genus of tropical insects of marked beauty, rivalling the Catocalas, to which they are closely allied.

Of the eight species of HYPOCALA described by Guenée (two of which are Fabrician species), five are from the East Indies, one from Africa, one from Honduras (N. Lat. 15°) and one from Hayti (N. Lat. 19°). That a species, typical of the genus, should occur in the State of New York (N. Lat. 42°), is a discovery of exceeding interest, adding, as it does, to our list of Noctuas, a peculiar and beautiful form, which, there is reason to believe, will long remain a rarity in our collections.

In consideration of the peculiar characters of these moths, Guenée, in his Noctuelites, Tome III, has arranged them in a separate family which he designates as Hypocalidæ, consisting of the single genus of HYPOCALA. The genus he defines as follows:

"Caterpillars unknown. Moths — Antennæ, medium, more or less pubescent in the ♂. Palpi very projecting, quite large, compressed, contiguous, with joints indistinct and ordinarily of triangular form—the last as scaly as the preceding. Tongue moderate. Eyes large and projecting. Frontal tuft elongated, carinated, thick and close (*serré*). Thorax oblong, scaly, stout. Abdomen long, swollen, not carinated above, somewhat hairy, yellow with black spots, bearing a small tuft at the base. Legs strong, slightly hairy. Wings subdentate; the superiors pulverulent, the subterminal line in part distinct: the inferiors yellow with a black border, having the nervule-independent [disco-central nervule — vein 5], inserted near the three others, opposite the 4th inferior [1st median nervule — vein 2]."

Guenée remarks: "The species of this genus are of medium size, and very similar to one another, so that their varieties would be absolutely confounding if the difference in the ciliation of their antennæ was not evident. The following is their general description:

"The superior wings are subdentate, of a powdery gray, bordering on a yellow, and usually dotted or striated with brown atoms. The orbicular is wanting, but the reniform is usually present, of an oval form and blackish. All the lines are indistinct except the subterminal, of which the inferior portion is always visible, blackish, slightly dentate, and followed by a contiguous, parallel, ferruginous line. The inferior wings are yellow, with a large cellular spot and a black border, irregular and interrupted near the anal angle by a spot of the color of the ground, as in CATOCALA. The under side of the same wings have the designs black and more distinct, and the costa gray. That of the superior is also yellow, with two black bands extending from the costa and terminating before the internal margin. The abdomen is not annulated with yellow and black as in HYBLÆA, but all yellow with some black bands occupying only the upper side of the segments, with the anus equally black above.

The Hypocalas inhabit India, Africa and America. They are not common in collections, where they are almost always found in a bad state."

The honor and credit of the discovery of this moth is due to Mr. W. W. Hill, of Albany, N. Y. I do not deem it an accidental discovery, but rather the direct consequence of so persistent and thorough a "working up" of a favorable locality by the aid of a greatly improved method of sugaring, that I believe I may venture the assertion that not even an approximation to it has hitherto been made in the annals of Lepidoptera collecting. While, therefore, I most earnestly deprecate the frequent introduction of names of individuals in our Entomological nomenclature,— often on no other ground than as a pretty compliment, an incentive to the enlargement of an amateur collection, or as a means of securing the favor of a collector, and while I would guard the honor as a just tribute, (valuable only from its rare bestowal), to those whose labors constitute a portion of the history of our science,— in the present instance, I have no hesitancy in proposing the name of the discoverer for association with the insect below described. The results obtained at Center during the year 1877, hereafter to be given to the public, will assuredly constitute an epoch in the collection of Noctuidæ, and an important chapter in their history.

Hypocala Hilli n. sp.

Antennæ scaled; under a microscope clothed with numerous short cilia on each joint, and with fascicles of longer ones, of the length on the basilar joints of about one-half of their diameter. In proceeding from the base toward the tip, these fascicles are less marked, until at about midway of the antennæ they have become changed into spinules, of which there may be three or more on each joint; toward the tip these spinules extend beyond the numerous short cilia to about once and a half the diameter of the joints, and are visible, as are also the fascicles, under an ordinary magnifier.

Palpi triangular, porrected, pointed, slightly beaked, pale brown sprinkled with black scales, as are also the head, prothoracic tufts and thorax: these latter tinged with ochraceous.

Abdomen yellow, the basal segments bearing some long dusky hairs. Segments 8–11, with a fuscous or black spot (black on the posterior ones) on their anterior two-thirds, approaching in form two mesially connected small segments of a circle. On the last segment a large black spot, triangular in front, extending on the sides and on the long anal tuft, of which the terminal portion is yellow, tipped with dusky. Sides of the abdomen with a row of five small black spots, and with its under surface pale yellow, sprinkled with black scales.

Wings: primaries sprinkled with black scales on a pale brown ground, which shows a distinct violet shade in a certain light, especially on their outer half, with a slight pale green reflection near the base on the inner portion of the wing. Costa delicately striated transversely with darker brown. Median lines and discal spots absent; the reniform traceable in a scarcely perceptible encircling line when sought for with a magnifier. Subterminal line ferruginous, bordered on each side with black, extending from opposite the cell just above vein 4, outwardly in two or three teeth to near the margin, then bending backward, and with some gentle curves, reaching the internal margin near the angle. The terminal line black, waved, and a yellowish one at the base of the fringe. The fringe brown, tipped with a darker shade. Secondaries of a luteous yellow, with their greater portion black, as follows: costal region blackish; apical portion broadly bordered with black, continued along the margin to vein 2, dividing at vein 3 where it runs inward, occupying somewhat the place of the median band in Catocala, bending downward to the internal angle, whence it runs along the internal margin to the base; within this, resting on the submedian, a broad black ray not reaching the band; a large black cellular spot, connected in its upper half with the black of the base and costal region, but disconnected from the submedian ray. Beneath, the black border and its inner continuation are better defined and narrower; the cellular spot is contracted to a narrow elongated subquadrangular black spot between veins 2 and 8; the submedian ray is wanting; costal region of both wings dotted with black. The under surface of the primaries is yellow, with a broad black straight band extending from below the costa to the submedian interspace, and another of the same length and of nearly twice the breadth, lying in the outer third of the wing, slightly indented on vein 5; apex and margin beyond this band, fuscous; costal slightly striated with black.

This species resembles *H. filicornis* Guenée, from Honduras, from which it is separable by its larger size, its non-striated primaries, etc. In its antennal structure, it is similar to *H. Pierreti* Guenée, from Hayti.

Expanse of wings, 1.90 in.; length of abdomen, .95 in.

From 1 ♀, taken at sugar at Center on Oct. 15, 1877. In the Collection of Mr. W. W. Hill.

Soon after the capture of the above, Mr. von Meske received from Texas, and has now in his Collection an example, which, notwithstanding some differences, should in all probability be referred to the above species. It was taken at Bastrop, Tex., (N. Lat. 30), by Mr. Heiligbrodt, on Sept. 2d, in perfect condition. It is of the same sex with the Center specimen — a fact deserving special mention, since all the descriptions of Guenée in which the sex is indicated (seven species), were drawn from ♂'s.

Both the reniform spot and the transverse lines are present in it: the former is conspicuously outlined in black, of an oval form, inclosing an oval ring of pale scales, of which the half toward the base of the wing is more distinct. Transverse lines single, brown, toothed: the anterior line moderately toothed, its general course nearly direct across the wing, except as strongly exserted outwardly below the median: the posterior line more prominently dentate, with three teeth between costa and vein 3, above which vein it is strongly bent inward so as to touch the lower end of the reniform and to extend beyond it for a space equal to its transverse diameter, thence curving outwardly, and with two angulations reaching the internal margin. There is a faint trace of a median shade running as a continuation of the lower half of the posterior transverse line. The thorax is concolorous with the wings, and together with the prothoracic tufts and head, is marked with a median brown line. The two basal segments have each a small black median tuft, a trace of one of which is visible in the Center example, the other having probably been lost in the partial denudation of the superior portion of the thorax. The antennal structure of the two examples is the same, and no material difference is noticeable in a comparison of the posterior wings.

XXI. ON THE IDENTITY OF HOMOPTERA LUNATA AND H. EDUSA.

Phalæna (Noctua) lunata DRURY. Illus. Nat. Hist., App. vol. ii. 1773.
" " *edusa* " Illus. Nat. Hist., App, vol. ii. 1773.
Noctua lunata WESTW.-DRURY. Illus. Exot. Entomol., v. i, p. 37, pl. 20. f. 3. 1837.
Erebus edusa, WESTW.-DRURY. Illus. Exot. Entomol., v. ii, p. 46, pl. 24, f. 4. 1837.
Homoptera lunata GUENÉE. Sp. Gen. Lep. Noct., vol. iii, p 12. 1852.
" *Edusa* " Sp. Gen. Lep. Noct., vol. iii, p. 14. 1852.
" *Saundersii* BETHUNE: in Proc. Ent. Soc. Phil., vol. ii, p. 215. 1865.
H. lunata and *edusa* BEAN: in Canad. Entomol., vol. ix, p. 174. 1877.

More than a century ago (in 1770), Drury, in the first volume of his admirable work cited above, illustrates a large and beautiful Homoptera from examples received from Virginia and Carolina, to which, in the appendix to the second volume, he applies the name of *Phalæna* (*Noctua*) *lunata*. In the second volume (in 1773), he describes and figures *Phalæna* (*Noctua*) *edusa*, a form from New York, differing from the preceding in having the subterminal space of the brown wings of a grayish or bluish white, which, on the superiors, is gathered in two lunulated spots.

From their wide distribution through several of the United States, their comparative abundance and their marked beauty, these two forms have found place in nearly every one of our collections of Lepidoptera, under the above names, and not unfrequently associated with their presumed companions of the opposite sex. Very recently, the interesting discovery has been made that the two constitute but a single species.

I was led to suspect the above relationship two years ago from the study of a few examples in my collection, and accordingly requested of some of my friends the careful inspection of their future captures, with a view of determining this point.

Since that time numerous examples of the two forms have come under our observation, in all of which the females are *lunata* and all the males "*Edusa*." As no other differences except sexual are perceptible, beyond the colorational features, there is no longer reason for questioning the identity of the two forms, and the necessity of henceforth dropping the name of "*edusa*" from our lists.

It is fortunate that the specific name which is to be retained will now indicate a marked feature of the species.

There has always been an annoying incongruity in designating as *lunata* that one of the two closely allied supposed species which was without the lunulated spots.

In consideration of the suggestion and careful observations of Mr. Bean, given in a late number of the *Canadian Entomologist* (ut cit.), I have included, with doubt, *H. Saundersii*, in the synonymy of *H. lunata*. My two examples of this form are both males, but with only these at my command (the form appears to be quite rare in the Albany district), it would not be proper to form an opinion as to its relations.

XXII. ON THE IDENTITY OF TWO FORMS OF HYPENIDÆ.

In the *Transactions of the American Entomological Society* Vol. IV, pp. 105, 106, Sept. 1872, Mr. Grote describes *Tortricodes bifidalis* and *T. indivisalis*, provisionally as two species. He indicates the principal difference between the two, to lie in a cleft in the outer margin of the primaries of *bifidalis*. Although designating them by different specific names, he remarks, "I am inclined to consider the two forms merely as sexes of one species, with the fore-wings cleft in the male. And with four specimens of *T. bifidalis* before me, and eight of *T. indivisalis*, I cannot but be sure that most, if not all, of my *T. bifidalis* are males, and of my *T. indivisalis*, females. * * * I shall then not be disappointed if the two should prove to be sexual forms of one species."

Subsequently, same vol., page 308, Jan. 1873, Mr. Grote writes: "Mr. J. A. Lintner informs me that he has both sexes of *Tortricodes bifidalis* with cleft primaries. I then refer *T. indivisalis* to HETEROGRAMMA *Guenée*, believing our species not to differ generically from the Brazilian species which M. Guenée uses for his type."

In accordance with the above reference, the two forms have been known up to the present as *Tortricodes bifidalis* and *Heterogramma indivisalis*.

In a recent study of a considerable number of examples of each of the above two forms, collected at sugar, during the months of June, July and August of 1875, I was surprised to find that among so many, all having the cleft wing were males, and all with the entire wing, females. Suspecting, from this discovery, that the two were but one and the same species, I examined my cabinet example of *T. bifidalis* "female," and found that I had been misled by an unnatural position which the frenulum had assumed, but that it was unquestionably simple, not yielding even to pressure after having been detached from the wing, and, therefore, indicating a male. There was then, no doubt of the specific identity of the two differing forms.

As the description of *Tortricodes bifidalis* precedes that of *H. indivisalis*, the latter must give way as a synonym.

In some of its features this species does not conform to the definition of the genus to which it has been referred. It will probably need a new genus for its reception.

That a single species has at the same time been given place in two genera, should not in the least degree reflect upon Mr. Grote. The author of the erroneous information conveyed to him, as above quoted, is alone responsible for the mistake. That two forms, so very unlike in construction of wing, should have been believed, at the first, to be identical, gives assurance that the difficult task of identification of species often imperfectly described, description of new forms, and systemization of the North American Heterocera, has been undertaken by one peculiarly fitted for the work.

XXIII. DESCRIPTIONS OF TWO NEW SPECIES OF PHALÆNIDÆ.

Acidalia lacteola n. sp.

Antennæ, vertex of head and thorax white. Abdomen white, sprinkled with pale brown scales and with bands of the same color.

Wings white, thinly sprinkled with pale brown scales, marked with sinuous dark brown bands and following shade lines, and with elongated black discal spots; inner margins with long white hairs, of which those opposite the bands are black tipped and spatulate.

Primaries with three transverse lines dividing the costa into four nearly equal parts — the two interior ones somewhat the shorter: the extrabasilar band runs outwardly from the costa to within the cell, where it is acutely reflected to the submedian fold, then with an outward angle on the submedian to the internal margin: the interior line is strongly reflected outwardly to the subcostal, thence, outside of the discal spot, with sharp angles and followed by brown scales to the inner margin: the outer line is more sharply defined, less sinuous, having but three prominent outward reflections, and is marked by a transverse black spot on each vein. The shade-line beyond this consists mainly of brown scales between the veins, arranged in a sagittate form, especially seen between the median nervules — veins, 2, 3 and 4. Terminal margin marked with brown scales between the veins (the fringes absent in the example).

Secondaries: the inner and outer band of the primaries are continued, and present much the same character; they divide the wing on the median vein in three nearly equal parts.

Expanse of wings, .87 in. Length of body, .25 in.

Described from one example, a female, not in very good condition, in the collection of Mr. Otto von Meske. Received from Mr. L. Heiligbrodt.

Habitat.— Bastrop, Texas.

Cidaria Packardata n. sp.

This name is proposed for the species described and figured by Dr. Packard on p. 124, Pl. viii, fig. 52, of the *Monograph of the Phalænidæ*. It is regarded by him as identical with the European *Phalæna populata* Linn., and a generic name applied to it and to several allied species, taken from the Tentamen list of Hübner's Stirps — PETROPHORA.

The two forms are so very different in contour of wings, markings and color, that it is difficult to understand how they could have been united. In *Packardata*, the primaries are quite excavated from the apex to vein 4, at which point, in one of the males before me, they are distinctly angulated; the secondaries are slightly angulated on vein 4. In *populata*, a scarcely perceptible excavation may be seen below the apex of the primaries, and the margin of the secondaries is regularly rounded.

In *Packardata*, the broad basilar band is composed of three lines, the first and second of which approximate, and are separated by pale yellow; the second and third are distant, with dark yellow between. In *populata*, the band consists of but two lines with pale yellow between, or a third line may be faintly seen *near the third.*

In our species, the whitish band between the basilar and mesial bands is more sharply angulated toward the costa above the median vein, and its outer border is more strongly toothed. The outer border of the mesial band differs very materially in its course from the same line in *populata*, in its running in a nearly direct line, with only a slight bending backward between veins 5 and 7, to near the outer margin of the wing, whence it proceeds to form two prominent produced teeth in cells 2 and 3. In *populata*, these teeth are comparatively improminent; they are at nearly twice the distance from the outer margin as compared with the other species; the excavation on vein 5 is deep and rounded, and thence, the course of the line to the costa is nearly parallel to the hind margin instead of quite oblique to it. The distinctly marked, lunulated, subterminal line of this species is barely indicated at its extremities in our form, in which there seems to be no place for it medially, from the approximation of the dentations of the mesial band to the outer margin. The subapical patch is large, uniformly dark, subsemicircular in outline, while in *populata* it is reduced to nearly a blackish oblique streak, behind which is the uniform shade of the terminal margin.

In color, *Packardata* is of a pale yellow, with great contrast between the ground color and the interior bordering of the two bands: *populata* is more uniform in its darker ochraceous shading, with the darker yellow tending to a diffusion over the whole of the mesial band. Our species is also of a larger size than the European, two examples measuring respectively 1.50 in. and 1.38 in., while the examples of the latter before me are but 1.25 in. in expanse.

Material under observation: One male of *C. Packardata*, from Collection of Otto von Meske, made at Sharon Springs, N. Y., August 14, 1875; one male from Collection of W. W. Hill, taken in Lewis county, N. Y., July 27, 1876. Of *C. populata*, two males and one female received from Dr. Speyer, of Germany, from Collections of Mr. von Meske, Mr. Hill and my own.

C. populata is very well represented on Pl. 22, fig. 590 of Wood's *Index Entomologicus*, in its characteristic shape of wings and disposition of bands. Its habitat and apparition are there given as "common in the north of England and in Scotland, in July." It is referred by Stephens to the genus ELECTRA.

XXIV. A NEW LOCALITY FOR BREPHOS INFANS MŒSCH.

This beautiful moth, possessing peculiar interest from its abnormal characters, its northern habitat, and from its close resemblance to the *B. Parthenias* of Europe, is rarely captured by the collector, and has been observed, so far as we know, in only few localities in the United States. Mr. Grote, in *Trans. Amer. Ent. Soc.*, I, p. 189, gives as its range, Labrador, southward through the Eastern States. In the *Canadian Entomologist* for 1875, VII, p. 40, we have the statement that "Mr. Kuetzing, of Montreal, has discovered a locality for *B. infans*, in a clump of white birch, north of the village of Hocheloga—the first record, it is believed, of its occurrence in this province [Quebec]." It was subsequently taken, in a number of examples, at Hyde Park, Mass., among white birch.

Its association in the above instances with the white birch, coupled with the knowledge that the European species *Parthenias*, *vidua* (notha), and *puella* (that of the Siberian *Middendorfii* not stated), also feed on birch, rendered it almost certain that our species would be found to have the same food-plant.

At the Center locality, and extending a mile or more in either direction, the N. Y. Central Railroad is bordered or has recently been, with a thick growth of white birch. It occurred to me that this would make a very fitting home for *B. infans* if its range extended to New York, and I accordingly suggested the probability of its presence there, to one of my entomological associates, Mr. W. W. Hill, who, I had reason to believe, would discover and capture it, if my surmises were correct. At about the time when it might be expected to appear, Mr. Hill visited the locality, found the moth, and was able to secure examples for his cabinet.

At my request, he has kindly furnished me with a statement of its discovery and capture, which I find to contain so many interesting particulars of the habits of the moth, as also of another extremly rare species (previous to 1876), that instead of extracting from it, I present it in full:

"In accordance with your suggestion that *Brephos infans* might be found at Center, on the 3d of April last [1876], I took the early train for that place. From 7. 30 A. M. to 12 M., I searched through the white birch swamps without meeting with one, but on emerging in a clearing on the west of the timber, I was at once brought in view of a half dozen or more, sailing around, at a height of from ten to twenty feet above the ground. In striving to capture a specimen, I observed that it would manage to keep from fifty to a hundred feet distant from me, except when flying swiftly by. They were evidently well aware of my presence. At 1.30 P. M. when moving very slowly and scanning closely the ground and brush, a *B. infans* rose from an open spot surrounded by some newly cut birches. To my great delight, I captured it. It proved to be a fresh and perfect specimen — taken, perhaps, in its first flight. Although they continued to fly for some time afterward, I failed to secure another.

April 17th, I again visited Center, but searched in vain for more of the *infans;* not a single example was to be seen. The day was colder, with more wind and but little sunshine. As the weather became more unfavorable, I left the birches and devoted the remainder of the day to examining the fences along the railroad, and was well rewarded by the capture of three examples of the rare *Xylina fagina.*

April 20th, visited the locality for the third time. At 1 P. M., I captured my second *infans* — this time resting on a small white birch — the wind blowing freshly toward me, with the sun bright and warm. I observed at least a dozen others, but did not succeed in taking another, although the chase was perseveringly continued for several hours. I repeatedly saw them alighting on small birch trees, but the moment a movement was made toward them, although at a distance of a hundred feet, it was noticed, and the moth at once took wing again. Later in the afternoon, the fine weather changed to cold and blustering, with snow, and as not a single *infans* was to be seen, I went in search of *Xylina fagina.* My examination of the fences was rewarded by the capture of nine specimens of the species — including those previously taken, twelve in all. Of these, ten were found at rest on the upper edge of the fence-boards (very singularly, none were on the upper board), the other two were on the more exposed north side of the fence. It seemed as if these little creatures had come out to enjoy the cold and driving snow-storm, which certainly failed to add to my comfort. At a late hour in the afternoon (5 P. M.) I also had the good fortune to secure a *Lobophora geminata* (Grote).

Neither of the above three species had before been found in this locality.

Subsequent trips gave me no additional examples of *B. infans*, although I saw it in flight — quite wild — so late as May 7th."

Guenée, in his Noctuelites, II, makes some interesting remarks on the habits of the European species of BREPHOS, which is here transcribed:

The caterpillars live on the tall trees, from which, letting themselves fall, they hang suspended by a thread, after the manner of many of the Geometers. They are found in autumn, in woods of considerable extent, and their moths fly in the first days of spring, or as might better be said, at the end of winter, about the leafless birches. Their flight is lively, jerky (*saccadé*) and rapid, but the sun is indispensable to draw them from their torpor. Hardly do its rays veil themselves, even for an instant, when the Brephos arrest their flight, to resume it as soon as it commences to shine. In these habits, they bear much resemblance to the Phalænidæ which like them fly in the early spring; and there is also a resemblance in the habits and forms of the caterpillars.

Mr. C. P. Whitney, of Milford, N. H., writes me in relation to the habits of *B. infans:* "I take it very early in the spring (this year [1877]) about the middle of April, only in the vicinity of the white birch (*Betula populifolia.*) Its season is about two weeks. Their flight is rapid and irregular, so that it is almost impossible to capture them except when resting on the ground, when almost every one discovered can be easily taken. Late in the season they fly high and alight on the twigs of the birches."

XXV. NOTES OF CAPTURE OF LEPIDOPTERA IN 1876.

RARE TO THE VICINITY OF ALBANY.

Argynnis Idalia Drury — quite rare in this vicinity — was captured at Center on September 16th.

Phyciodes Batesii Reak., was abundant at Center, on the 17th of June, when twenty-five examples were collected in a short time — the sexes in about equal number.

This is apparantly quite a local species, and so far as known, occurring in limited localities in New York, Pennsylvania and Virginia. It may usually be found in Center during the first half of June — my earliest recorded observation of it being May 31st, and latest June 22d. It has also been taken by me in limited numbers at Schoharie, N. Y.

Several specimens of *Pholisora Catullus* were collected in Bethlehem on the 12th and 13th of July. This species had been so diligently sought for in former years that it is difficult to see how it could have escaped detection if it had been a habitant of the locality for any considerable length of time. Its ascribed food-plant, *Monarda punctata*, is not known to the State Botanist to occur in this portion of the State. Mr. Scudder records *P. Catullus* as "very rare in New England, found in southern portions." (*Proc. Ess. Ins.*, III, p. 170.)

Eudamus Lycidas (Sm.-Abb.), was taken by Mr. W. W. Hill, at Center, on June 15th. It had previously been found by Mr. Edwards at Newburgh, N. Y., ninety miles south of Albany.

Sphinx luscitiosa Clem.— One of the most rare of our New York sphinges — was captured in perfect condition, at Center, on the 3d of June.

Euchœtes Oregonensis Stretch, was not at all rare at Center during the month of May — a single collector taking thirty-three examples of it, viz: eight examples on the 10th, four on the 13th, twelve on the 16th, four on the 19th, three on the 20th, and two on the 24th. Mention of the collection of two examples of the species at the same locality is made in the 26th *Report of the N. Y. St. Mus. of Nat. Hist.*

Spilosoma latipennis Stretch (see Report ut cit., p. 144), occurred in a single example at Center, on June 10th.

Nephelodes violans Guen., in forty examples were taken at Bath-on-the-Hudson, between August 30th and September 15th, by Mr. Hill. It had not been observed in this vicinity for several preceding years.

Among the above are several examples which conform to the description given of *N. minians*, but its separation seems to rest on no other ground than depth of color, and in these, all the intermediate grades are to be found.

It is worthy of mention, that not a perfect example of *N. violans* has thus far been captured, or is present in any of the Albany collections.

Mr. R. Thaxter, in his list of Noctuidæ, taken at Newton, Mass. (Psyche, II, p. 36), states the fact, that the examples of *N. minians* (probably the same above recorded as *N. violans*), were almost always in poor condition.

Mr. William Grey, a collector of Lepidoptera for many years, has been rewarded for the additional attention which he has recently given to Entomology, by the capture of four species of CATOCALA, never before taken in the vicinity of Albany, viz.: *C. Epione* West., *C. flebilis* Grote, *C. Robinsoni* Grote and *C. coccinata* Grote — a single example of each species.

That graceful and beautiful Sphinx, *Chærocampa tersa* (Linn.), not previously taken so far north as this, it is believed, is also one of the rarities captured by Mr. Grey.

The above were taken upon the grounds of Mr. Erastus Corning, at Kenwood, near Albany.

COLLECTIONS IN 1877.

In advance of a more extended notice of the extensive and interesting collections recently made near Albany, the following species may be mentioned, as of special interest from their rarity. Not one of the number had previously been taken in the Albany district, and less than half the number were represented in the Albany collections, from other localities. With few exceptions they are from the Center locality.

Thyatira pudens *Guen.*
Charadra propinquilinea *Grote.*
Diphthera fallax *H.-S.*
Acronycta grisea (*Barn.*).
Acronycta tritona (*Hubn.*).
Acronycta lobeliæ *Guen.*
Acronycta Radcliffei (*Harvey*).
Acronycta dactylina *Grote.*
Acronycta luteicoma *Grote.*
Acronycta subochrea *Grote.*
Acronycta ovata *Grote.*
Acronycta albarufa *Grote.*
Acronycta exilis *Grote.*
Acronycta lithospila *Grote.*
Agrotis turris *Grote.*
Agrotis albipennis *Grote.*
Agrotis perpolita *Morr.*
Agrotis opacifrons *Grote.*
Agrotis dilucida *Morr.*
Agrotis badicollis *Morr.*
Agrotis janualis *Grote.*
Agrotis phyllophora *Grote.*
Agrotis mimallonis *Grote.*
Agrotis Bostoniensis *Grote.*
Agrotis fumalis *Grote.*
Agrotis cupida *Grote.*
Agrotis Hilliana *Grote.*
Agrotis pressa *Grote.*
Mamestra lubens *Grote.*
Mamestra lilacina *Harvey.*
Mamestra assimilis *Morr.*
Mamestra vindemialis (*Guen.*).
Mamestra congermana (*Morr.*).
Dianthœcia modesta *Morr.*
Hadena leucoscelis (*Grote*).
Hadena arna (*Guen.*).
Hadena algeus *Grote.*
Polia diffusilis *Harvey.*
Dryobata stigmata *Grote.*
Morrisonia evicta *Grote.*
Morrisonia vomerina *Grote.*
Conservula anodonta *Grote.*
Phlogophora v-brunneum *Grote.*
Gortyna appassionata *Harvey.*
Gortyna nebris *Guen.*
Gortyna purpurifascia *Gr.-Rob.*
Caradrina bilunata *Grote.*
Crocigrapha Normani *Grote.*
Orthosia euroa (*Gr.-Rob.*).
Glæa tremula *Harvey.*
Glæa apiata *Grote.*
Glæa venustula *Grote.*
Glæa pastillicans *Morr.*
Xanthia togato (*Esp.*).
Scopelosoma ceromatica *Grote.*
Scopelosoma Græfiana *Grote.*
Scopelosoma devia *Grote.*
Scopelosoma vinulenta *Grote.*
Scopelosoma tristigmata *Grote.*
Xylina ferrealis (*Grote*).
Xylina signosa *Walk.*
Xylina semiusta (*Grote*).
Xylina Georgii (*Grote*).
Xylina unimoda *Lintn.*
Xylina Thaxteri (*Grote*).*
Xylina tepida (*Grote*).
Xylina pexata (*Grote*).
Xylina Baileyi (*Grote*).
Xylina querquera (*Grote*).
Xylina lepida *Lintn.*
Xylina capax (*Gr.-Rob.*)
Calocampa nupera *Lintn.*
Calocampa cineritia *Grote.*
Calocampa curvimacula *Morr.*
Lithomia germana (*Morr.*).
Xylomiges tabulata *Grote.*
Abrostola ovalis *Guen.*
Plusia formosa (*Grote*).
Lygranthœcia brevis *Grote*
Hypocala Hilli *Lintn.*
Catocala tristis *Edw.*
Catocala minuta *Edw.*
Homoptera unilineata *Grote.*
Homoptera obliqua *Guen.*
Homoptera benesignata *Harvey.*
Homoptera Woodii *Grote.*

*** Stated by Dr. Speyer to be a well marked variety of *Xylina lambda* Fabr.**

XXVI. NOTES ON SOME LEPIDOPTERA.

Grapta Satyrus Edw.: in Trans. Amer. Ent. Soc., II, p. 374.

An interesting discovery, and addition to our New York Fauna, is the recognition of the above species, among the collections made by Mr. W. W. Hill, at Fenton's, Lewis Co., (Adirondack region) during the season of 1877.

The type specimens, described as above cited, were from the Rocky Mountains in Colorado. In 1871, the butterfly was reared by Mr. Henry Edwards, from examples of the larvæ found by him on a species of Urtica, at Congress Springs, Santa Clara Co., Cal. The larva is described and figured, from a MS. and drawings by Mr. R. H. Stretch, in Edwards' *Butterflies of North America* I, p. 120, pl. 40. On the same plate, the butterfly is beautifully represented in both surfaces and sexes. Mr. Edwards also credits the species to New Mexico, Oregon and British America.*

The occurrence of this species in Northern New York is believed to be but the second instance of its collection east of the Rocky Mountains. The larvæ were found by Mr. C. W. Pearson, of Montreal, Quebec, feeding on nettle, at Chateaugay Basin, about fifteen miles south of Montreal. The butterflies emerged after a pupation of ten or twelve days. †

Lycæna Lucia Kirby and **L. pseudargiolus** Bd.-Lec.

In consideration of the interest attaching to these species, from the identity shown by Mr. Edwards, of *L. violacea*, *L. pseudargiolus* and *L. neglecta*, (see *Canadian Entomologist*, vol. x, pp. 9, 10), and the suggestion by the same author, that *L. Lucia* may prove to be but a northern form of the same, I give below some memoranda kindly furnished to me by E. C. Howe, M. D., of Yonkers, N. Y., of observations made at that locality :

---

* Catalogue of the Diurnal Lepidoptera of North America, p. 28. 1877.

† Canadian Entomologist, vol. vii, p. 216 1875.

"*Lycæna Lucia* was first observed on April 17th [a month earlier than its Albany record — see p. 55], and lasted until May 17th. *L. neglecta* first appeared May 21st, and continued throughout June, quite common. It afterward appeared in September.

"During the latter part of August, two specimens of *L. Lucia* were taken, unless they were the var. *violacea*. The markings on the underside of the wings were dark and heavy, exactly like those collected in the spring. It was not observed again."

The occurrence of *L. Lucia* or the form *violacea* in the month of August has never before been recorded, and it would seem at variance with the idea entertained of their being but spring forms. Mr. Scudder records that one example of *violacea* has been taken at Walpole, N. H., so late as the 7th of July, but it has ordinarily been confined, in this latitude, to the months of May and June.

Do not the observations of Dr. Howe of the August examples rather indicate that they were an exceptional second appearance of *L. Lucia*, and lend additional confirmation to the belief hitherto entertained of its being distinct from *L. pseudargiolus*.

Agrotis nigricans Linn., var. **maizii** Fitch.

Of this moth, figured and described at considerable length in Dr. Fitch's Ninth Annual Report (pl. 4, figs. 2 and 3, pp. 237–249, Sixth–Ninth Reports: 1865), there are five examples, labelled as above by Dr. Fitch, in the collection of the Museum of the New York State Agricultural Society, and bearing also the additional popular name of the corn dart moth.

No one else has recognized *A. nigricans* in any of our American forms, nor, very strangely, has this determination of Dr. Fitch been referred as a synonym to any other species.*

A critical examination of the examples above mentioned has enabled me to refer them unhesitatingly, although much faded from ten or more years' exposure to strong sunlight in an exposed table-case, to the typical form of *A. tessellata* Harris, as recognized in the collections at Buffalo and Albany. The specific name "*A. tessellata*" embraces at present a variety of forms, which seem to me to vary too much among

* Since the above has been in type, the reference of "*A. nigricans*" to the *A. tessellata* of Harris, made by Prof Grote in the *Canadian Entomologist*, vol. vi, 118 (1874), has come to my notice.

themselves to really constitute a single species. I do not doubt but that, from these, other species will eventually be separated, and it is probable that ample material has already been accumulated for the purpose, in the two hundred or more "tessellata forms" in my possession which are awaiting a convenient time for their critical study.

Agrotis perpolita *Morr.*

From a cursory inspection of the female of this species, it would probably be referred to *A. velleripennis*, with which it is compared in the original description. A closer examination, however, would disclose several points of difference, the more prominent of which are the following:

The head, wings and thorax are nearly black, having only a brown reflection: in *A. velleripennis*, although described as coal-black, they are of a dark brown shade in all the examples before me, some of which are quite fresh.

The usual transverse lines in velvety-black, which are distinctly to be seen in *A. velleripennis*, are in this entirely wanting. Its orbicular spot differs from the ordinary shape, in that it is elongated in the direction of the cell, and, in two of the three examples examined, pyriform in outline, having the contracted portion directed toward the base of the wing: in its elongation, it resembles that of *A. clandestina*, but the normal shape of the spot in that species is acutely ellipsoidal.

The legs and spines, which in *A. velleripennis* are distinctly annulated with pale scales, are much less conspicuously ringed in this.

The males of the two species — from the contrast presented in their secondaries — are not liable to be confounded; those of *A. velleripennis* being white, subhyaline, with a median line of nervular spots, more or less distinct, and a very narrow terminal bordering of brown, and in *A. perpolita* smoky brown, with a broad darker border. There is also a marked difference in their antennal structure — the serrations in the latter species being not half so long or so strong as in the former.

This species seems to be quite rare. The original specimen was from Orono, Maine. It was unknown to Prof. Grote, previous to his identification, from the published description, of the examples from Mr. Hill's collection. Three examples were taken by Mr. Hill at Center, N. Y., on August 16th, 30th and 31st. No other examples of it have occurred among the very large collections made at this locality.

Agrotis cupida Grote.*

A correspondent of the *Country Gentleman*, from Erie Co., O., has recently discovered the food-plant of the larva of this species. From his communication we obtain the following facts: In 1874, $200 worth of grape-buds were destroyed by some unseen enemy. Repeated and careful examinations for many days throughout the season failed to bring to light the depredator. The following year, upon examining the vine at night with a lantern, a caterpillar was seen crawling along the vine, and to stop at a bud and commence eating it. After this discovery, searches were made each night, by six or eight persons bearing lanterns, during the continuance of the caterpillar, and two thousand caterpillars were taken and destroyed. It was calculated, on the basis of one caterpillar eating a single bud in a night, that the buds destroyed by the two thousand which were killed, might have produced eight tons of grapes.

As one bud a night would be a very small allowance for a half-grown cut-worm, it would be safe, we think, to double the above estimate of possible resulting damages.

Nothing is stated of the appearance or habits of the larva, except that it is of the color of the vine, and commences its depredations as soon as the buds begin to start.

During my sugaring operations at Schenectady, larvæ were occasionally seen upon the grape-trellis at night, feeding upon the bait, but from my recollection of them, they had not the aspect of such of the Agrotis forms as I have seen.

A single example of the moth occurred among my collections at Schenectady, on the 7th of August, 1876. It had been a rare species in this vicinity, until the remarkable Center collections of 1877 made it a common form. It occurred abundantly at this locality during the latter part of August and through most of September.

Mr. George Norman (*Canadian Entomologist*, 7, p. 5) records it frequent at sugar, at St. Catharines, Ont., from 17th July to August. Mr. Westscott (op. cit., 8, p. 12), notes a

---

* *Noctua cupida* Grote: in Proc. Ent. Soc. Phila., vol. iii, p. 525, pl. 5, fig. 7. 1864.

Agrotis cupida Grote: in Trans. Amer. Ent. Soc., vol. ii, p. 309. 1869.

A new grape-insect: The Cultivator and Country Gentleman, vol. xliii, p. 166. Albany, N. Y., March 14, 1878.

single example at Maywood, Ill., on September 15. Mr. Thaxter, in his List (*Psyche*, II, p. 36), mentions it as rare, at light, in August.

The moth is very variable in its features, in the color of its wings, form and coloring of its discal spots, etc., or different species are included under that name. The varieties which Mr. Grote has referred to this species, seem to me to differ too much to really constitute a single species. Another season's collections may enable us to determine the value of these differences.

Mr. Grote has referred, with some doubt, an Agrotis received from California, to this species.

Agrotis brunneicollis Grote. (*Noctua*) Proc. Ent. Soc. Phila., III, p. 524, pl. 5, f. 5. 1864.

This species has been captured for the first time in this district, by Mr. S. C. Waterman, at Ten Eyck's woods, near Albany, on the 15th of June, 1877. An example has also been taken at Kenwood during the same month. Mr. Thaxter records it among his Newton (Mass.) collections, as rather rare, at light, in August.

Cucullia intermedia Speyer.

Several examples of the larva of this species were taken at Center, N. Y., during the last of June and first of July, in their second, third and fourth stages, feeding on *Mulgedium leucophæum*, popularly known as false or blue lettuce.

The young larva is striped laterally and dorsally, and bears a strong resemblance to that of *C. lucifuga* in its early stages, as described by Dr. Speyer in the *Stettiner Entomologische Zeitung* for 1870, and in *23rd Rep. St. Mus. Nat. Hist.*, p. 222.*

I regret that I have only the following brief notes of the appearance of the larva — a part of what was intended to be a detailed description. "The lateral stripe is white, of about one-third the diameter of the body, traversed by a yellow stripe of about one-half the breadth of the white, on which are the

* To the kindness of Dr. Speyer I owe it, that the two specimens of the *lucifuga* larvæ prepared by Mr. O. Schreiner of Weimar, Prussia, from which his descriptions above referred to are drawn, have place in my collection. The specimens show the life-like appearance which may be imparted to larvæ when prepared by inflation (see Scudder in *American Naturalist*, vol. viii, p. 321), and also illustrate the service which they may render in the comparison of forms from widely separated localities and countries.

stigmata and a number of black spots, as follows: on the first segment, a black spot precedes the stigma; on segments 4–9 and 11, a black spot behind the stigma, bearing a black hair; on segments 4–9 each, on the white portion of the band, a larger, rounded black spot above, and a little in advance of, the stigma, having within it a longer black hair."

It is probable that the larva may again be found the coming season, to afford the means of completing the above description, which is quite desirable in consideration of its close resemblance to that of the European *C. lucifuga*. The mature forms taken at Center bore a closer resemblance to *C. lucifuga* than did my Schoharie examples. Dr. Speyer (*St. Mus. Rep.* ut cit., p. 222), in indicating the difference between the two, states "the first [*intermedia*] has thirteen dorsal spots, the other twenty-five." The examples collected as above, show that the number, shape and division of these spots are not constant, and that therefore, from such numerical or geometrical features, no reliable specific characters are to be drawn. I am able to recall my enumeration of eighteen dorsal orange spots on one of the larvæ.

In some notes on this larva, published by me in 1873, in the *23rd Rep. St. Mus. Nat. Hist.*, p. 213, its probable food-plant is given as the common burdock (*Lappa officinalis*), as inferred from finding several examples near this plant, and some of the larvæ having fed upon it in confinement. I now incline to the belief that their eating the burdock was under the provocation of hunger, and that their range of food under natural conditions would not extend to a plant so different in character from that of the blue lettuce. It is an interesting fact that in its systematic arrangement, the genus Mulgedium stands between Lactuca and Sonchus, the two genera which embrace, in Europe, the food-plants of *C. lucifuga* (see loc. cit., p. 214). As two of these plants, *Sonchus oleraceus* and *S. arvensis* have been introduced in this country from Europe, and the former has become quite common around dwellings and by roadsides in New York, it is not improbable that the *C. intermedia* larva may hereafter be discovered upon them. I have repeatedly examined the burdock for the larvæ, but without success.

Xylina lambda (Fabr.), var. **Thaxteri** Grote.

The reference of *X. Thaxteri* as a variety of *X. lambda*, by Dr. Speyer, has been noted on page 120. From the examination of the photograph of *X. Thaxteri* given in the Grote Check List, Pt. ii, fig. 3, Dr. Speyer was led to express his opinion* that it was but a variety of the European *X. lambda*. Mr. Grote, in its description (*Bul. Buf. Soc. Nat. Sci.*, ii, p. 196), had instituted a comparison between it and *Zinckenii*, under which name *X. lambda* was formerly known.

The recent examination by Dr. Speyer of a number of examples of *X. Thaxteri* has confirmed the opinion above expressed. He finds it to "correspond with the European species in all particulars except in that its primaries are a little broader, and its secondaries somewhat more excavated on vein 5. *X. lambda* had heretofore been known under three varieties; *Zinckenii* Treits., *rufescens* Mén., and *somniculosa* Hering. *Thaxteri* is now the fourth well marked variety; in its sharp markings, it resembles *Zinckenii* var., and in its color, *rufescens* var."

Hypena humuli Fitch.

In Harris' *Report on the Insects of Massachusetts*, p. 345 (Edition of 1841), the author, after describing some caterpillars infesting the hop-vine, says of the moths proceeding from them, that "they have been named [by him] the *Hypena humuli*, on the supposition that they are distinct from the *Hypena rostralis* or hop-vine snout-moth of Europe."

"These moths are readily known by their long, wide, and flattened feelers, which are held close together, and project horizontally from the fore part of the head in the manner of a snout. The antennæ in both sexes are naked and bristle-formed. The wings vary in color, being sometimes dusky or blackish brown, and sometimes of a much lighter rusty brown color. The fore-wings are marbled with gray beyond the middle, and have a distinct oblique gray spot on the tip; they are crossed by two wavy blackish lines, one near the middle, and the other near the outer hind margin; these lines are formed by little elevated black tufts, and there are also two similar tufts on the middle of the wing. The hind wings are dusky brown or light brown, with a paler fringe, and are without bands or spots. The wings expand about one inch and a quarter."

* Stettiner Entomologische Zeitung, for 1876, page 203.

The above description is inapplicable to the *Hypena humuli* of our collections ; it applies fully to *Hypena scabra* Linn., and to no other species with which we are acquainted. In *H. scabra*, the fore-wings are marbled with gray beyond the middle" — in *H. humuli*, not: "the two wavy blackish lines, one near the middle, and the other near the outer hind margin, are formed by little elevated black tufts ;" in *humuli* these lines do not consist of elevated scales or tufts, even in examples just emerged from pupæ. The "two similar tufts on the middle of the wing" of *scabra*, are replaced in *humuli* by four. Of the former species, "the wings expand one inch and a quarter ;" of the latter, no specimen of large numbers under my observation, have equaled that expanse.

From the above it seems evident that the description of Harris was drawn from examples of *H. scabra* before him: from the general resemblance of the two species, *scabra* examples may have accidentally replaced those which he had reared from the hop, and which he intended to describe.

An earlier reference to the species is made by Harris in his *Catalogues of the Animals and Plants of Massachusetts*, 1835, page 74, where it is included among the Tineidæ as *Crambus humuli.*

The description of "*Hypena humuli*" is also given in the subsequent editions of Dr. Harris' Report.*

In 1855, in the *Transactions of the N. Y. State Agricultural Society*, vol. xv, pp. 555–558, pl. 1 fig., 1, Dr. Fitch describes and figures the moth obtained from the hop-vine larvæ as the *Hypena humuli* of Harris, in the event of its not proving identical with the European *H. rostralis* L. This being the first description of the species, the name of Dr. Fitch will hereafter have to be associated with it. The above is also to be found in the volume of the *First and Second Report on the Noxious Insects of New York*, pp. 323–326, pl. 1, fig. 1, printed in 1856.

In the *Annals of the Lyceum of Natural History of New York*, ix, p. 311 (1870), Mr. Coleman T. Robinson describes a form as *Hypena evanidalis.* This form proves to be the female of *H. humuli* — the sexes differing so much that for a long time they were regarded as distinct species.

---

* Insects of New England Injurious to Vegetation, p. 373. 1852. Insects Injurious to Vegetation, p. 477, f. 273. 1862.

Depressaria Le Contella Clem., etc.

Through some misunderstanding, the captures of the following named species, which were added to the proof of the short list of Microlepidoptera given on p. 51 of these papers, were omitted in the printing:

Depressaria Le Contella *Clem*	Aug. 10
Acrobasis nebulo (*Walsh*)	July 8
Tortrix furvana *Rob*	June 27
Tortrix limitata *Rob*	July 10
Tortrix Pettitana *Gr.-Rob*	July 1
Sericoris campestrana *Zeller*	June 19
Ditula blandana *Clem*	June 26

The following species of New York Tortricidæ have also been taken by me, at the date stated, but were not captured at sugar, as were the preceding species. With a few exceptions they were collected at Schoharie, Schenectady and Center. Most of the determinations were made for me by Prof. Fernald:

Tortrix gurgitana *Rob*	July 14
furvidana *Clem*	July 15
fumosa *Rob*	July 4
discopunctana *Clem*	June 30
incertana *Clem*	July 18
lævigana *S.-V.*	June 15
nigridia *Rob*	June 21
puritana *Rob*	Aug. 5
sulfureana *Clem*	Aug. 21
Penthina albeolana *Zell*	June 18
chionosema *Zell*	June 13
nimbatana *Clem*	Aug. 18
Steganoptycha flavocellana *Clem*	Aug. 15
Phoxopteris mediofasciana *Clem*	May 16
nubeculana *Clem*	June 10
spireæfoliana *Clem*	June 13
Pædisca dorsisignatana *Clem*	Sept. 14
otiosana *Clem*	July 12
Sericoris coruscana *Clem*	June 12
constellatana *Zell*	June 18
Exartema fasciatanum *Clem*	July 30

XXVII. ON SOME SPECIES OF COSSUS.

Cossus reticulatus n. sp.

Allied to *C.robiniæ* in shape of wings and markings, having the stronger scales and reticulated ornamentation of that species, in which it differs from the minute and sparse scales and transverse lines of *C. querciperda* and *C. Centerensis.*

Primaries reticulated with black on a pale ash ground, the wings lighter than n *C. robiniæ*, from the absence of the conspicuous intranervular black spots and streaks which characterise that species, and are well represented in fig. 205, p. 413, of Harris' *Insects Injurious to Vegetation.* In this species, only between the internal, submedian and 1st median nervule (veins 1a, 1b, and 2), at the outer third of the wing, do the reticulations coalesce so as almost to form spots. In the terminal and subterminal portions of the wing, the small ash spots (sometimes ocellated with a black dot or line) for the greater part rest upon the veins; between 2 and 5, there are other spots intermediate to these venular ones — elsewhere, with a few exceptions, the spots are venular, forming two intranervular rows. The costal region is pale ash, traversed by black lines rather than reticulated. The median portion of the wing is imperfectly reticulated. The terminal margin, and the unicolorous fringe is conspicuously marked with a black spot on each vein.

Secondaries thinly clothed with fuscous hairs, permitting the reticulations of the lower surface to be seen in transparency, except between the margin and costal nerve, where it is scaled in pale ash as the primaries. Terminal margin and the pale fringe, black spotted as the primaries.

Beneath: primaries much as above, with these differences: There is an accumulation of blackish elongated scales in the basilar region, in the interspaces of the internal, submedian and median nerves; also between veins 2 and 3 at their origin, and a triangular spot having its apex on the costa at its outer third. Costa with about fourteen black lines between the base and vein 12, and three lines between veins 12 and 11, of which the outer, at the junction of vein 11 with the costa, is broad and extends inwardly to vein 10; another broad, black costal line each at the junction of veins 10 and 9 with the costa — these last three spots (also shown on the upper surface of the wings) are the equivalents of the three anteapical costal white dots of many of the Noctuidæ, designated by Guenée as the virgular spots (*traits virgulaires*).

Secondaries, from inner margin to vein 1 b, clothed with fuscous hairs as above; thence to costal margin with ash scales, nearly plain between 1b. and 2; reticulated between the median nervules (veins 2–5) and to vein 7 as on the primaries above; thence to costal margin, the interspaces barred by black lines, of which about eighteen are seen on the margin. A fuscous cloud borders the median nerve from the base to within its branches. Thorax above and beneath covered with pale ash scales. Tibiæ and tarsi ash, annulated with black. Abdomen above, apparently (the body of the example is much greased) concolorous with the secondaries above. Antennæ black, strongly bipectinate, like those in *C. robiniæ*.

Expanse of wings, 3.35 inches; length of body, 1.75 inch.

Habitat, etc., Texas, Rio Grande. Described from one female, in the collection of Mr. B. Neumœgen, New York city

Cossus undosus n. sp.

A Cossus was taken by Mr. S. H. Scudder, at Green River Station, Union Pacific Railroad, Wyoming, resting on a "cotton-wood" (*Populus balsamifera*, probably), together with a pupa case projecting from the same trunk. Through the kindness of Mr. Scudder, these have been placed in my possession. The moth unfortunately is a wreck, in no condition for accurate description, having lost its antennæ, one-half of one pair of wings, and one-third (the apical portion) of the other.

It differs from any of our known species in markings and squammation. Both pair of wings in their ground color are white, and are crossed by numerous, narrow, black, transverse lines. Of these the most prominent one crosses the outer third of the cell, with an outward inflection from the costa to the subcostal nervure, an inward inflection in the cell to the first median nervule which it follows for a short distance, and thence proceeds in a double curve to the internal margin. Another black line, less heavily marked, runs irregularly from the costa to the internal margin, passing over the middle of the cell. Between the stronger transverse lines are fainter ones, which sometimes reticulate with the former. The thorax, abdomen, basal and internal portion of the hind wings, are thickly clothed with pale gray hairs or elongated scales: the remaining portion of the hind wings (the portion preserved in the example) is as thickly scaled as the primaries, and nearly as distinctly lined; beneath they are stronger lined than above. Palpi barely extending beyond the eyes, clothed with white scales interspersed with narrow black ones. Thorax beneath, with long gray hairs. Legs similarly clothed, with their tibiæ and tarsi banded with black.

Length of body, with extruded ovipositor, 1.50 inch. Expanse of wings, entire, unknown; from one discal cross-vein to the opposite, 1.85 inch.

The pupa-case projecting from the tree, was that of a moth differing in sex from the captured example, but presumably of the same species. It measures 1.40 in. in length and 0.3 in. in its broadest diameter. Its color is about that of the ♀ of *C. robiniæ*, but of a paler brown than the ♂ of that species. Its terminal segment and rows of teeth on the segments are darker brown, approaching fuscous, but presenting quite a contrast with the black terminal segment and wing-cases of *C. Centerensis*. Its armature (transverse rows of teeth) is much stronger than in the ♀ *C. robiniæ*, and a little more so than in the ♂. It is stronger than in *C. Centerensis* (in which the armature is nearly equal in the sexes), having the teeth longer, although not so broad at their base: on the 8th, 9th and 10th segments, the teeth continue quite prominent in their extension below the stigmata, where in *C. Centerensis* they are weak.

The armature of the pupa-cases of Cossus, unquestionably presents excellent specific characters. From its study, I am able to announce the existence, in the State of New York, of another species of Cossus, boring in the white birch (*Betula populifolia*), the imago of which has not yet been detected.

From near the base of a prostrate birch at Center which had been extensively mined, I took, in 1876, a pupa-case, clearly differing from any known species. Unfortunately the specimen has been mislaid, or I should not hesitate, from the characters it presented to describe it and give the species a name. Other trunks of birch have been observed by me, similarly mined, and evidently by the same Cossus.

The species above described as *Cossus undosus*, may possibly be the *C. populi* of Walker (*Cat. Lep. Br. Mus.*, vii, p. 1515), from Hudson's Bay, which has not yet been identified. The very general terms in which its brief description is given, will not admit of its separation from allied forms, and unless the type is preserved in the British Museum, and cómparison be made, it must be handed over to the long list of undetermined and indeterminable species of Walker.

It cannot be the *Cossus nanus* of Strecker, from its non-resemblance to *Cossus ligniperdi*, which *C. nanus* is said closely to resemble* ; and from its differing markedly from a Colorado example of a ♀ Cossus which I refer to the ♂ *named* by Mr. Strecker but unfortunately accompanied by the mention of only a few specific features.

Cossus plagiatus Walker.

This species, briefly indicated, loc. cit., p. 1515 (1856), is another unknown species. Mr. Grote in his *List of the North American Platypterices, Attaci*, etc., p. 8 (1874),† refers it as a synonym of *Mac Murtrei* (Boisd.), Icon. Règne An., pl. 85, f. 2,— marking it, however, as an unrecognized species. It does not appear why this reference is made, and we may presume that it is based on a citation of Dr. Packard, in his *Synopsis of the Bombycidæ of the United States*,‡ where under the synonymy of *Xyleutes plagiatus*, he quotes from the *Systematic List of Canadian Lepidoptera* by W. S. M. D'Urban, § the following :

**Cossus nanus* n. sp.—Expands 1⅛ inches. Has the appearance of a minature *Cossus ligniperda*, is gray, of lighter and darker shades, and reticulated with black lines which are most noticeable across the disk and on the terminal part of wing. Secondaries uniform grayish. Beneath grayish, faintly reticulated.—Hab. Colorado. Proc. Acad. Nat. Sci. Phila., 1876. p. 151.

† Read before the American Philosophical Society, Nov. 20th, 1874.

‡ Proc. Ent. Soc. Phila., vol. iii, p. 390. 1864.

§ Can. Nat. and Geol., Aug. 1860, p. 247

"*Cossus plagiatus* Walk. Rare, July.

"In 1857, Mr. T. R. Peale, of the U. S. Patent Office, named this species *Cossus McMurtrici* [sic], and informed me that it was common south of Pennsylvania, but rare in the Middle States."

Cossus crepera Harris.

This name appears in Dr. Harris' *Catalogues of Animals and Plants of Massachusetts*, p. 72. 1835, but is not continued in his subsequent reports. In 1839, Doubleday, having suspected its true relationship, writes to Dr. Harris of this species: "There is a true *Cossus* with mottled upper wings, and yellow under wings, black at the base and inner margin, —*Robiniæ* ♂?" It is decribed by Dr. Packard, loc. cit., p. 388, as *Xyleutes crepera*, and catalogued by Grote in his List above cited, as an unrecognized species, under the new generic name proposed by him of Xystus. It is now known to be but the ♂ form of *C. robiniæ*, from the ♀ of which it differs so greatly in the angulated form of its posterior wings and their yellow color, as to have been mistaken for another species.

Cossus querciperda Fitch.

The species described under this name by Dr. Packard, loc. cit., p. 389, is not the one so named and briefly described by Dr. Fitch,* but some other form — possibly *C. Centerensis*.† The types are the only pair, so far as known, in existence, and are in my Collection. A male and female were taken in copulation, June 27, 1857, at Schoharie, on the trunk of a young black oak (*Quercus tinctoria*), four inches in diameter. A second male was taken at the time, from the same tree, a short distance from the attached pair. One of these, together with the female, it is believed, were subsequently given to Mr. J. W. Weidemeyer, of New York, with other duplicates from my boxes, without statement of their rarity, which at the time was not known. As Mr. Weidemeyer's Collection is no longer in his hands, the examples have probably been destroyed.

In March following the capture of the above examples, two Cossus larvæ were found in burrows in some pieces of black

* Trans. N. Y. State Agricul. Soc., vol. xviii, p. 790. 1859. Fifth Report of the Insects of New York, p. 10 (section 294 of the volume of the Third, Fourth and Fifth Reports).

† Canadian Entomologist, vol. ix, p. 129. 1877.

oak which had been prepared for fuel. They were both frozen rigidly when discovered. One was lying in a cell, in its burrow, formed by some slight threads in which its cuttings had been thinly woven above and below it: the other had constructed a cell of about the capacity of its body, branching off from its main burrow—its entrance closed by a thin wall of the cuttings.

The smaller of the two larvæ measured one inch and a half in length. It was of a pale green color, with a darker green dorsal stripe, bordered faintly with yellow. Head flat, subtriangular, dark brown, clouded with black. First segment with two brown spots extending across it, narrowed laterally, and of nearly the length of the segment medially, where they unite to inclose on the dorsal line an elongate-elliptical green spot. The anterior segments are flattened, and broader than the following, which gradually diminish in breadth toward the posterior end. The segments are marked dorsally with four rose-colored elevated points — the trapezoidal spots of Guenée; on the 10th and 11th segments they form a square: a similar spot is present above each stigma, a smaller one below, and another in front—each of these bearing a short brown hair. The stigmata are oval, orange colored, centered with dark brown. The legs are tipped with chestnut brown, and the prolegs armed with brown plantæ.

One of the larvæ escaped from its burrow by gnawing through the stick of wood in which it was inclosed and its paper box, and was found some weeks thereafter, dead within a roll of clothing. The other disclosed a perfect imago on the 29th of April—the female type of the rare *C. querciperda* of my Collection. The larva had constructed within its burrow a very slight cocoon of delicate silk. The long ovipositor of the moth was a marked feature of it, when alive, measuring in its full extrusion, three-tenths of an inch. It displayed a tenacity of life remarkable even in a gravid Bombycid, as it lived for twenty-four hours after a strong solution of cyanide of potassium had been pricked in its thorax.

The most interesting character of this species is not referred to in its description, viz.: the great disparity in the size of the sexes. The ♂ measures in length of body 0.55 in., and in expanse of wings, 1.23 in. The body of the ♀, exclusive of its ovipositor, is 1.25 in. long, and the expanse of wings is 2.62 in. Their comparative weight is as one to four, even after the removal of the eggs and viscera from the female—its weight being twelve grains, and that of the male three grains. Were it not for the capture of a second male at the same time, of the same diminutive size, it might have been supposed that the example was a dwarfed individual. This disproportion in size is the more interesting, from the fact that in a congeneric species—*Cossus Centerensis*, the size affords no indication of sex, for the males of this species are often larger than the females.

Dr. Fitch has erred in representing the hind wings of the male *querciperda* as colorless. They are pale yellow on the disc beyond the fold at vein 1a—the color of the delicate scales showing more plainly when their surface is viewed in line with the eye. Within vein 1a to the inner margin, the wing is covered with black hairs.

The yellow coloring of these wings, together with the angle on vein 1, and the antennal structure, ally this species to *C. robiniæ*.

INDEX TO ENTOMOLOGICAL CONTRIBUTIONS.